T0341146

Chemical Thermodynamics

Theory and Applications

Chemical Thermodynamics

Theory and Applications

W. John Rankin

CRC Press
Taylor & Francis Group
Boca Raton London New York

CRC Press is an imprint of the
Taylor & Francis Group, an **Informa** business

CRC Press
Taylor & Francis Group
6000 Broken Sound Parkway NW, Suite 300
Boca Raton, FL 33487-2742

Library of Congress Cataloging-in-Publication Data

Names: Rankin, W. J., author.
Title: Chemical thermodynamics : theory and applications / authored by W.J. Rankin.
Description: First edition. | Boca Raton, FL : CRC Press/Taylor & Francis Group, 2019. | Includes bibliographical references and index. | Summary: "This book develops the theory of chemical thermodynamics from first principles, demonstrates its relevance across scientific and engineering disciplines, and shows how thermodynamics can be used as a practical tool for understanding natural phenomenon and developing and improving technologies and products. Concepts such as internal energy, enthalpy, entropy, and Gibbs energy are explained using ideas and experiences familiar to students and arealistic examples are given so the usefulness and pervasiveness of thermodynamics becomes apparent. The worked examples illustrate key ideas and demonstrate important types of calculations, and the problems at the end of chapters are designed to reinforce important concepts and show the broad range of applications. Most can be solved using digitized data from open access databases and a spreadsheet. Answers are provided for the numerical problems. A particular theme of the book is the calculation of the equilibrium composition of systems, both reactive and non-reactive, and this includes the principles of Gibbs energy minimization. The overall approach leads to the intelligent use of thermodynamic software packages but, while these are discussed and their use demonstrated, they are not the focus of the book, the aim being to provide the necessary foundations. Another unique aspect is the inclusion of three applications chapters: heat and energy aspects of processing; the thermodynamics of metal production and recycling; and applications of electrochemistry. This book is aimed primarily at students of chemistry, chemical engineering, applied science, materials science and metallurgy, though it will be also useful for students undertaking courses in geology and environmental science"-- Provided by publisher.
Identifiers: LCCN 2019026322 (print) | LCCN 2019026323 (ebook) | ISBN 9780367222475 (hardback ; acid-free paper) | ISBN 9780429277252 (ebook)
Subjects: LCSH: Thermochemistry. | Thermodynamics.
Classification: LCC QD511 .R28 2019 (print) | LCC QD511 (ebook) | DDC 541/.36--dc23
LC record available at https://lccn.loc.gov/2019026322
LC ebook record available at https://lccn.loc.gov/2019026323

Visit the Taylor & Francis Web site at
http://www.taylorandfrancis.com

and the CRC Press Web site at
http://www.crcpress.com

Contents

Preface

This book is aimed primarily at students of chemistry, chemical engineering, applied science, materials science and metallurgy, though it will be useful also for students undertaking courses in geology and environmental science.

My aim has been to produce a book that rigorously develops the theory of thermodynamics and demonstrates its relevance across scientific and engineering disciplines and to show that thermodynamics, far from being an esoteric and abstract discipline, is a practical tool for understanding natural phenomenon and developing and improving technologies and products. The author has been a user of thermodynamics in process development applications, an experimental thermodynamicist and teacher of the subject for over 30 years, and it is hoped that some of this experience is reflected in the content, style and approach adopted. It is likely students will have previously encountered elementary aspects of thermodynamics in other subjects, but this book assumes no prior exposure to the subject. Only a basic knowledge of calculus is assumed.

Many textbooks of thermodynamics follow the historical development of the discipline. Many also concentrate on the mathematical aspects of the subject and inadequately link these to real phenomena and processes. In more applied texts, the approach often is to identify the key equations and show how to use them to perform routine calculations with little theoretical understanding. None of these approaches is ideal. At one extreme, thermodynamics becomes almost a branch of applied mathematics; at the other, it becomes little more than teaching the student to pick the right equation for a particular calculation. In both cases, the fundamental principles of thermodynamics, its over-arching relevance and the inherent beauty of the subject, are either obscured by detail or not revealed at all.

Thermodynamics has been revolutionised over the past several decades by the conversion of thermodynamic data into digitised formats amenable to manipulation using spreadsheets and by the development of software products for performing thermodynamic calculations. This requires newer ways of presenting the subject. This book adopts a modern approach.

- Abstract concepts such as internal energy, enthalpy, entropy, Gibbs energy, etc., are explained using ideas and experiences that are familiar to students so that these have real meaning.
- Mathematical concepts are developed fully, and intermediate steps are shown so students can easily follow the logical development of the subject but without overwhelming the student with mathematics.
- Realistic and familiar examples are given so the usefulness and pervasiveness of thermodynamics become apparent. Worked examples are designed to illustrate important ideas and demonstrate important types of calculations.
- Modern forms of data, including databases, are used.
- The problems provided at the ends of chapters are designed to reinforce important concepts and/or to show the broad range of applications of thermodynamics. Most can be solved using digitised data and a spreadsheet. Answers are provided for most of the numerical problems.
- The overall approach adopted leads naturally to the intelligent use of thermodynamic software packages.
- SI units and IUPAC nomenclature are used throughout.

Many thermodynamic computational software products are available. These are largely black boxes which produce solutions to a range of thermodynamic problems. These have great power and are very useful when used by skilled practitioners. Indeed, practitioners now routinely use these, and the tedious hand calculations, still so frequently used in undergraduate courses, are never performed in practice today. However, when users have little theoretical understanding of thermodynamics, these products often produce results which are either wrong or which are misinterpreted. It is important that students are aware of these packages and their capabilities, and these are discussed in this book, but I believe they should have a complementary role in introductory courses of thermodynamics and not be the focus of the course.

Many modern texts are overloaded with detail and extraneous information which, even if interesting, makes it difficult for the student to identify the key messages and points. This book is purposely succinct, focussed and easy to read. Each chapter starts with a brief statement of the scope and the key learning objectives. The text of each chapter begins with an introductory paragraph to set the context, and key points and key equations are emphasised in the text.

The applications chapters (Chapters 6, 11 and 17) are more extensive than in most introductory books, and the examples have been chosen to show the wide-ranging applications of thermodynamics. These chapters could be omitted for less applied courses (or only a few examples selected).

Many people, directly and indirectly, have made this book possible. Dr Anil K. Biswas first introduced me as an undergraduate student at the University of Queensland to the wonders of thermodynamics. Many colleagues over the subsequent years helped sustain my interest and enthusiasm for the subject. More recently, Drs R. J. Sinclair, D. E. Langberg, D. R. Swinbourne, G. Senanayake and J. B. See read drafts of some or all of the chapters and gave useful feedback. David Langberg also worked many of the chapter problems. All errors in the text of course are mine alone. Finally, I wish to express my gratitude to Marsha Rankin for her support and for enduring my preoccupation during the writing of the book over an extensive period.

A note to students

It's been said that the first time we are exposed to thermodynamics we don't understand very much of it; the second time we think we understand it all; the third time we realise we don't understand it at all. This is not meant to put you off the study of thermodynamics but simply to emphasise that some thermodynamic concepts are difficult to fully comprehend; indeed, most practitioners and teachers continue to develop new insights into thermodynamics throughout their working lives. Hopefully, this book will lead some of you onto that path.

HOW TO USE THIS BOOK

For each chapter, the following steps might be found useful.

- Read the chapter scope and introductory paragraph to understand the aim of the chapter.
- Scan the chapter to get an idea of the contents – this includes understanding the structure of the chapter, the main headings and terms.
- Read the chapter from beginning to end with a view to understanding it fully – this may require rereading all or parts of the chapter.
- Review the learning objectives at the beginning of the chapter.
- Attempt some or all of the problems.

TACKLING THE PROBLEMS

The problems aim to reinforce the learnings of the chapter and illustrate some ways thermodynamics can be applied to real situations. A systematic approach to solving the problems involves the following steps:

- Understand what the question requires and decide the calculation procedure.
- Identify and list any required assumptions.
- Collect the necessary data and enter them into a spreadsheet. Use of a spreadsheet rather than a calculator to perform calculations makes it easy to alter/correct equations, change input values, etc.
- Convert all the data to SI units (if not already in SI units). All calculations should be done in SI units – this makes life so much easier than trying to use mixed units!
- Perform the necessary calculations. If you have performed all the calculations in SI units, the answer will also be in SI units.
- Finally, try to assess the reasonableness of the answer and select the appropriate number of significant figures.

Author

W. John Rankin has a BSc and PhD from the University of Queensland, Australia. He worked initially for MINTEK in South Africa, then lectured in extractive metallurgy and chemical engineering at the University of Stellenbosch, South Africa, the Royal Melbourne Institute of Technology, Australia, and the University of Waterloo, Canada. During the 1990s, he was a Professorial Research Fellow and Director of the G.K. Williams Cooperative Research Centre for Extractive Metallurgy at the University of Melbourne, Australia. He then accepted a position in CSIRO (Australia's national science agency) and held the role of Chief Scientist of the Division of Process Science and Engineering. His research interests are in the fields of high temperature thermodynamics, pyrometallurgy and the implications of sustainability for the minerals industry, and he has published over 130 research papers. He authored the book *Minerals Metals and Sustainability: Meeting Future Material Needs* (CRC Press) and is co-editor of the journal *Mineral Processing and Extractive Metallurgy*. He is currently an Adjunct Professor at the Swinburne University of Technology, Melbourne, and an Honorary Fellow of CSIRO.

Units

SI units* are used throughout the text. There are seven base SI units, and all other units are derived from these. Derived units are expressed algebraically in terms of base units or other derived units and are obtained by the mathematical operations of multiplication and division. A few derived units have been given special names and symbols. For completeness all base units are listed here although candela is not used in this book.

SI BASE UNITS

Quantity	Unit	Symbol
length	metre	m
mass	kilogram	kg
time	second	s
electric current	ampere	A
temperature	kelvin	K
amount of substance	mole	mol
luminous intensity	candela	cd

SOME DERIVED UNITS

Quantity	Unit	Symbol
area	square metre	m^2
volume	cubic metre	m^3
velocity	metre per second	ms^{-1}
acceleration	metre per second squared	ms^{-2}
density, mass density	kilogram per cubic metre	$kg\ m^{-3}$
specific volume	cubic metre per kilogram	$m^3\ kg^{-1}$
concentration	mole per cubic metre	$mol\ m^{-3}$

DERIVED UNITS WITH SPECIAL NAMES AND SYMBOLS

Derived Quantity	Name	Symbol	Expression in Terms of Other SI Units	Expression in Terms of SI Base Units
force	newton	N		$m\ kg\ s^{-2}$
pressure	pascal	Pa	$N\ m^{-2}$	$m^{-1}\ kg\ s^{-2}$
energy, work, heat	joule	J	$N\ m$	$m^2\ kg\ s^{-2}$
power	watt	W	$J\ s^{-1}$	$m^2\ kg\ s^{-3}$
electrical resistance	ohm	Ω	$V\ A^{-1}$	$m^2\ kg\ s^{-3}\ A^{-2}$
electric charge	coulomb	C		$s\ A$

* *Système international*, in English the International System of Units (SI), is the most widely used system of measurement. *The Bureau International des Poids et Mesures* (International Bureau of Weights and Measures), under the supervision of the International Committee for Weights and Measures, has responsibility for the system.

potential difference	volt	V	$W A^{-1}$	$m^2 kg\ s^{-3} A^{-1}$
celsius temperature	degree Celcius	°C		K
conductance (electrical)	Siemens	S	Ohm^{-1}	$m^2 kg^{-1} s^3 A^2$

NON–SI UNITS ACCEPTED FOR USE WITH THE SI

Unit	Symbol	Value in SI Units
litre	L	$1 L = 1\ dm^3 = 10^{-3} m^3$
tonne (or metric ton)	t	$1\ t = 10^3\ kg$
bar	bar	$1\ bar = 10^5\ Pa$
kilowatt hour	kW h	$1\ kW\ h = 3600\ kJ$

COMMONLY USED SI PREFIXES

Factor	Prefix	Symbol
10^{15}	peta	P
10^{12}	tera	T
10^9	giga	G
10^6	mega	M
10^3	kilo	k
10^2	hecto	h
10^1	deka	da
10^{-1}	deci	d
10^{-2}	centi	c
10^{-3}	milli	m
10^{-6}	micro	μ
10^{-9}	nano	n
10^{-12}	pico	p

SOME USEFUL CONVERSIONS TO SI UNITS

$1\ atm = 101\ 325\ Pa = 1.013\ 25\ bar$
$1\ L = 0.001\ m^3$
Celsius to Kelvin: $+273.15$
$1\ cal = 4.184\ J$
$1\ kW\ h = 3600\ kJ$

Nomenclature

The following symbols and abbreviations are used in the text. The IUPAC[*] naming convention has been followed (with a couple of exceptions). Source: Cohen, E. R. et al. 2008. *Quantities, units and symbols in physical chemistry, IUPAC Green Book*, 3rd edition, Cambridge: IUPAC & RSC Publishing.

Symbol	Name	SI unit
A	Helmholtz energy	J
a	activity	–
A_r	relative atomic mass; atomic weight	–
C	number of components in a system	–
C_p	heat capacity at constant pressure	J mol^{-1} K^{-1}
C_V	heat capacity at constant volume	J mol^{-1} K^{-1}
c	concentration	mol m^{-3}; g L^{-1}; etc.
E	electromotive force	V
ΔE	electric potential difference	V
E_a	activation energy	J mol^{-1}
F	degrees of freedom in a system; also called variance	–
F	Faraday constant	C mol^{-1}
f	fugacity	–
G	Gibbs energy	J
G_B^0	standard molar Gibbs energy of substance B	J mol^{-1}
G_B	partial molar Gibbs energy, or chemical potential, of substance B	J mol^{-1}
H	enthalpy	J
H_B^0	standard molar enthalpy of substance B	J mol^{-1}
H_B	partial molar enthalpy of substance B	J mol^{-1}
I	electric current	A
K	equilibrium constant	–
k	Boltzmann constant	J K^{-1}
L	Avogadro constant	mol^{-1}
m	mass	kg
m_B	molality of substance B	mol kg^{-1}
M_B	molar mass of substance B	kg mol^{-1}
M_r	relative molecular mass; molecular weight	–
N	number of entities (for example, atoms, molecules, ions)	–
N_B	number of entities of substance B (for example, atoms, molecules, ions)	–
n_B	amount of substance B	mol
n	number of moles; total number of moles	mol
P	number of phases in a system	–
P	power	Watt
p	pressure; total pressure	Pa, bar
p_B	partial pressure of substance B	Pa, bar
q	heat	J
Q	reaction quotient	–

[*] The International Union of Pure and Applied Chemistry (IUPAC) is the international authority on chemical nomenclature and terminology.

Q	electric charge; quantity of electricity	C
R	universal gas constant	J mol^{-1} K^{-1}
R	electric resistance	Ω
r	number of independent chemical reactions in a system	–
S	entropy	J K^{-1}
S_B^0	standard molar entropy of substance B	J K^{-1}
S_B	partial molar entropy of substance B	J K^{-1}
s	number of species in a system	–
T	thermodynamic temperature	K
U	internal energy	J
V	volume; total volume	m^3
V_B^0	standard molar volume of substance B	m^3 mol^{-1}
V_B	partial molar volume of substance B	m^3 mol^{-1}
w	mass fraction	–
w	work	J
x_B	mole fraction of substance B	–
X_m	a molar quantity	$\lvert X \rvert$ mol^{-1}
X_B	partial molar quantity of substance B	$\lvert X \rvert$ mol^{-1}
X_B^X	excess molar quantity of substance B	$\lvert X \rvert$ mol^{-1}
Z	compressibility factor	–
z	electron number of an electrochemical cell	–
z	number of stoichiometric and mass balance restrictions on a reactive system at equilibrium	–
γ_B	activity coefficient of B; pure B standard state	–
$\gamma_{c,B}$	activity coefficient of B; infinitely dilute solution standard state	–
μ	chemical potential (partial molar Gibbs energy)	J mol^{-1}
μ_B^0	standard chemical potential of substance B	J mol^{-1}
$\Delta_r G^0$	standard reaction Gibbs energy	J mol^{-1}
$\Delta_r H^0$	standard reaction enthalpy	J mol^{-1}
$\Delta_r S^0$	standard reaction entropy	J mol^{-1}
$\Delta_f G^0$	standard Gibbs energy of formation	J mol^{-1}
$\Delta_f H^0$	standard enthalpy of formation	J mol^{-1}
$\Delta_f S^0$	standard entropy of formation	J mol^{-1}
Λ_B	molar conductivity	S m^2 mol^{-1}
κ	conductivity (electrical)	S m^{-1}
θ	Celsius temperature	°C
ρ	density; mass density	kg m^{-3}
ϕ	volume fraction	–
φ	fugacity coefficient	–

Constants

F	Faraday constant	$96\,485.332\ \text{C mol}^{-1}$
K	Boltzmann constant	$1.380\,648\,8 \times 10^{-23}\ \text{J K}^{-1}$
L	Avogadro constant	$6.022\,140\,857(74) \times 10^{23}\ \text{mol}^{-1}$
R	universal gas constant	$8.314\,459\,8(48)\ \text{J mol}^{-1}\ \text{K}^{-1}$

1 An overview of thermodynamics

A theory is the more impressive the greater the simplicity of its premises is, the more different kinds of things it relates, and the more extended is its area of applicability. Therefore the deep impression which classical thermodynamics made upon me. It is the only physical theory of universal content concerning which I am convinced that within the framework of the applicability of its basic concepts, it will never be overthrown.

Albert Einstein*

SCOPE

This chapter provides a brief historical introduction to the discipline of thermodynamics and a qualitative introduction to the laws of thermodynamics.

LEARNING OBJECTIVES

1. *Gain an appreciation of the historical context of thermodynamics and its development, particularly the branch known as chemical thermodynamics.*
2. *Understand, qualitatively, the concepts of internal energy, enthalpy, entropy and Gibbs energy.*
3. *Gain a simple understanding of the four laws of thermodynamics.*

1.1 WHAT IS THERMODYNAMICS?

In broad terms thermodynamics is the study of the behaviour of matter in an environment where it loses or gains energy. How our understanding of this has developed has been a long and often circuitous process. The physical behaviour of matter was well understood by the beginning of the 19th century, due in great measure to the pioneering work of Isaac Newton, and it was known that it can be described in terms of its motion (its kinetic energy) and position (its potential energy). However, it was also understood that its behaviour could be influenced by other factors such as heat and pressure. The realisation that kinetic and potential energy, heat and pressure could be converted from one form to another was a major advance in understanding which ultimately led to the development of a general description of the behaviour of matter. It also led to the development of steam and other heat engines in which thermal energy is converted to mechanical energy.

Thus, as originally conceived, thermodynamics was the study of the relationship between heat (and the related concept of temperature), energy and work. This is reflected in the name: Thermo (meaning heat) and dynamics (meaning motion). However, as understanding developed, the discipline of thermodynamics expanded to encompass all the influences that affect the behaviour of matter – chemical, electrical and magnetic effects, in addition to mechanical and thermal.

The discipline of thermodynamics is relevant to all branches of science and engineering because it enables us to determine whether a process or change of interest is feasible and, if it is feasible,

---

* A. Einstein, Autobiographical Notes, in Schilpp, P. A. 1949. *Albert Einstein: Philosopher–Scientist*, New York: Tudor Publishing Company, p. 33.

1

the maximum energy that can be obtained from the process or, if the process is not feasible, the minimum energy required to make it occur. This is true whether the process belongs to the realm of physics, chemistry, biology, geology, astrophysics or any other branch of science. Accordingly, thermodynamics has great explanatory power and is useful in developing understanding of existing processes and phenomena, both natural and artificial. Equally importantly, it has great predictive capabilities which are useful in developing new materials, processes and products and improving existing materials, processes and products. It has practical application in fields such as the development and application of materials, energy production, storage and consumption, geochemistry, the chemical and metallurgical industries and the environment.

1.2 A BRIEF HISTORY

The beginnings of thermodynamics as a scientific discipline can be traced back almost 400 years. In 1650, in Magdeburg in modern day Germany, Otto von Guericke, a natural philosopher and politician, designed and built a pump and demonstrated that a vacuum can exist, thereby disproving Aristotle's long-held assertion that a vacuum is physically impossible to achieve – 'nature abhors a vacuum'. In England, the physicist and chemist Robert Boyle read of Guericke's designs and, in 1656 with scientist Robert Hooke, built a vacuum (or air) pump. Using this pump, Boyle and Hooke discovered the correlation between pressure and volume now known as Boyle's law.

In 1679, Denis Papin who was collaborating with Boyle built a steam digester, an early form of pressure cooker consisting of a closed vessel with a tightly fitting lid that confined the steam to create a high pressure. His designs included a pressure release valve that kept the vessel from accidentally exploding. By watching the valve move rhythmically up and down to maintain a constant pressure in the vessel, Papin conceived the idea of using a piston moving backwards and forwards in a cylinder as the basis of a steam-driven engine. After moving to Marburg (in Germany), Papin built the first piston steam engine in 1690. This relied on creating a partial vacuum in a vessel by condensing steam, then allowing atmospheric pressure to drive water into the vessel. In 1697 Thomas Savery built a steam-operated water pump based on Papin's designs. In 1695 Papin moved to Kassel (also in Germany) where, in 1705, he developed a second steam engine with the help of Gottfried Leibniz, based on Savery's design but utilising steam pressure rather than atmospheric pressure. Details of the engine were published in 1707.

Around 1712 in England, Thomas Newcomen, an ironmonger by trade, combined the ideas of Savery and Papin to make the first practical steam engine.* He replaced Savery's receiving vessel (in which the steam was condensed) with a cylinder containing a piston. Instead of the vacuum drawing in water, it drew down the piston. This was used to rock a large wooden beam supported on a central fulcrum. On the other side of the beam was a chain attached to a pump located at the bottom of a mine so that water could be removed from the mine. In each cycle, the cylinder was filled with steam as the piston was pushed outwards; then a small amount of water was injected into the cylinder to condense the steam and create a low pressure, allowing the piston to move back inwards.

The concepts of heat capacity (or specific heat) and latent heat, necessary for the development of thermodynamics, were developed around 1761 by Joseph Black, a professor at the University of Glasgow. James Watt was employed as an instrument maker at the university where he met Black who encouraged him to improve the efficiency of Newcomen's steam engine. While working from time-to-time over the period 1763 to 1775, Watt conceived the idea of a condenser located external to the cylinder. In this way, steam could be condensed without cooling the piston and cylinder walls as did the internal spray in Newcomen's engine. The efficiency of Watt's engine was more than double that of the Newcomen engine. Watt's engine was commercialised in 1776 and became one

---

* An animation showing the operation of Newcomen's engine can be found at: https://en.wikipedia.org/wiki/Thomas_New comen

of the major driving forces of the industrial revolution, making possible the replacement of water power by steam power.

Drawing on previous work, in 1824 Sadi Carnot, a French military engineer and physicist, published the book *Reflections on the Motive Power of Fire* which outlined the basic energy relations between heat engines and motive power. It anticipated the second law of thermodynamics and marked the start of thermodynamics as a modern science. In 1843 James Joule, an English physicist from a wealthy brewing family, published a paper *Mechanical Equivalent of Heat* which anticipated the first law of thermodynamics. The first and second laws of thermodynamics emerged in a formal sense in the 1850s primarily out of the works of William Rankine (a professor of civil and mechanical engineering at the University of Glasgow), Rudolf Clausius (a German physicist) and William Thomson (later Lord Kelvin). The first textbook of thermodynamics, published in 1859, was written by William Rankine.

Thus, by the middle of the 19th century, the nature of energy was well understood. Its transformations from one form to another had been well studied, and both the principle of its conservation (the first law of thermodynamics) and the basis for determining the direction of energy-driven changes (the second law) had been established. All of this knowledge had been derived largely from the desire to improve the efficiency of steam engines. Thermodynamics was thus founded on the basis of turning heat from burning fuel into work in the form of mechanical motion.

In 1865, Rudolf Clausius suggested that the principles of thermodynamics could be applied to chemical reactions. However, it was mainly through the work of the American mathematical physicist Josiah Willard Gibbs, a professor at Yale College (now Yale University), that it became clear that the principles developed for steam power applied equally well in other situations. Between 1873 and 1876 Gibbs published a series of three papers, the most famous being *On the Equilibrium of Heterogeneous Substances*. In these papers, Gibbs built on the work of Clausius and showed, through graphical and mathematical means, how the first and second laws of thermodynamics could be used to determine the thermodynamic equilibrium of chemical processes as well as their tendencies to occur. However, the significance of Gibbs' work was not fully appreciated for several more decades.

In the early decades of the 20th century, two major books were written which showed how the principles developed by Gibbs could be applied to chemical processes. These books established the foundation of the science of chemical thermodynamics: *Thermodynamics and the Free Energy of Chemical Substances* by Gilbert N. Lewis and Merle Randall (at the University of California, Berkeley), published in 1923, and *Modern Thermodynamics by the Methods of Willard Gibbs* by Edward A. Guggenheim (at the University of Reading), published in 1933. Lewis, Randall and Guggenheim are considered the founders of modern chemical thermodynamics because of the major contribution of their books in unifying the application of thermodynamics to chemistry.

1.3 THE LAWS OF THERMODYNAMICS

The discipline of thermodynamics is based on four laws, the *laws of thermodynamics*. These are simple, universal statements of conclusions drawn from observations and measurements of phenomena that occur on Earth and throughout the universe; that is, they are empirical laws based on observations and experimental results produced over time and across all areas of science and found to be repeatable and internally consistent.

The zeroth law. The zeroth law of thermodynamics is concerned with the relationship between temperature and heat flow. It states that if two systems are each in thermal equilibrium with a third, then all three are in thermal equilibrium with each other.

The first law. The first law of thermodynamics deals with the conversion of thermal energy into other forms of energy, enabling us to calculate how much heat may be obtained from, or is required to carry out, a given process. In its simplest form, the first law states that the total energy of an isolated system* is constant; energy can be transformed from one form to another but cannot be created or destroyed.

The second law. The second law of thermodynamics tells about the direction in which natural processes occur and allows answers to questions such as:

- Will a system change from State 1 to State 2 under given conditions? If not, how can the conditions be altered in order to make the change occur?
- Under what conditions will the system change from State 1 to State 2 spontaneously, that is, without any external help?

The second law deals only with the feasibility of change, not with the rate of change. The study of rate of change (kinetics) is not the domain of thermodynamics.

The third law. The third law of thermodynamics states that there is an absolute zero of temperature (−273.15°C), though in practice it can never be achieved. At this temperature, systems have their greatest order at the molecular level. The third law is used in the evaluation of a quantity called entropy.

These four laws lead directly to the definition of a number of important thermodynamic quantities, as briefly discussed below.

From the first law, since energy cannot be created or destroyed the energy within a system, its *internal energy, U* (or total energy or intrinsic energy), must increase or decrease if the system takes in or gives out energy, respectively. In a closed system,† there are two ways in which this energy exchange can take place: Through work done on or by the system and through heat transfer to or from the system. Because many processes of interest are open to the atmosphere, it is convenient to focus on the exchange of thermal energy that occurs at constant pressure. This property is called *enthalpy, H.* Working with enthalpy rather than internal energy has the advantage that it bypasses the need to keep track of the work that is done whenever the volume of a system changes through expansion or contraction.

A spontaneous change in a system is one that occurs naturally, without any external energy. The direction of such a change is determined by the natural tendency of matter and energy to disperse in a random manner. The measure of this dispersal is called the *entropy, S.* The second law of thermodynamics states that in an isolated system undergoing a spontaneous change the entropy of the system always increases. This law accounts for changes in an entirely general way, whatever the process might be and regardless of whether physical or chemical transformations are involved.

The criterion for spontaneity of a process is that the total entropy of the universe must increase. To apply this to the common case of a closed system, rather than an isolated system, it is necessary to establish the entropy change both in the system and in its surroundings. When the process occurs at constant temperature and pressure, it is possible to express the entropy change of the surroundings in terms of the change in the system's enthalpy. In this way, the overall entropy change of the system plus its surroundings brought about by a process can be described in terms of the properties of the system alone (*H, S* and *T*). This combined measure of the total entropy change from the perspective of the system alone is called the *Gibbs energy, G,* of the system. At constant temperature and pressure, every chemical reaction and every physical process that occurs of its own accord (that is, without the input of external energy) is driven by a lowering of the Gibbs energy.

We explore these and other concepts in greater detail in the remainder of this book and show how they are used to understand, predict and control physico-chemical processes.

---

* An isolated system is one that has a boundary that prevents the transfer of both matter and energy; no matter or energy is exchanged with its surroundings.

† A closed system is one that allows the exchange of energy but not matter with its surroundings.

2 Fundamental concepts

'When I use a word', Humpty Dumpty said in rather a scornful tone, 'it means what I choose it to mean – neither more nor less'.

Lewis Carroll*

SCOPE

This chapter introduces fundamental concepts and terms essential for the subsequent development of thermodynamic theory and its applications.

LEARNING OBJECTIVES

Understand and be able to give examples of the following concepts and terms:

- *states of matter*
- *systems, the types of systems, the composition of systems*
- *microscopic and macroscopic properties; intensive and extensive properties*
- *equilibrium of a system*
- *reversible and irreversible processes; spontaneous processes*
- *state functions and path functions*
- *energy, heat, work and temperature*

2.1 INTRODUCTION

Thermodynamics, like all branches of science, has its own language. Many of the terms commonly used such as work, energy and heat are already familiar and are used in everyday conversation. However, the scientific meaning of these words is more precise than their everyday meaning, and a good grasp of what these and other terms mean (and imply) is the key to understanding thermodynamic concepts and applying these to solve practical problems. This chapter introduces important concepts and terms that will be encountered frequently in subsequent chapters.

2.2 SUBSTANCES AND THE STATES OF MATTER

As noted in Chapter 1, thermodynamics is the study of the behaviour of matter when it loses or gains energy. All matter is composed of substances, which in turn are composed of atoms. The term substance has a specific meaning in chemistry and thermodynamics. A *substance* (or pure substance) is a form of matter that has a definite composition and specific properties. A substance cannot be separated into components by physical methods, that is, without breaking chemical bonds. Substances can be elements, compounds, ions or alloys. Thus, water is a substance, but a salt solution is not. In geology, naturally occurring substances are called minerals, while physical mixtures (aggregates) of minerals are called rocks. Substances can exist as solids, liquids, gases or plasmas.[†] These are the *states of matter*. The state of a substance may change with changes in temperature or pressure;

---

* Lewis Carroll (1871) *Through the Looking-Glass, and What Alice Found There*. London, Macmillan.
[†] A plasma is a gaseous mixture of electrons and positive ions formed by heating a gas or by subjecting a gas to a strong electromagnetic field. Unlike the other three states of matter, plasma does not exist naturally on Earth.

for example, ice (the solid state) can melt to form water (the liquid state), and water can evaporate (the gaseous state). Substances may combine with, or be converted to, other substances by means of chemical reactions.

A *gas* is a substance that fills any container it occupies. The atoms or molecules of a gas are free to move independently of one another, and thus a gas will expand or contract to fill the available space. The pressure of a gas is due to the impact of the atoms or molecules on the walls of the container. In *liquids*, the molecules have less freedom of movement. They are free to move relative to one another but maintain their mutual attraction sufficiently to have a fixed volume at a particular temperature. Hence, they take the shape of the container but maintain a fixed volume. In *solids*, the movement of atoms or molecules is even more restricted. In this state the attractive forces are strong enough to bind the atoms or molecules into a regular arrangement. Solids, therefore, have a fixed volume and shape. If the atoms or molecules of a substance are arranged in a definite pattern or lattice it is said to be *crystalline*. If the atoms or molecules are not so arranged the substance is said to be *amorphous*.

In liquids, the molecules (or atoms or ions) comprising it are in constant random motion. The molecules have a distribution of kinetic energy, and some that are near the surface and travelling towards it will have sufficient kinetic energy to overcome the attractive forces of the surface molecules and escape and diffuse away. These form a *vapour* and the process of forming a vapour is called *evaporation* or *vapourisation* (Figure 2.1a). Evaporation of water and gasoline are common everyday examples. Evaporation also occurs with solids. Common examples are the evaporation of naphthalene (moth balls), as evidenced over time by the decreasing size of the piece of naphthalene and the associated odour of the vapour; the shrinkage over time of ice cubes in an ice cube tray in a freezer and of 'dry' ice (frozen carbon dioxide); and the eventual failure of the filament in incandescent light bulbs. In solids the vibration of the lattice molecules (or atoms or ions) is responsible for dislodging surface molecules with the highest kinetic energy. These enter the vapour state directly from the solid state. Such evaporation is called *sublimation*.

If evaporation of a solid or liquid occurs into an empty, enclosed space (Figure 2.1b), the accumulation of vapour molecules will result in an increase in the number of molecules returning to the liquid or solid since the molecules in the vapour are moving randomly and colliding. When the state is reached at which the rate of return of molecules equals the rate of escape, the vapour is said to be *saturated* with respect to the evaporating entity, and a state of dynamic equilibrium is achieved. The *saturated vapour pressure*, or more commonly *vapour pressure*, of a substance is defined as the pressure exerted by its vapour in equilibrium with the solid or liquid state of the substance at a particular temperature. The stronger the mutual attraction of the atoms or molecules of a substance the lower will be its vapour pressure. The vapour pressure of substances increases with temperature because the average kinetic energy of the molecules (or atoms or ions) comprising it increases with

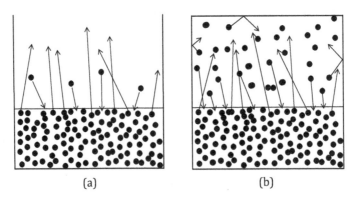

(a)   (b)

FIGURE 2.1 Schematic illustrating the evaporation of a solid or liquid at the microscopic level: (a) free evaporation (open system), (b) evaporation in a closed system.

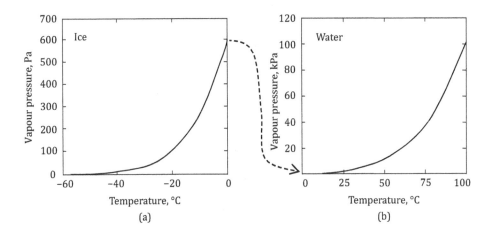

FIGURE 2.2 The variation of the vapour pressure of H_2O with temperature: (a) solid ice, (b) liquid water. Note the different pressure scales (a) Pa, (b) kPa.

increasing temperature. This is illustrated in Figure 2.2 which shows the variation with temperature of the vapour pressure of water in solid and liquid form.

In an open system, when the vapour pressure of a liquid becomes equal to the external pressure vapourisation can then occur throughout the liquid (rather than just at the surface), and the vapour can expand into the surroundings (rather than just diffuse into the surroundings). Vapourisation becomes rapid, and the substance is said to boil. The temperature at which this occurs is called the *boiling point*. For example, the boiling point of water is 99.97°C at a pressure of 1 atm (1.01325 bar) and 99.63°C at 1 bar pressure. Further addition of heat does not raise the temperature but increases the rate of evaporation.

2.2.1 AMOUNT OF SUBSTANCE

In everyday usage, we express the amount of a substance in terms of mass in units such as kilograms or grams. However, in chemistry and the related discipline of thermodynamics the unit of the amount of a substance is the mole. The *mole* is defined as the amount of a substance which contains as many elementary entities (atoms, molecules, ions, etc.) as are contained in 12 g of ^{12}C. This number is approximately $6.022\ 14 \times 10^{23}$ (Avogadro's constant, L). The abbreviation for mole is mol. To avoid confusion, it is necessary to specify the entity when reporting the amount of a substance. Thus, the statement '1 mol nitrogen' is misleading. It is necessary to specify either 1 mol N (14.00 g) or 1 mol N_2 (28.00 g). The mole is the unit for reporting the thermodynamic quantity called *amount of substance, n*. For example, we could say the amount of substance FeO in a sample is 2.62 moles. However, it is more usual to simply say the *amount* of FeO is 2.62 mol or the number of moles of FeO is 2.62. The amount of substance is proportional to the number (N_B) of elementary entities of the substance. The proportionality constant is the reciprocal of the Avogadro constant, that is,

$$n_B = \frac{N_B}{L}$$

The amount of substance is related to the mass of substance by the relation:

$$n_B = \frac{m_B}{M_B} \tag{2.1}$$

where m_B is the mass (kg) and M_B is molar mass (kg mol^{-1}) of substance B, respectively. The *molar mass* of an element is numerically equal to the standard relative atomic mass of the element, and the molar mass of a compound or ion is numerically equal to the sum of the standard relative atomic mass of the atoms which form the compound or ion.

EXAMPLE 2.1   Amount of substance

i. 1 mol of H_2 contains approximately 6.022×10^{23} molecules of H_2 and 12.044×10^{23} atoms of H.

ii. 1 mol of NaCl has a mass of $23.00 + 35.45 = 58.45$ g.

iii. 2 mol of $CuSO_4$ has a mass of $2 \times (63.55 + 32.06 + 4 \times 16.00) = 143.61$ g.

iv. 100 g Fe: $n_{Fe} = \dfrac{100}{55.85} = 1.79$ mol.

v. 255 g FeS_2: $n_{FeS_2} = \dfrac{255}{55.84 + (2 \times 32.06)} = 2.13$ mol.

2.3   SYSTEMS

We obtain information about phenomena by carrying out experiments and making observations. When doing this it is necessary to define exactly what is being studied. It is in this context that a *system* is defined as that part of the universe which is under consideration. The rest of the universe is called the *surroundings*. The system is separated from the surroundings by a *boundary,* which may be an actual physical boundary or a conceptual boundary drawn around the system of interest. The concept is illustrated in Figure 2.3. If we are investigating a solution contained in a glass beaker, for example, the boundary of the system could be defined as the extremities of the solution being studied or the walls of the beaker containing it. A chemical reaction can be considered as a system in which case the system consists of specified amounts of the reactants and products.

Systems can be classified according to the nature of their boundary (Figure 2.4). An *isolated system* has a boundary that prevents the transfer of both matter and energy. If the forms of energy are restricted to work and heat, terms that will be defined in Section 2.6, then a sealed vacuum flask approximates an isolated system – the seal prevents the transfer of matter, the rigid structure prevents work being performed by expansion or contraction against an external pressure, and the mirrored walls and vacuum minimise the exchange of heat. A *closed system* has a boundary that prevents the transfer of matter but allows exchange of energy. The matter may be contained within

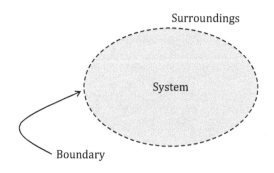

FIGURE 2.3   A system and its surroundings.

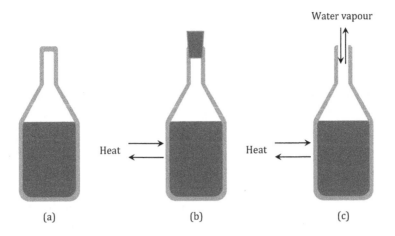

FIGURE 2.4 A system consisting of water: (a) an isolated system – no transfer of matter or energy, (b) a closed system – transfer of energy but not matter, (c) an open system – transfer of both matter and energy.

a rigid physical boundary, such as the walls of a closed vessel. Alternatively, it may be contained within an imaginary boundary defined by a closed volume in space or simply by a boundary defined as the extremities of the matter itself, as is the case when the components in a chemical reaction are defined as the system. An *open system* has a boundary that permits free exchange of both matter and energy. This may be contained within a rigid physical boundary, such as the walls of an open vessel, or it may be contained within fixed dimensions in space, such as a specified volume (for example, 5 L) but with no physical boundaries. In practice, isolated systems are of little importance, and most real processes occur in either closed or open systems.

Each physically or chemically distinct, uniform and physically (or mechanically) separable part of a system is called a phase. In particular, a *phase* is a system or a part of the system in which the properties are the same throughout. A phase may be solid, liquid or gaseous. Distilled water is a phase, as is water containing dissolved salts. An example of a system containing two phases is water in contact with its vapour. Other examples are droplets of water in air (fog, mist) and an emulsion of oil droplets in water. Even though in both latter cases there are many droplets there are still only two phases. A *homogeneous system* contains only a single phase, and a *heterogeneous system* contains two or more phases.

A *component* is a chemically independent constituent of a system. The components of a system define the composition of the system. For thermodynamic purposes, the choice of components is often arbitrary, though the number is not. For example, a sulfuric acid solution consists of two components, usually designated as H_2SO_4 and H_2O. But SO_3 and H_2O could also be considered as the components since SO_3, H_2SO_4 and H_2O are related stoichiometrically through the chemical equation

$$SO_3 + H_2O = H_2SO_4$$

so if the amounts of SO_3 and H_2O are known, the amount of H_2SO_4 can be calculated.

A *solution* is a *single-phase* system *consisting of two or more components*. A solution can be solid, liquid or gaseous. For example, sea water is a solution of water containing a range of dissolved salts, and air is a solution of nitrogen, oxygen and small amounts of other gases. A *mixture* is a physical combination of two or more substances, without chemical bonding or other chemical change, in which the identities of the substances are retained. Mixtures, therefore, are heterogeneous systems. For example, suspensions and colloids are mixtures.

The concept of *ideal systems* is often invoked in thermodynamics. An ideal system is a simplification which allows it to be treated theoretically in a rigorous manner. The behaviour of ideal systems simulates the behaviour of real systems under certain limiting conditions and may approximate its real behaviour over some range of conditions. Ideal systems can provide a convenient reference against which the behaviour of real systems may be compared. For example, the pressure–temperature–volume properties of most gases are described quite well by the ideal gas law over a wide range of conditions. Another commonly invoked ideal system is the ideal solution. This is one in which the atoms or molecules comprising it have equal attraction to each other so there is no preferential associations of the atoms or molecules – they are randomly mixed. While very few actual solutions are ideal, some approach ideal behaviour while others deviate from ideal behaviour considerably. In any case, ideal solutions provide a basis for comparing and assessing the thermodynamic behaviour of real solutions.

2.3.1 COMPOSITION OF SYSTEMS

When more than one component is present in a system the state of the system is not completely specified unless the relative amounts of the components are known. Of the various methods possible – mass percent, volume percent, kilograms per cubic metre, etc. – in most cases mole fraction is the most useful for expressing composition for thermodynamic purposes. Consider a system consisting of components A, B, … . The *mole fraction* of component B is defined as:

$$x_B = \frac{n_B}{n} \tag{2.2}$$

where n is the total number of moles of components and is given by

$$n = n_A + n_B + \cdots$$

For the system A, B, …

$$x_A + x_B + \cdots = \frac{n_A}{n} + \frac{n_B}{n} + \cdots = \frac{n_A + n_B + \cdots}{n} = 1 \tag{2.3}$$

Frequently, the composition of solid and liquid systems is given in mass percent for everyday purposes, and it is necessary to convert this to mole fraction for thermodynamic purposes. This is achieved using the following relation:

$$x_B = \frac{w_B / M_B}{w_A / M_A + w_B / M_B + \cdots} \tag{2.4}$$

Conversion back from mole fraction to mass fraction, if required, is accomplished using the equation:

$$w_B = \frac{x_B M_B}{x_A M_A + x_B M_B + \cdots} \tag{2.5}$$

The algebraic derivation of these relations is straightforward and is left as an exercise for the reader.

The composition of gases is most frequently expressed in terms of volume percent or volume fraction. However, as will be seen in Section 3.3.4, for ideal gas mixtures the volume fraction is identical to the mole fraction, that is,

$$x_B = \phi_B \tag{3.4}$$

where ϕ_B … is the volume fraction of B in the gas mixture.

The concentration of dilute solutions may be expressed in a variety of ways: The mass percent or parts per million (ppm) of the solute.* For thermodynamic calculations involving dilute aqueous solutions, however, composition is always expressed in terms of *molality, m,* defined as the number of moles of solute per kilogram of solvent, that is,

$$m_B = \frac{n_B}{m_s} \tag{2.6}$$

If the molar mass of the solvent is M_s the molality of the solute B is related to its mole fraction by:

$$m_B = \frac{x_B}{x_s M_s} \tag{2.7}$$

Molality should not be confused with molarity which is the number of moles of solute per litre of solution. The molality of a solution is independent of temperature, but molarity decreases with increasing temperature (because water expands with increasing temperature). The molality and molarity of a dilute aqueous solution are nearly the same at ambient temperatures since 1 kg of water (solvent) has a volume of very nearly 1 L, and the presence of a small amount of solute has little effect on the volume.

EXAMPLE 2.2 Calculate the composition of a system in terms of mole fraction

i. A mixture consists of 500 g SiO_2, 400 g Al_2O_3 and 200 g CaO. Express its composition in terms of mole fraction of the components.

SOLUTION

$$n_{SiO_2} = \frac{500}{60.1} = 8.319 \text{ moles}$$

$$n_{Al_2O_3} = \frac{400}{102.0} = 3.922 \text{ moles}$$

$$n_{CaO} = \frac{200}{56.1} = 3.565 \text{ moles}$$

$$n = 8.319 + 3.922 + 3.565 = 15.806 \text{ moles}$$

Therefore,

$$x_{SiO_2} = \frac{8.319}{15.806} = 0.526;$$

$$x_{Al_2O_3} = \frac{3.922}{15.806} = 0.248;$$

$$x_{CaO} = \frac{3.565}{15.806} = 0.226$$

As expected, the sum of the mole fractions is 1.000.

ii. A mixture consists of 50 mass% SiO_2, 30 mass% Al_2O_3 and 20 mass% CaO. Express its composition in terms of mole fraction of the components.

---

* 1 ppm is equal to $\frac{1}{1\,000\,000} \times 100 = 0.0001 \text{ mass\%}$.

SOLUTION

Consider 100 g of the mixture, then

$$n_{SiO_2} = \frac{50}{60.1} = 0.832 \text{ moles}$$

$$n_{Al_2O_3} = \frac{30}{102.0} = 0.294 \text{ moles}$$

$$n_{CaO} = \frac{20}{56.1} = 0.357 \text{ moles}$$

$$n = 0.832 + 0.294 + 0.357 = 1.483 \text{ moles}$$

Therefore,

$$x_{SiO_2} = \frac{0.832}{1.483} = 0.561;$$

$$x_{Al_2O_3} = \frac{0.294}{1.483} = 0.198;$$

$$x_{CaO} = \frac{0.357}{1.483} = 0.240$$

Again, as expected, the sum of the mole fractions is 1.000.

As a check, the mole fractions can be converted back to mass fractions using Equation 2.5:

$$w_{SiO_2} = \frac{x_{SiO_2} \, M_{SiO_2}}{x_{SiO_2} \, M_{SiO_2} + x_{Al_2O_3} \, M_{Al_2O_3} + x_{CaO} \, M_{CaO}}$$

$$= \frac{0.561 \times 60.1}{(0.561 \times 60.1) + (0.198 \times 102.0) + (0.240 \times 56.1)} = \frac{33.725}{67.450} = 0.500$$

Similarly,

$$w_{Al_2O_3} = \frac{0.198 \times 102.0}{67.450} = 0.300$$

$$w_{CaO} = \frac{0.240 \times 56.1}{67.450} = 0.200$$

iii. A gas mixture consists of 30 vol% O_2, 25 vol% N_2 and 45 vol% H_2. Express its composition in terms of mole fraction of the components.

SOLUTION

From Equation 3.4, $x_B = \phi_B$,

$$x_{O_2} = 0.30; \quad x_{N_2} = 0.25; \quad x_{H_2} = 0.45$$

2.3.2 Macroscopic and microscopic properties

Consider a closed system consisting of a gas contained in a sealed balloon in which the gas is the system and the fabric of the balloon is the boundary. A complete description of this system could involve knowing the position and velocity of every gas molecule at any given time. While this information does characterise the system exactly, the amount of data required is enormous. Alternatively, we could describe the system in terms of average or system properties such as mass, pressure, volume, temperature and other properties such as enthalpy, entropy and Gibbs energy which are less familiar (but which form the basis of much of this book). Thus a system may be studied from either of two perspectives: The *microscopic* and the *macroscopic*. The microscopic approach considers the behaviour of every molecule by using statistical methods. This is the basis of the discipline of *statistical mechanics*. The macroscopic approach is concerned with the gross or average effects of many molecules and their interactions. The latter approach, known as *classical thermodynamics*, greatly reduces the complexity of the problem. In the classical thermodynamic approach it is not necessary to know the nature of the matter (its atomic or molecular level structure), only what phenomena are possible. This is the approach adopted in this book although microscopic aspects are discussed where these can provide greater insights into particular concepts.

Macroscopic properties of systems are of two types:

Extensive properties. These are additive properties. The value of an extensive property of a system is directly proportional to the amount of material in the system. Mass and volume are extensive properties. For example, if a bar of copper is cut in two both the mass and the volume of a piece are directly proportional to the amount that is left after cutting the bar. Enthalpy, entropy and Gibbs energy are also extensive properties.

Intensive properties. These are bulk properties. The value of an intensive property of a system does not depend on the size of the system or the amount of material in the system. Examples include temperature, pressure, refractive index, density and hardness of a substance. For example, if a bar of copper is cut in two, each piece has the density and hardness of copper. The ratio of two extensive properties is itself an intensive property. For example, the ratio of the extensive properties mass and volume, the density, is an intensive property.

To distinguish between the actual value of an extensive property and its value based on 1 mole we adopt the following conventions, where X is an extensive property:

Convention: X refers to the total value for the system.
               X_m refers to the value for 1 mole (the molar value).

X_m is called a *molar quantity*. For example, the molar volume of water, $V_m(H_2O)$, at 277 K is 18.016 mL. For a system consisting of components A, B, ... , the relationship between the total quantity and molar quantity is:

$$X_m = \frac{X}{n_A + n_B + \cdots} \tag{2.8}$$

2.3.3 The concept of equilibrium

An *equilibrium state* is one in which the macroscopic properties of a system have unique values which do not change with time. In other words, it is sufficient to specify a single value for the pressure, another single value for the volume, temperature and so on, in order to fully characterise the system. More precisely, a system is in equilibrium when it is in simultaneous mechanical

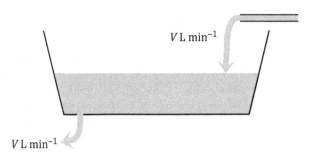

FIGURE 2.5   A steady-state process. Water flows into the bath at the same rate it flows from the plug hole so the water level remains constant over time. The temperature may vary along the length of the bath and over the depth of the water but remains constant with time at each point in the bath.

equilibrium (no imbalance of forces), thermal equilibrium (no temperature gradients) and chemical equilibrium (no concentration gradients or on-going chemical reactions). In essence then, equilibrium is a concept associated with the absence of any tendency for spontaneous change when the system is isolated.

It is possible for systems to be unchanging, and exhibit no tendency to change, but not be at equilibrium. Steady-state and metastable systems are of this type. A *steady-state* system is one in which all properties are constant despite ongoing processes that strive to change them. For a system to be at steady state there must be a flow of mass or energy through the system. To illustrate a steady-state system, think of a bathtub with the tap open and with the bottom outlet open, as in Figure 2.5. If water flows in and out at the same rate so the water level remains constant this is a steady-state system. In a steady-state system, the values of intensive properties (temperature, density, etc.) may vary from point to point but will remain unchanged with time at any given point. Many natural and industrial processes (those that operate with an approximately constant flow of inputs and outputs) can be considered to be steady-state for some purposes.

If a system is not in its most stable state and there are no flows of any kind, it is said to be *metastable*. For example, diamond is a metastable form of carbon at ambient* temperature and pressure. It can be converted to graphite, the stable form, but only after overcoming an activation energy barrier. Metastability occurs as a result of a local minimum in a system's energy. A more familiar example is super-heated water, that is, water heated above 100°C at atmospheric pressure but without boiling. Such water is in an unstable state, and something which causes water vapour bubbles to nucleate within the liquid will result in the water starting to boil, even in the absence of additional heat. This is sometimes encountered when water is heated in a container in a microwave oven. When the vessel containing the water is removed or the water is stirred nucleation of water vapour bubbles may occur, and the water begins to boil, sometimes with catastrophic results.

2.4   PROCESSES

In thermodynamics, a *process* is said to occur when a system changes from one state of equilibrium to another, for example, an ice cube in a glass (state 1) melting to form water in the glass (state 2) or a piece of zinc reacting with sulfuric acid to form hydrogen and zinc sulfate. Processes are classified according to the conditions under which they occur as follows:

- An *isothermal process* is one that occurs at constant temperature.
- An *adiabatic process* is one that occurs with no heat exchange with the surroundings.

---

* The word ambient means 'relating to the immediate surroundings'. Ambient temperature is usually taken as 25°C and ambient pressure as 1 atm.

- An *isobaric process* is one that occurs at constant pressure.
- An *isochoric process* is one that occurs at constant volume.

Processes either take place of their own accord (that is, they occur naturally) or occur because of some external influence. Changes that have a natural tendency to occur are called *spontaneous changes*. A *spontaneous process* is defined as a change in a system the cause of which cannot be traced to a phenomenon originating in the surroundings, that is, one acting on the system at the boundary. It occurs by itself due to causes wholly within the system itself. For example, a cup of hot coffee will cool naturally (spontaneously); a piece of zinc metal placed in dilute sulfuric acid will react with the acid. All processes occurring within an isolated system are spontaneous since, by definition, any change that occurs cannot be due to an external influence. Since the universe can be viewed as an isolated system, all naturally occurring processes within the universe are spontaneous. On the other hand, *non-spontaneous changes* must be made to occur by the external application of energy. For example, a cup of cold coffee can be reheated by the external application of heat; the zinc dissolved in sulfuric acid can be recovered and the acid regenerated by electrolysis involving the application of electrical energy.

The *path* of a process is the series of states through which a system passes during the process. Consider the expansion of a gas in a closed system in which the gas (the system) is contained in a perfectly insulated cylinder fitted with a frictionless piston, as shown in Figure 2.6. Let the initial equilibrium state of the system be 1 and the final expanded equilibrium state be 3. Now consider state 2, one of the intermediate states which occur just after the partial withdrawal of the piston. The region just behind the piston is subjected to reduced pressure (because of the increased volume), and because of this the pressure within the system (the gas) is no longer uniform and cannot be represented by a single number. State 2 is thus a non-equilibrium state. By the same reasoning, all the states which make up the path of the process are non-equilibrium, and the path is thus a non-equilibrium path. This example illustrates an important principle: *All real processes proceed via paths consisting of non-equilibrium states.*

Now consider what happens when the path taken by a process involves a sequence of infinitesimally small steps so each step can occur at equilibrium. In such a process, an infinitesimal change

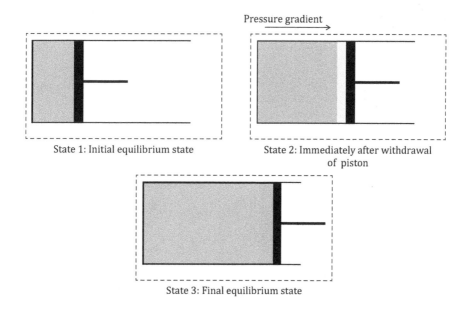

FIGURE 2.6 Schematic illustration of the expansion of a gas in an enclosed space.

in conditions at the system boundary can cause the process to return to its previous state. Such processes are said to be *reversible*. There are no actual reversible processes; a reversible process would be infinitely slow. In contrast, all real processes proceed at a finite rate and are said to be *irreversible. Since spontaneous processes occur at a finite rate, it follows that spontaneous processes are thermodynamically irreversible.*

Chemical reactions can be made to go in the forward and reverse directions, but a chemical reaction is reversible in the thermodynamic sense only if at all times the reaction remains at equilibrium, which would require an infinitely long time. Reversibility is a hypothetical concept used to examine processes carried out under equilibrium conditions.

2.5   STATE FUNCTIONS AND PATH FUNCTIONS

Real processes proceed via a pathway of non-equilibrium states, and we can't describe such states by single values of temperature, pressure, etc., in the same way that we can if the pathway is made up of a series of equilibrium states since we usually don't know much about the pathways taken by real processes. State functions allow us to analyse the changes that occur during real processes without the need to know the path taken.

The concept of state function can be illustrated using a simple analogy. Suppose the process of interest is that of climbing from point 1 to point 2 on a hill (Figure 2.7). Regardless of the path chosen (a, b, c or any other), once point 2 is reached the change in height (ΔX) is always constant, that is,

$$\Delta X = X_2 - X_1 \tag{2.9}$$

where X_1 and X_2 are the heights of points 1 and 2 above sea level, respectively. In this sense, X is analogous to a state function. *The symbol Δ (Greek letter capital delta) is used to represent a change in value of a function.* In contrast, consider the actual distance travelled while proceeding up the hill. This value is neither a characteristic of point 1 nor point 2 and will depend on the path chosen. Such functions are called *path functions*. In thermodynamic systems, heat and work are path functions since their value depends on how the system changes from the initial to the final state. Thus, the change in a path function cannot be determined without knowledge of the path, but the change in a state function is constant (for a given initial and final state) regardless of the path.

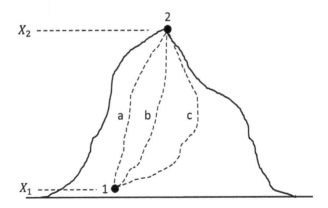

FIGURE 2.7   The concept of state function. The value of $\Delta X = X_2 - X_1$ is always the same irrespective of the path taken from X_1 to X_2.

State functions (also called *state variables*) are functions that describe the state of a system at equilibrium. They are independent of how the system arrived at the state, and their values depend only on the prevailing system conditions, that is, for the state function X,

$$X = X(T, p, n_A, n_B, \ldots) \tag{2.10}$$

where T and p are the temperature and pressure of the system and n_A, n_B, ... are the moles of components A, B, ... When a system undergoes a process by which it changes from equilibrium state 1 to equilibrium state 2, the change in a given state function is constant regardless of how the process is carried out (that is, the path taken). Many thermodynamic properties are functions of the state of the system, the most important being:

p pressure
T temperature
V volume
U internal energy
H enthalpy $(= U + PV)$
S Entropy
A Helmholtz energy $(= U - TS)$
G Gibbs energy $(= H - TS)$

If two observations are made on a system, and if any state function has a different value on each observation, then that system is said to have undergone a *change of state*. If X_1 is the value of a state function (X) in state 1 and X_2 is the value of the same function in state 2, then the change in the value of the state function when the system changes from equilibrium state 1 to equilibrium state 2 is:

$$\Delta X = X_2 - X_1 = \int_{X_1}^{X_2} dX \tag{2.11}$$

This means that if we sum all the incremental changes in X (that is, dX) in going from state 1 to state 2 the sum will be ΔX, and it will have the same value no matter what path is taken. Mathematically, since the value of X is independent of the path, dX is said to be an *exact differential*. The differential of a path function, on the other hand, is an *inexact differential*. To understand the difference, consider a quantity of heat (a path function) transferred to a system. The total amount of heat transferred is the sum of all the incremental contributions along the heating path, that is,

$$q = \int_{q_1}^{q_2} f(q) dq$$

where $f(q)$ is a function describing the heating path. Different functions may describe the path between q_1 and q_2. Mathematically, dq is said to be an inexact differential. To distinguish inexact differentials from exact differentials inexact differentials are written as δq, δw, etc.

Summary

- *Changes in the values of state functions are obtained by subtracting the value in the initial state from the value in the final state. Their difference is represented by ΔX.*
- *The usefulness of state functions is that differences between them can be evaluated and used without any knowledge of the way in which that difference was brought about.*
- *The value of path functions depends on the path taken and can be evaluated only if the path is known.*

2.5.1 THE STANDARD STATE

The numerical value of state functions for a substance depends on the *state of aggregation* of the substance. The most common forms of aggregation and the symbols used are listed in Table 2.1. The state of aggregation must be specified for the value of a state function to have any meaning, and this is indicated by appending the appropriate symbol in parentheses after the chemical symbol or formula for the substance, for example Fe(s), SiO_2(cr), Al(l), O_2(g), NaOH(aq).

The state of aggregation of a substance used for thermodynamic purposes is the *standard state*. Because any state could be chosen as the standard state, the standard state must be defined in the context of use. A physical standard state is one that exists for, at least, time that is sufficient to allow measurement of its properties. The most common physical standard state is the form of the substance that is stable at the temperature of interest at a pressure of 1 bar.* Thus, liquid water at 1 bar would normally be chosen as the standard state for water at temperatures between 0 and 100°C and water vapour at 1 bar at temperatures greater than 100°C. A non-physical standard state is one for which properties are obtained by extrapolation from a physical state – for example, a liquid supercooled below its melting point or an ideal gas at a condition where the real gas is non-ideal.

In practice, the International Union of Pure and Applied Chemistry (IUPAC) recognises four standard states. These are briefly summarised below; their significance and applications will be discussed in subsequent chapters.

Standard state for gases. The standard state is the (hypothetical) state of the pure substance in the gaseous phase at a pressure of 1 bar, assuming ideal behaviour.

Standard state for solid or liquid substances, mixtures of substances and for solvents. The standard state of a pure substance or solvent is the state of the substance or solvent at a pressure of 1 bar.

Standard state for solutes in solution. The standard state is the (hypothetical) state of the solute at a molality of 1 mol kg^{-1} of solvent and pressure of 1 bar and exhibiting infinitely dilute solution behaviour. Alternatively, other measures of concentration, such as mass percent or mole fraction can be used instead of molality.

Biochemical standard state. The standard state is chosen to be $[H^+] = 10^{-7}$ mol L^{-1}. The concentration of the solutes may be grouped together, for example, the total phosphate concentration rather than the concentration of each component.

Standard conditions of temperature and pressure (abbreviated to STP) are sometimes used to permit comparisons to be made between sets of data. These should not be confused with standard state as defined above as they serve a different purpose. The most used standards for comparison purposes are those of IUPAC and the US National Institute of Standards and Technology (NIST), although these are not universally accepted, and other organisations have alternative definitions.

TABLE 2.1

The symbols used to represent the states of aggregation of chemical species

a, ads	species adsorbed on a surface	l	liquid
am	amorphous solid	lc	liquid crystal
aq	aqueous solution	mon	monomeric form
aq, ∞	aqueous solution at infinite dilution	n	nematic phase
cd	condensed phase (that is, solid or liquid)	pol	polymeric form
cr	crystalline	s	solid
f	fluid phase (that is, gas or liquid)	sln	solution
g	gas or vapour	vit	vitreous substance

---

* The bar is not an SI unit but is accepted for use as an SI unit. One bar is defined as being equal to 10^5 Pa.

The IUPAC definition of STP is: 273.15 K (0°C) and an absolute pressure of 1 atm (1.013 25×10⁵ Pa). NIST uses a temperature of 20°C (293.15 K) and an absolute pressure of 1 atm. This standard is also called normal temperature and pressure (NTP). *Use of the terms STP and NTP, without an accompanying definition, should be avoided.* Many technical publications simply state standard conditions without specifying them, often leading to confusion and errors.

2.6 ENERGY, WORK, HEAT AND TEMPERATURE

The everyday meaning of the words energy, work, heat and temperature will be well known to all readers, but in scientific literature they have very specific meanings. An understanding of these is essential to developing an understanding of thermodynamics. We treat these terms together because they are closely related.

2.6.1 ENERGY AND WORK

The *energy* of a system is its capacity to do work. *Work w* is done when a force f acts upon an object to cause a displacement s of the object and is defined as

$$\delta w = f(s)\,ds$$

where $f(s)$ is a function describing the change of force along the path taken. Therefore,

$$w = \int_{s_1}^{s_2} f(s)\,ds$$

where w is the work done in moving from point 1 to point 2. Work is a path function – the work done in moving from point 1 to point 2 depends on the path taken, that is, on the function $f(s)$. In the simple case where the force applied is constant and movement is in the direction of the force (as, for example in lifting a mass vertically),

$$w = fs$$

The unit of energy (and work) is the Joule (J), which is defined as the energy expended in applying a force of one Newton through a distance of one metre. The Newton (N) is defined as the force required to accelerate a mass of one kilogram at a rate of one metre per second per second. *Power* is the rate of expending energy (or doing work) and is given by the relation

$$p = \frac{w}{t}$$

Power has the unit Joules per second, called the Watt. The kilowatt hour (kW h) is the unit of energy most commonly used for electrical energy. It is the billing unit for energy delivered to customers by electric utilities. One kW h is the energy required to operate a 1 kW device (for example, a motor or heater) for a period of 1 hour. One kW h, therefore, is equal to 1000×60×60 Joules, or 3.6 MJ.

Energy can exist in two fundamental forms – potential energy and kinetic energy. Other forms of energy such as thermal energy, chemical energy and nuclear energy are manifestations of kinetic or potential energy.

Potential energy is the energy stored in a body or a system due to its position in a force field (gravitational, electrostatic or magnetic) or its configuration (both microscopic and macroscopic). For example, when a coiled spring is stretched in one direction, it exerts a force in the opposite direction so as to return to its original state. Similarly, when a mass is raised vertically, the force of gravity acts to bring it back to its original position. The action of stretching the spring or lifting the mass requires energy. The energy that went into lifting the mass is stored in its position in the gravitational field, and the energy required to stretch the spring is stored in the metal. This energy is released when the mass or spring returns to its original state. At the microscopic level, chemical energy is the energy stored in the bonds between atoms and nuclear energy is the energy stored in the nucleus of atoms. This energy can be released through chemical reactions and nuclear fission or fusion, respectively.

Kinetic energy is the energy of an object due to its motion. It is equal to the work required to accelerate a body of given mass from rest to its current velocity. The body maintains this kinetic energy until its velocity changes. The same amount of work is required to decelerate the body from its current velocity to a state of rest. The kinetic energy of a non-rotating body of mass m travelling at a velocity v is equal to $\frac{1}{2} mv^2$. The microscopic kinetic energy associated with the movement of atoms and molecules comprising a substance manifests as thermal energy (Section 2.6.2). Electrical energy is a form of kinetic energy, being the manifestation of the movement of electrons through a substance.

Absolute values of energy cannot be determined; rather all energies are assigned values relative to a reference point. For example, for the macroscopic kinetic energy of a system the velocity would usually be expressed relative to the earth, which for many applications can be considered as being stationary. The gravitational potential energy of a body is usually expressed relative to its height above a point on the Earth.

Energy can be converted from one form to another in various ways. A vehicle rolling down a hill gains kinetic energy as its velocity increases but loses potential energy as its height decreases – its potential energy is being converted into kinetic energy. Electrical energy can be produced by many kinds of devices which convert other forms of energy into electrical energy, for example, photovoltaic cells (thermal energy), turbines driven by running (kinetic energy) or falling water (gravitational potential energy), fuel-burning internal combustion engines and gas turbines (chemical energy) and batteries and fuel cells (chemical energy).

2.6.2 HEAT

One way of increasing or decreasing the energy of a system is by the transfer of energy as heat. *Heat can be thought of as energy in the process of being transferred from one object to another due to a difference in temperature.* The energy flows from the hotter body to the cooler body. Heat is the macroscopic manifestation of the *total* kinetic energy of the atoms and molecules making up a body or a system. Atoms and molecules are in constant motion in all substances, and the sum of the kinetic energies of all the atoms and molecules making up a system determines the thermal energy (or heat content) of the system.

2.6.3 TEMPERATURE

Temperature is the degree of hotness of a system. It is the macroscopic manifestation of the kinetic energy of the individual atoms and molecules in the system. Except at extremely low temperatures, the temperature of a substance is directly proportional to the average kinetic energy of the translational motion of the particles. The faster these move, the greater their average kinetic energy and the higher the temperature. Thus, heat and temperature are related but different concepts.

The scientific concept of temperature is based on the empirical observation that if bodies A and B are each in thermal equilibrium with a third body, then they are in thermal equilibrium with each other and at the same temperature. This relationship is known as the *zeroth law of thermodynamics*. It is also an empirical observation that when two bodies are brought into contact without mechanical, chemical, electrical or magnetic interactions and without any other change, the temperatures of the bodies converge to the same value. This is a consequence of the second law of thermodynamics. Together, these laws provide the basis for measuring temperature using a device called a *thermometer.* Thermometers make use of a material with a measurable property that varies in a consistent way with its degree of hotness. A scale against which to measure this variation is also required. This has two components: A zero point and a magnitude of the temperature interval (often called the *degree*). Thus, if we measure the value of a property for a chosen material when it is in contact with a second body, we may determine the temperature of the second body using the scale (Figure 2.8). The main types of thermometers are listed in Table 2.2.

The most common temperature scale, the Celcius* scale, was established originally by arbitrarily letting the temperature of water in equilibrium with ice at atmospheric pressure be 0°C and the temperature of boiling water (in equilibrium with steam) at atmospheric pressure be 100°C. The interval was then divided into 100 equal units, called degrees. However, a more precise definition is

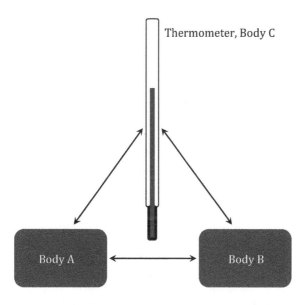

FIGURE 2.8 The principle of the thermometer. If thermometer C is placed in contact with body A it will come into thermal equilibrium with it, and the temperature of A can be measured. If the thermometer is then placed in contact with body B, it will come into thermal equilibrium with it. If the measured temperatures of A and B are the same, then bodies A and B are in thermal equilibrium with each other.

TABLE 2.2
Various types of devices for measuring temperature

Thermometer	Thermometric Property
Liquid-in-glass	expansion
Gas (constant V)	pressure
Electrical resistor	electrical resistance
Thermocouple	potential difference
Paramagnetic salt	magnetic susceptibility
Black body radiation	spectral radiance

* Anders Celsius (1701–1744) was a Swedish astronomer, physicist and mathematician and a professor at Uppsala University. He proposed the temperature scale which now bears his name in a paper to the Royal Society of Sciences in Uppsala in 1742.

used today. At sufficiently low pressures, the volumes of all gases held at a particular constant pressure change in a consistent way with change in temperature – the volume increases with increasing temperature and decreases with decreasing temperature. This behaviour predicts a zero volume for all gases at −273.15°C. Since negative volumes are not possible (atoms occupy space!), this temperature must represent an absolute value of zero and not an arbitrary one as for the Celsius scale. The scale based on absolute zero is called the *thermodynamic temperature scale*. This scale employs a unit, called the Kelvin, which has been given the same magnitude as the Celcius degree. The scale was proposed by William Thompson (later Lord Kelvin)* from calculations based on the second law of thermodynamics. Kelvin's theory gives an absolute measure of temperature since it is totally independent of any variation in thermometer type.

In SI units, the Celsius degree unit and the Celsius scale are defined by two points: Absolute zero, defined as being exactly −273.15°C, and the triple point† of specially prepared water (Vienna Standard Mean Ocean Water or VSMOW), defined as being 0.01°C. Similarly, these points are used to define the thermodynamic temperature scale. Absolute zero is defined as exactly 0 K (−273.15°C), and the triple point of water is defined as being exactly 273.16 K (0.01°C). It follows that the relationship between Kelvin and Celsius temperatures is given by:

$$T(\text{K}) = \theta(°\text{C}) + 273.15 \qquad (2.12)$$

In thermodynamic equations (and calculations) the temperature is always expressed in Kelvin.

Note, temperatures in Kelvin are indicated without using the term degree, either as the word or the symbol. So, we write, for example, 273.15 K (with a space between the temperature value and the unit) and say 271.15 kay. For Celcius temperature we write, for example, 25°C and say 25 degrees Celcius.

The values of temperatures measured by thermometers depend on the type of thermometer used and how its particular property varies with temperature. Small variations occur between readings from different types of thermometers. This has been overcome by the development of a standard temperature scale against which all other temperature measuring devices can be calibrated. This scale, the *International Temperature Scale (ITS−90)*, established in 1990, has been designed to reproduce the thermodynamic temperature scale as closely as possible throughout its range. Many different thermometer designs are required to cover the entire range, including helium vapour pressure thermometers, helium gas thermometers, standard platinum resistance thermometers and monochromatic radiation thermometers. For all practical purposes the International Temperature Scale is identical to the thermodynamic temperature scale.

PROBLEMS

2.1   Which of the following properties of a system are extensive properties, and which are intensive? Explain.
   i.   The density of iron.
   ii.   The gravitational potential energy of a system.
   iii. The volume of air in a room.

---

* William Thomson, first Baron Kelvin (1824–1907), was a British mathematical physicist and engineer. At the University of Glasgow he worked on the mathematical analysis of electricity and formulation of the first and second laws of thermodynamics and did much to unify the emerging discipline of physics in its modern form.
† The triple point of a substance (Section 13.4) is the temperature and pressure at which the solid, liquid and gaseous phases of that substance coexist in thermodynamic equilibrium. The triple point of water is exactly 273.16 K (0.01°C) and pressure of 611.657 Pa.

    iv. The concentration of HCl in an aqueous solution.

    v. The number of moles in a piece of copper.

    vi. The heat absorbed by bringing a beaker of water to boiling point.

2.2   Discuss each of the following situations, and assess whether it approximately represents an isolated, a closed or an open system.

    i. A river.

    ii. The interior of an unopened can of soda water.

    iii. The interior of a closed refrigerator that is switched on.

    iv. The interior of a closed refrigerator that is switched off.

    v. The Earth.

2.3   Calculate the number of moles in 100 g of each of the following substances: O_2, HCl, $CuSO_4$, $CuSO_4.5H_2O$, $PbCl_2$, C_2H_5OH.

2.4   Convert the following mass quantities to moles: 350 g of $FeSO_4$, 25 g of NH_3, 1.5 kg $Ca(OH)_2$; 1000 mL water; 100 mL C_2H_5OH (ethyl alcohol), $\rho = 790$ kg m^3.

2.5   How many molecules are in (a) 125 g of methane (CH_4), (b) 35 g of ethyl alcohol (C_2H_5OH)?

2.6   Express the composition of each of the following mixtures in (a) mass percent, (b) mole fraction:

    i. 25 g SiO_2 and 75 g $Ca(OH)_2$

    ii. 25 g SiO_2, 75 g $Ca(OH)_2$ and 30 g $MgCO_3$

    iii. 25 g SiO_2, 75 g $Ca(OH)_2$, 30 g $MgCO_3$ and 30 g $BaSO_4$

2.7   Which of the examples below represents a change in a state function? Explain.

    i. The work done in climbing from the bottom to the top of a staircase.

    ii. The change in the gravitational potential energy of an object when it is carried from the bottom to the top of a building.

    iii. The change in density of water in a pot when it is heated from 25 to 50°C.

    iv. The amount of heat released from a gas burner used to heat a beaker of water at 25 to 50°C.

2.8   How many, and what, phases are present in the following systems?

    i. A solution of salt in water.

    ii. A mixture of sand and water.

    iii. A beaker containing a solution of salt and water.

    iv. The Earth's atmosphere (assuming no solid or liquid particles in suspension)

    v. The contents of a bottle of soda water.

    vi. A drinking glass containing the recently poured contents of a bottle of soda water.

3 Gases

SCOPE

This chapter examines the physical behaviour of gases, first by assuming gases behave ideally, then considering the behaviour of real gases and how they differ from ideal gases.

LEARNING OBJECTIVES

1. *Understand the nature of gas pressure and the p−V−T relationship for ideal gases.*
2. *Understand the ideal gas law and be able to use it in calculations.*
3. *Understand why real gases deviate from ideal behaviour and how their properties are described quantitatively.*

3.1 INTRODUCTION

We begin the study of the thermodynamics of systems by considering gases, the simplest of the states of matter and the easiest to describe thermodynamically. This chapter looks at the physical behaviour of gases and gas mixtures. A *pure gas* is one made up of individual atoms (such as an inert gas like argon), molecules made from one type of atom (such as O_2 and N_2) or compound molecules made from a variety of atoms (such as water vapour, H_2O, and carbon dioxide, CO_2). A *gas mixture* is a mixture of two or more pure gases. Air, for example, is a gas mixture. The thermodynamic description of gases is developed in subsequent chapters, and it forms the basis for examining the thermodynamics of the other states of matter.

3.2 GAS PRESSURE

At the microscopic level, what distinguishes a gas from liquids and solids is the vast separation of the individual gas particles. These particles are in constant random movement with velocities that increase as the temperature increases. The force exerted by these particles on the walls of a container gives rise to the property of pressure of a gas. *Pressure p* is defined as the force applied perpendicular to the surface of an object per unit area over which that force is distributed: $p = F/A$. The SI unit of pressure is the Pascal (Pa), defined as a force of 1 Newton acting over an area of 1 square metre: $1\ Pa = 1\ N\ m^{-2}$. However, several other units are also used in thermodynamics, in particular the bar and the atmosphere (atm):

1 bar = 100 000 Pa = 100 kPa (by definition)
1 atm = 101 325 Pa (by definition)

The bar is used as the standard pressure for reporting values of thermodynamic data. For many practical purposes, atmospheric and bar pressures are interchangeable because the difference between them is small.

Pressure, as for temperature, can be expressed in absolute or relative terms. Scales for *absolute pressure* are based on a value of 0 Pa for a perfect vacuum. Scales for *relative pressure* set 0 Pa at a finite pressure, usually 1 atm. Gauges for measuring pressure are often calibrated to read zero at ambient pressure, so the absolute pressure of the gas is the measured (relative) pressure plus the ambient pressure (Figure 3.1). This pressure is often referred to as the gauge-pressure. Gauges for

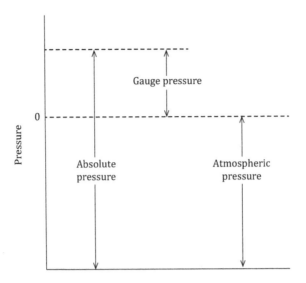

FIGURE 3.1   The relationship between absolute and gauge pressure of a gas.

measuring pressures less than atmospheric pressure, called vacuum gauges, however, are normally calibrated in absolute pressure units to avoid the need to use negative values.

Only absolute pressure has any significance in thermodynamics, and relative pressure values must be converted to absolute pressure for thermodynamic calculations.

3.3   IDEAL GASES

An *ideal gas* is a hypothetical gas defined as one in which all collisions between atoms or molecules are perfectly elastic and in which there are no intermolecular forces. It can be visualised as a set of perfectly hard spheres which can collide but which otherwise do not interact with each other. All the internal energy[*] of such a gas is in the form of kinetic energy (due to the movement of atoms or molecules), and any change in internal energy is thus accompanied by a change in temperature. The ideal gas concept is useful because ideal gases obey the ideal gas law and are amenable to analysis using statistical mechanics. Gases that deviate from ideal behaviour are known as *real gases*. Though the ideal gas is a hypothetical concept, in many situations many common gases approach ideal behaviour, and ideal behaviour can be assumed in many situations without introducing significant errors.

3.3.1   AVOGADRO'S LAW AND AVOGADRO'S CONSTANT

Avogadro's law states that, under the same conditions of temperature and pressure, equal volumes of all gases contain the same number of molecules. The law is named after Amedeo Avogadro who, in 1811, first hypothesised the relationship. The volume of an ideal gas is directly proportional to the amount of the gas, if the temperature and pressure are constant. Therefore, at constant temperature and pressure,

$$\frac{V}{n} = \text{constant} \qquad\qquad (3.1)$$

---

[*] The internal energy of a substance is the total energy of the system at the microscopic level and is the sum of the microscopic kinetic energy and microscopic chemical potential energy (this is discussed further in Section 4.2.1).

where V is the volume of the gas, n is the amount of gas (moles). In practice, real gases deviate to varying degrees from ideal behaviour, and, in this case, the law holds only approximately. It is now known that 1 mole of every substance, not only gases, contains the same number of atoms or molecules. This number is called the *Avogadro constant L* and is approximately 6.02214×10^{23}.

3.3.2 THE COMBINED GAS LAW

The combined gas law is a law that combines Boyle's law (1662), Gay-Lussac's law (~1700) and Charles's law (1787). These laws each relate one thermodynamic variable (temperature, pressure and volume) to another mathematically while holding the third variable and the amount of gas constant.

> *Boyle's law*: Pressure and volume are inversely proportional to each other at fixed temperature.
> *Gay-Lussac's law*: Pressure and temperature are directly proportional to each other at fixed volume.
> *Charles' law*: Volume and temperature are directly proportional to each other at fixed pressure.

The inter-dependence of pressure, volume and temperature is shown in the *combined gas law*, which, for a fixed amount of an ideal gas, is expressed mathematically as follows

$$\frac{pV}{T} = c \tag{3.2}$$

where c is a constant.

3.3.3 THE IDEAL GAS LAW

The magnitude of the constant in Equation 3.2 is proportional to the amount of gas n and can be expressed as the product of the number of moles of gas and a proportionality constant. Substituting, and rearranging, results in the equation:

$$pV = nRT \tag{3.3}$$

where R is the *ideal gas constant (or universal gas constant)* and is equal to 8.314 J K^{-1} mol^{-1}. This equation is known as the *ideal gas law*. It can be shown by statistical mechanics that R is equal to the product of Boltzmann's constant k and Avogadro's constant ($R = kL$).

EXAMPLE 3.1   The ideal gas law

Calculate the volume of 1 mole of an ideal gas at 0°C and 1 atmosphere pressure.

SOLUTION

Converting data to SI units: 0°C = 273.15 K; 1 atm = 101325 Pa. $n = 1$
From Equation 3.3,

$$V = \frac{nRT}{p} = \frac{1 \times 8.314 \times 273.15}{101325} = 0.0224 \text{ m}^3 = 22.4 \text{ L}$$

3.3.4   GAS MIXTURES

The composition of gas mixtures is usually expressed in volume percent; for example, the composition of air (ignoring minor components) is approximately 21 vol% O_2 and 79 vol% N_2. However, for thermodynamic calculations we need the composition in mole fraction. The conversion is quite simple. For a gas mixture of components A, B, ... the mole fraction of B in the mixture is (Equation 2.2):

$$x_B = \frac{n_B}{n_A + n_B + \cdots}$$

From the ideal gas law, at constant volume and temperature,

$$n_A = c\, V_A; \quad n_B = c\, V_B; \ldots \text{ where } c = \frac{p}{RT}$$

Therefore

$$x_B = \frac{c\, V_B}{c\, V_A + c\, V_B + \cdots} = \frac{V_B}{V_A + V_B + \cdots}$$

Therefore

$$x_B = \phi_B \tag{3.4}$$

where ϕ_B is the volume fraction of B in the mixture. Thus:

The composition of an ideal gas mixture in terms of mole fraction is the same as the volumetric composition expressed in volume fractions.

The composition of air therefore is: $x_{O_2} = 0.21$; $x_{N_2} = 0.79$.

Dalton's law of partial pressures

Dalton's law of partial pressures states that the total pressure exerted by a mixture of non-reactive ideal gases is equal to the sum of the pressures that each individual gas would exert if it were present alone. The individual pressure of a gas within a mixture of gases is called the *partial pressure* of the gas. This law was observed experimentally by John Dalton in 1801. The pressure of a mixture of ideal gases, therefore, is the sum of the partial pressures of the gases making up the mixture:

$$p = p_A + p_B + \cdots \tag{3.5}$$

where p_A, p_B ... are the partial pressure of components A, B, ... in a gaseous mixture. The partial pressures of the individual gases are related to the total pressure by the relation

$$p_B = p\, x_B \tag{3.6}$$

where x_B is the mole fraction of component B in the total mixture. Combining Equation 3.4 with Equation 3.6 yields

$$p_B = p\, \phi_B \tag{3.7}$$

EXAMPLE 3.2 Dalton's law of partial pressures

A mixture of gases in a closed vessel at a total pressure of 2 atm consists of 4 moles of hydrogen, 8 moles of oxygen, 12 moles of helium and 6 moles of nitrogen. What is the partial pressure of each gas? What is the volume percent of each gas?

SOLUTION

$n = 4 + 8 + 12 + 6 = 30; p = 2$ atm

$$x_{H_2} = \frac{4}{30} = 0.133; \quad x_{O_2} = \frac{8}{30} = 0.267; \quad x_{He} = \frac{12}{30} = 0.400; \quad n_{N_2} = \frac{6}{30} = 0.200$$

From Equation 3.6,

$$p_{H_2} = 2 \times 0.133 = 0.267 \text{ atm}; \quad p_{O_2} = 2 \times 0.267 = 0.533 \text{ atm};$$

$$p_{He} = 2 \times 0.400 = 0.800 \text{ atm}; \quad p_{N_2} = 2 \times 0.200 = 0.400 \text{ atm}$$

Note, the sum of the partial pressures is equal to the total pressure (2 atm).
From Equation 3.4, and converting fractions to percentages,

$$\text{vol\%H}_2 = 13.3; \text{vol\%O}_2 = 26.7; \text{vol\%He} = 40.0; \text{vol\%N}_2 = 20.0$$

3.4 REAL GASES

The ideal gas law can be derived from first principles from the kinetic theory of gases in which simplifying assumptions are made. Chief among these are that the molecules or atoms of gas have mass but no volume, are sufficiently far apart that they do not interact with each other and undergo only perfectly elastic collisions with each other and the walls of the container. Real gases deviate from ideal behaviour for two main reasons:

- As pressure increases, the volume of a gas becomes smaller and approaches zero. However, while volume does approach a small number it can never be zero because molecules occupy space.
- Intermolecular forces do exist in real gases, and these become increasingly important under certain conditions.

The ideal gas law tends to fail at low temperatures and high pressures where intermolecular forces and molecular size become important. It also fails for most dense gases (for example, many refrigerants) and for gases with strong intermolecular forces, notably water vapour. At high pressures, the volume of a real gas is often considerably greater than that of an equal quantity of ideal gas. At low temperatures, the pressure of a real gas is often considerably less than that of an equal quantity of ideal gas. Furthermore, at some combination of low temperature and high pressure, real gases undergo a phase transition to a liquid or a solid.

3.4.1 THE p–V–T RELATIONSHIP

For an ideal gas the relationship between pressure and volume at constant temperature is hyperbolic ($p = RT/V$). For a real gas, below a certain critical temperature compression will eventually lead to the gas condensing to the liquid state with a resulting change in volume of the system. As the gas condenses its volume will change at constant pressure until all the gas has converted to liquid. Then the pressure will rise markedly with any further compression of the liquid. This is illustrated in Figure 3.2. The situation for a series of temperatures is illustrated in Figure 3.3. The locus of the end points of the horizontal sections of the curves defines the boundary of gas–liquid mixtures (shown as the dashed curve). Points along the horizontal sections represent mixtures of gas and liquid, with the conditions for each phase defined by where the horizontal section crosses the boundary.

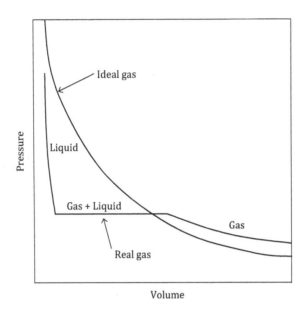

FIGURE 3.2 The $p-V$ relationship for an ideal gas and a real gas at constant temperature.

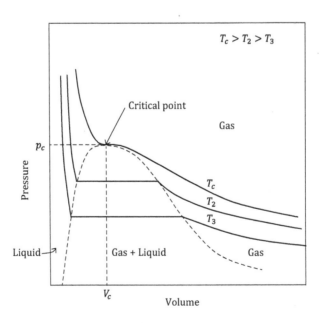

FIGURE 3.3 The effect of temperature on the $p-V$ relationship for an ideal gas.

The boundary curve in Figure 3.3 has a maximum value, called the *critical point*, which gives the temperature, pressure and molar volume beyond which only a single phase, called a *supercritical fluid*, is present. The critical point can be understood as follows. Imagine a liquid being heated in a closed container. The vapour pressure of the liquid in the space above the liquid increases as the temperature is raised, and its density therefore increases. A temperature will be reached at which the density of the gas becomes equal to the density of the liquid. The interface between the liquid and gas then disappears, and only one phase is present. This occurs at the *critical temperature*. The critical point of water is at 647 K (374°C) and 22.064 MPa (218 atm).

3.4.2 COMPRESSIBILITY

Empirically, the deviation from ideal gas behaviour can be described by a dimensionless quantity, the *compressibility factor Z*, defined as:

$$Z = \frac{pV}{nRT} \qquad (3.8)$$

The compressibility factor is determined experimentally by measurements of p, T and V or estimated using mathematical models. For ideal gases, $Z = 1$, and Equation 3.8 reduces to the ideal gas law, but for real gases Z is either greater than or less than one. It is difficult to generalise at what pressures or temperatures the deviation of gases from ideal behaviour becomes important. As a rule of thumb, most gases obey the ideal gas law reasonably accurately up to a pressure of about 2 bar and even higher for gases composed of small non-associating molecules. For example, methyl chloride (CH_3Cl), with a highly polar molecule and therefore a gas with significant intermolecular forces, has a compressibility factor of 0.9152 at a pressure of 10 bar and temperature of 100°C. For air (consisting of small non-polar molecules) at similar conditions the compressibility factor is 1.0025.

Typical trends in compressibility factor are shown in Figure 3.4. At very low pressures the molecules have large separations and interaction between them is negligible; therefore, $Z = 1$. At

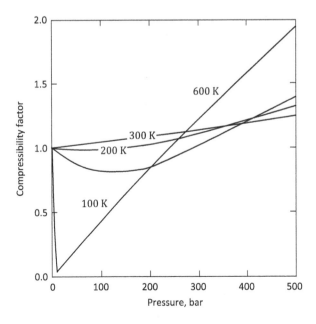

FIGURE 3.4   Variation of the compressibility factor for air with pressure and temperature.

moderate pressures the molecules are close and attractive forces are significant. The gas in this case occupies less volume than predicted by the ideal gas law, and, therefore, $Z < 1$. At high pressures the molecules are compressed closely and repulsive forces dominate. In this case the gas volume barely decreases with increase in pressure, and $Z > 1$.

It has been found that all real gases have similar properties if they are expressed as a ratio of those properties at the critical point. These are termed *reduced properties*. Thus,

$$T_r = \frac{T}{T_c}; \quad V_r = \frac{V}{V_c}; \quad p_r = \frac{p}{p_c}$$

Applying the compressibility concept to reduced conditions enables a common set of compressibility factors to be developed for all real gases as a function of T_r and p_r:

$$p_r V_r = ZRT_r$$

Figure 3.5 is an example of a generalised compressibility factor graph derived from hundreds of experimental $p - V - T$ data points for ten gases.

3.4.3 EQUATIONS OF STATE FOR REAL GASES

Various equations have been proposed to express the equation of state for gases that deviate significantly from ideality. One of the more common is the *van der Waals* equation which was derived to take account of the fact that the particles of a gas do occupy a finite space and that their interaction increases with increasing pressure:

$$p = \frac{nRT}{V - nb} - a\frac{n^2}{V^2}$$

where a and b are empirical constants, unique for each gas. Another equation of state is the *virial equation*. This is in the form of a series in powers of the variable n/V and has the form

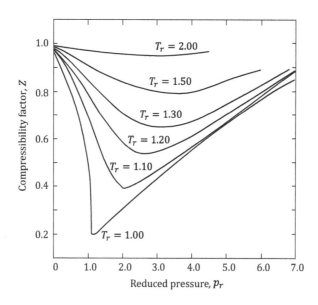

FIGURE 3.5 The variation of compressibility factor with reduced pressure at several reduced temperatures (based on experimental data for methane, ethane, ethylene, propane, n-butane, i-pentane, n-hexane, nitrogen, carbon dioxide and steam).

$$Z = \frac{pV}{nRT} = 1 + B\frac{n}{V} + C\frac{n^2}{V^2} + \cdots$$

The *virial coefficients* B, C, ... are functions of temperature. In practice, terms above the third coefficient are rarely used. Although the van der Waals and virial equations are better than the ideal gas law they are also limiting laws and become increasingly inaccurate as the pressure increases.

PROBLEMS

Assume ideal behaviour unless otherwise stated.

3.1   Calculate the volume of a container needed to store 0.05 moles of helium gas at 200 kPa and 200°C.

3.2   Calculate the pressure exerted by 50 g of hydrogen in a 10 L capacity gas cylinder at 25°C.

3.3   A 50 L cylinder is filled with argon at atmospheric pressure at 35°C. How many moles of argon are in the cylinder?

3.4   To what temperature would a 500 mL cylinder containing 1.0 g helium have to be cooled so that the pressure is 255 kPa?

3.5   The density of ethane (C_2H_6) is 1.264 kg m^{-3} at 20°C and 1 atmosphere pressure. Calculate the molecular weight of ethane. Compare this to the value calculated from its formula, and suggest reasons for the difference.

3.6   A gas mixture in a closed container consists of 4 mol H_2, 6 mol O_2, 11 mol of Ar and 5 mol N_2. What is the mol fraction and volume percent of each gas in the mixture? If the total pressure is 1.8 atm what is the partial pressure of each gas in the mixture?

3.7   The mole fraction of nitrogen in air is 0.21. What is the mass of nitrogen in a sealed 10 L container of air at 1 bar pressure and 25°C?

3.8   A sample of gas is introduced into the space above a layer of water in a closed vessel. What is the pressure of the gas if the total pressure is 1.1 bar and the water vapour pressure is 0.6 bar?

3.9   25.0 L of nitrogen at 2 bar and 15.0 L of oxygen gas at 1 bar are added to a closed 10 L container at 25°C. Calculate the total pressure and the partial pressures of nitrogen and oxygen.

3.10 A 5 L sealed vessel contains 50 g of hydrogen, 20 g of nitrogen and 30 g of carbon dioxide. Calculate the partial pressure of each gas and the total pressure of the gas mixture if the vessel temperature is 25°C.

3.11 Calculate the pressure exerted by 2 g of helium in a 300 mL container at −25°C using (a) the ideal gas law, (b) van der Waals equation. Suggest the reason for the difference between the non-ideal and ideal cases. Data: a = 3.46×10^{-3} Pa m^6; b = 23.71×10^{-6} m^3 mol^{-1}. Using the van der Waals value of pressure calculate the compressibility factor of helium at these conditions.

3.12 Using van der Waals equation, calculate the temperature of 25.0 moles of helium in a 10 L gas cylinder at 150 bar pressure. Compare this value with the temperature calculated using the ideal gas equation. Calculate the compressibility factor of helium at these conditions.

4 The first law

In this house, we obey the laws of thermodynamics!

Homer Simpson ('The PTA Disbands' episode of *The Simpsons* TV series)

SCOPE

This chapter explains the first law of thermodynamics and its implications for chemical thermodynamics.

LEARNING OBJECTIVES

1. *Understand quantitatively the meaning of the first law of thermodynamics.*
2. *Understand the concepts of the internal energy, enthalpy and heat capacity of substances.*
3. *Be able to calculate the enthalpy of a substance from its heat capacity.*
4. *Be able to calculate the enthalpy change of a system when substances react.*
5. *Understand how heat capacity and enthalpy are determined experimentally.*

4.1 INTRODUCTION

Historically, the discipline of thermodynamics developed in order to better understand the relationships between heat and mechanical energy, particularly in steam engines. It was only later that the principles so developed were applied to chemical systems, in particular to the relationships between heat, temperature, pressure and the composition of systems. This is the domain of chemical thermodynamics. There are two main aspects to chemical thermodynamics. The first is concerned with the thermal effects associated with changes in temperature, pressure and chemical composition. These relations are derived from the first law of thermodynamics and are the subject of this chapter. The second aspect is concerned with the tendency for a change in a system (for example, a chemical reaction or physical transformation) to take place, the extent to which it will take place, the equilibrium composition of a system and the effects of temperature and pressure on the equilibrium position. These relations are derived from the second law of thermodynamics and are discussed in Chapter 7 onwards.

4.2 THE FIRST LAW

Development of the understanding that led to the first law of thermodynamics occurred over many years. An early important development was the realisation that heat and mechanical energy are related. In 1798, at the arsenal in Munich, Benjamin Thompson, later Count Rumford,* investigated the heat generated by friction while the barrels of cannons were being bored. He immersed a cannon in a drum of water and, using a blunt boring tool, found that the water came to the boil after about two and a half hours and that the heat generated was seemingly inexhaustible as long as the boring continued. He confirmed that no physical change had taken place in the material of the cannon by comparing the material machined away to that remaining. He published an account of his experiments but did not attempt to quantify the amount of heat generated.

---

* Benjamin Thompson, Count Rumford (1753–1814) was an American-born British physicist and inventor who challenged the established caloric theory of heat.

The next major step was made by James Prescott Joule* in 1850. Joule found that the temperature of water in an insulated (adiabatic) container increased by a fixed amount when work was performed on it irrespective of how the work was performed – by rotating a paddle wheel immersed in water; passing an electric current through a coil immersed in water; compressing a gas in a cylinder immersed in water; and rubbing together two metal blocks immersed in water. In all cases, within experimental error, the same amount of work was required to cause a given rise in temperature. The most accurate experiment involved using a paddle wheel operated by falling weights. The results, expressed in SI units, showed that about 4.3 kJ of mechanical work was required to raise the temperature of 1 kg of water by 1 K. Historically, this is referred to as the mechanical equivalent of heat. It is now known from more accurate measurements that the value (to four significant figures) is 4.184 kJ.

Joule's experiments are a demonstration of the *first law of thermodynamics* which, in its most general form, may be stated as follows:

First law: In any non-nuclear process, energy can neither be created nor destroyed.

This is also known as the *law of conservation of energy*. The only exception to the law is nuclear reactions, in which matter is converted into energy through the destruction of protons and neutrons.[†] We will see later how this law applies to chemical systems, but to do that we first need to introduce the concept of internal energy.

4.2.1 INTERNAL ENERGY

All systems contain energy in different forms. The system as a whole may be moving, in which case it possesses macroscopic kinetic energy. It may contain macroscopic gravitational, electrostatic or magnetic potential energy by virtue of its position relative to the Earth's surface or its position in an electric or magnetic field. It contains chemical potential energy by virtue of the structural arrangements of the atoms or molecules comprising it and thermal energy by virtue of the kinetic energy of its atoms and molecules. The *total energy* of the system is the sum of all the energies.

The energy of interest for a thermodynamic system is usually its internal energy, the energy contained within the system, rather than its macroscopic kinetic and potential energy. Consider, for example, a football resting on a table that is moving with the Earth's rotation and orbit. Thermodynamic studies would focus on the energy associated with the gas inside the football (the system) and ignore the kinetic energy associated with movement of the football (its macroscopic kinetic energy) and its gravitational potential energy.

The *internal energy* of a system, U, is the total energy of the system at the atomic and molecular (microscopic) scale and has two components: Kinetic energy and chemical potential energy. The microscopic kinetic energy is due to the motion of the system's atoms and molecules (translations, rotations, vibrations). The microscopic chemical potential energy is the energy required to break bonds between atoms within molecules or crystals and relates to the electrostatic forces resulting from the sharing of electrons between bonded atoms. The internal energy of a system is the sum of these two components. For practical purposes, it is rarely necessary, convenient or even possible to consider all the energies contributing to the internal energy of a system. Indeed, for most systems it is not possible to calculate the total internal energy. It is the change in the internal energy of a system as it moves from one state to another that is of most interest in thermodynamics.

---

* James Prescott Joule (1818–1889) was an English physicist and brewer. The SI derived unit of energy, the Joule, is named after him. Joule worked with Kelvin to develop the absolute scale of temperature.

† During nuclear fission and fusion processes, mass is converted into energy. The relationship is given by Einstein's equation $E = mc^2$ where m is the mass loss (kg), E is its energy equivalent (J) and c is a constant equal to the speed of light (299 792 458 m s^{-1}).

Internal energy is an extensive property of a system and has the unit of Joules. Internal energy is a state function, as may be illustrated as follows. As noted previously, 4.184 kJ of energy is required to raise the temperature of 1 kg of water in an adiabatic container by 1 K. This value is independent of the device used to perform the work (for example, mechanical agitation or electrical resistance heating). Since the container is adiabatic, no heat is exchanged with the surroundings, and, therefore, any change in internal energy is due to work done on or by the system. Since a given amount of work always produces the same change of state (that is, the same increase in temperature) it follows that a change in U is always associated with the same change in state.

4.2.2 MATHEMATICAL STATEMENT OF THE FIRST LAW

If the forms of energy considered are restricted to thermal energy and work then for an infinitesimally small change in a system the mathematical statement of the first law is

$$dU = \delta q - \delta w \qquad (4.1)$$

where dU is the change in the internal energy of the system due to absorption of a quantity of heat δq from the surroundings and the performance of an amount of work δw on the surroundings.* Note, here we have adopted the following arbitrary conventions:

Conventions: Work done *by* the system on the surroundings is taken to be positive.
Heat flow *into* the system from the surroundings is taken to be positive.

If a system absorbs heat and converts this entirely to work without a change of state, then $dU=0$ and $\delta q = w$.

Because it is usually not possible to determine the absolute internal energy of a system, the first law of thermodynamics is formulated so as to be a statement for the difference in internal energy between one system state and another ($\Delta U = U_2 - U_1$). Since U is a state function and dU is an exact differential which can be integrated independently of the path between two states,

$$\Delta U = U_2 - U_1 = \int_{U_1}^{U_2} dU \qquad (4.2)$$

where U_1 and U_2 are the internal energies of the system in the initial and final states, respectively.

Work and heat are not state functions since their values depend on the path a system takes in changing from one state to another, and δq and δw are inexact differentials. Work done by a system on the surroundings may be of different kinds, and δw is the sum of these:

$$\delta w = \delta w_{volume} + \delta w_{electrical} + \delta w_{mechanical} + \cdots \qquad (4.3)$$

Chemical reactions perform work by changing the volume of the reacting system against an external pressure and, less commonly, by producing electrical energy. The work done by a system expanding against a constant external pressure can be evaluated by considering a piston moving within a

---

* Other energy terms can be added to the left-hand side of Equation 4.1 if required by the situation. For example, the system may have kinetic energy (E_k) by virtue of its motion and gravitational potential energy (E_p) by virtue of its height above a reference plane. In this case the statement of the first law would be: $dU + dE_k + dE_p = \delta q - \delta w$. However, in most systems of interest in chemical thermodynamics the systems are at rest relative to the Earth, and, hence, dE_k and dE_p are zero and Equation 4.1 is the appropriate form to use.

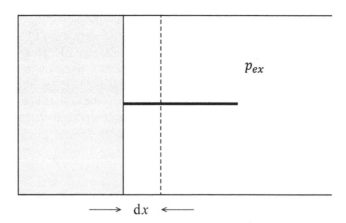

$$\longrightarrow \quad dx \quad \longleftarrow$$

FIGURE 4.1   A piston moving within a cylinder against a constant external pressure.

cylinder as illustrated in Figure 4.1. The piston is assumed to be weightless and frictionless. The work done by the piston in moving a distance dx against an external pressure p_{ex} is:

$$\delta w = p_{ex} \, A \, dx$$

where A is the cross-sectional area of the piston. The volume displaced by the piston dV is:

$$dV = A \, dx$$

Therefore,

$$\delta w = p_{ex} \, dV \tag{4.4}$$

Now suppose the cylinder contains a substance, which could be a solid, liquid or gas, and the expansion of this substance causes the piston to move. If the pressure p exerted by the substance is just infinitesimally greater than p_{ex}, then the substance expands reversibly as the piston moves. In this case, pdV is equal to the work done by the reversible expansion of the substance. Note, we have replaced p_{ex} in Equation 4.4 by p, the pressure of the substance in the cylinder since for reversible expansion $p_{ex}=p$ at each stage of the expansion. If the substance expanded irreversibly (at a finite rate), the external pressure would be less than p and the work done would be less than pdV. The work done by a reversible process is always greater than the work done by the same process carried out irreversibly and is, therefore, the *maximum work* for the process.

It follows from Equation 4.4 that for a reversible process,

$$w = \int_{V_2}^{V_1} p \, dV \tag{4.5}$$

Since p is a state function, from Equation 2.11 at constant pressure,

$$w = p(V_2 - V_1) = p \, \Delta V \tag{4.6}$$

Substituting for δw_{volume} in Equation 4.3 and then into Equation 4.1 yields

$$dU = \delta q - p \, dV - \delta w' \tag{4.7}$$

where $\delta w'$ is the work done on the surroundings other than by expansion, given by:

$$\delta w' = \delta w_{\text{electrical}} + \delta w_{\text{mechanical}} + \cdots$$

If the volume of the system remains constant, $dV = 0$ and if $\delta w' = 0$,

$$dU = \delta q \qquad\qquad (4.8)$$

Integrating

$$\int_{U_1}^{U_2} dU = U_2 - U_1 = \Delta U = \Delta q$$

The heat absorbed or evolved by a process occurring at constant volume is equal to the change in internal energy. It is a unique quantity since it is equal to the change in a state property.

EXAMPLE 4.1   The work of expansion due to chemical reaction

i. Calculate the work of expansion when 100 g of iron reacts at 25°C with sulphuric acid in (a) a closed vessel and (b) an open beaker.

SOLUTION

The reaction is: $Fe(s) + H_2SO_4(aq) = FeSO_4(aq) + H_2(g)$

In case (a), the volume doesn't change. Since $w = p\,\Delta V$ and $\Delta V = 0$, therefore $w = 0$.

In case (b), the gas evolved by the reaction pushes against the atmosphere (constant pressure), and the work done is given by:

$$w = p(V_2 - V_1)$$

The initial volume of the system is very small compared to the final volume (due to the evolution of gas) so $V_1 \approx 0$. Therefore,

$$w = pV_2$$

Since 1 mole of H_2 is produced when 1 mole of Fe is consumed,

$$n_{H_2} = n_{Fe} = 100 / 55.85 = 1.7905$$

For an ideal gas, $pV = nRT$. Therefore,

$$w = pV_2 = nRT = 1.7905 \times 8.314 \times 298.15 = 4438 \text{ Joules}$$

ii. Calculate the work of expansion when iron filings and sulphur react to form iron sulphide.

SOLUTION

The reaction is: $Fe + S = FeS$

The densities of the reactants and products are Fe: 7860 kg m^{-3}; FeS: 4740 kg m^{-3}; S: 2070 kg m^{-3}. The molar volumes, therefore, are:

$$V_m(Fe) = \frac{55.85}{7\ 860\ 000} = 7.1056 \times 10^{-6} \text{ m}^3$$

$$V_m(\mathrm{S}) = \frac{32.06}{4740000} = 1.5488 \times 10^{-5} \text{ m}^3$$

$$V_m(\mathrm{FeS}) = \frac{87.91}{2070000} = 1.8546 \times 10^{-5} \text{ m}^3$$

Since 1 mole of Fe reacts with 1 mole of S to form 1 mole of FeS, the change in volume is:

$$\Delta V = 1.8546 \times 10^{-5} - \left(7.106 \times 10^{-6} + 1.549 \times 10^{-5}\right) = -4.0471 \times 10^{-6} \text{ m}^3$$

The negative sign means the volume of the system decreases as a result of the reaction. At 1 atmosphere pressure, the work of expansion (actually contraction) by the reaction is:

$$w = p(V_2 - V_1) = p\,\Delta V = 101325 \times \left(-4.0471 \times 10^{-6}\right) = -0.4101 \text{ J}$$

For 100 g of iron reacting (the same basis as in Case 1 above),

$$w = -0.4101 \times \frac{100}{55.85} = -0.7432 \text{ J}$$

The above examples show that the work of expansion due to a chemical reaction when a gas phase is involved is likely to be very much greater than when only solid or liquid phases are involved.

4.3 ENTHALPY

Processes occurring at constant volume are much less common than processes at constant pressure since, in practice, many systems are open to the atmosphere – the pressure is constant but the volume is not. Equation 4.7 applies to processes occurring at constant pressure and, since U and V are state functions, for a system changing from state 1 to state 2 it can be rewritten as

$$\Delta q = (U_2 - U_1) + p(V_2 - V_1) + \Delta w' = (U_2 + pV_2) - (U_1 + pV_1) + \Delta w' \qquad (4.9)$$

Since p is also a state function it follows that the quantity Δq has a unique value at every state of the system at constant pressure.

The term $(U+pV)$ frequently arises in the study of constant pressure processes, and it has been found to be useful to define a function, *enthalpy* H*, as follows:

$$H = U + pV \qquad (4.10)$$

Enthalpy is a state function; that is,

$$H = H(T, p, n_1, n_2, \ldots)$$

Since enthalpy is the sum of two energy terms its unit is the Joule (J). Enthalpy is an extensive property, and for dealing with pure substances it is useful to define the *molar enthalpy* $H_m(\mathrm{B})$ as the enthalpy of 1 mole of substance B; it has the unit J mol^{-1}.

---

* From the Greek enthalpein (ἐνθάλπειν) meaning 'to warm in'. The concept of enthalpy was first discussed in 1875 by Josiah Willard Gibbs who introduced a 'heat function for constant pressure'. However, Gibbs did not use the word enthalpy, this being used for the first time in scientific literature in 1909.

The relation between enthalpy and absorbed heat is obtained by differentiating Equation 4.10,

$$dH = dU + p\,dV + V\,dp \tag{4.11}$$

and substituting for dU from Equation 4.7

$$dH = \delta q + V\,dp - \delta w' \tag{4.12}$$

At constant pressure, $dp = 0$, and if no work other than reversible volume work is performed, $\delta w' = 0$, then

$$dH = \delta q \tag{4.13}$$

Integrating

$$\int_{H_1}^{H_2} dH = H_2 - H_1 = \Delta H = \Delta q \tag{4.14}$$

The heat absorbed or evolved by a process occurring at constant pressure is equal to the change in enthalpy. It is a unique quantity since it is equal to the change in a state property.

4.3.1 THE NATURE OF ENTHALPY

Conceptually, enthalpy is the sum of the energy required to create a system (U) and the amount of energy required to create space for it by displacing its surroundings (pV). Enthalpy, therefore, is a measure of the total energy of a system, that is:

The difference in enthalpy between two states of a system is the energy required to transform the system from one state to the other at constant temperature and pressure.

For example, when a system consisting of a gas of volume V is created or brought to its present state from absolute zero at constant pressure p and temperature, energy must be supplied equal to its internal energy plus pV, where pV is the work done in pushing against the external (often atmospheric) pressure to create the space for the system. When the system consists of a condensed phase (solid or liquid), V will be small (compared to the volume of an equivalent amount of the substance in the gaseous form), and the pV term will be small (see Example 4.1) relative to the internal energy of the substance, and, accordingly, the difference between the enthalpy of the substance and its internal energy will be small at normal pressures. It is only at very high pressures that the pV term is no longer negligible for solids and liquids.

4.4 THE ENTHALPY OF MIXING

Mixtures are physical combinations of two or more substances without chemical bonding or other chemical change in which the separate identities of the substances are retained. It follows that when two or more unreactive substances are mixed at constant pressure there is no change in enthalpy of the system; that is,

$$\Delta_{\text{mix}} H_m = H_m - H_m^0 = 0 \tag{4.15}$$

where $\Delta_{\text{mix}} H_m$ is the *molar enthalpy of mixing* and is equal to the difference between the enthalpy of the mixture H_m and the sum of the enthalpy of the components of the mixture prior to mixing H_m^0. This is unexceptional and accords with everyday experience. However, as shown in later chapters, the entropy and Gibbs energy of unreactive systems do change when the components of the systems are mixed.

4.5 THE ENTHALPY OF PHASE CHANGES

When thermal energy is added to a pure solid or liquid substance at constant pressure, its temperature increases until a temperature is reached at which a phase change occurs (for example, a melting point or boiling point). With the further addition of heat, the temperature remains constant while the phase change takes place. The amount of substance that transforms is a function of the amount of heat added. After the change is complete, the addition of more heat increases the temperature of the substance.

At a phase change temperature, at constant pressure, the two phases are in equilibrium (for example at the melting point, the solid and the liquid are in equilibrium) and the temperature of the system and the surroundings is the same. Thus the heat is transferred reversibly between the system and the surroundings. The heat transferred q is equal to the enthalpy change: $q = \Delta H$ (Equation 4.14). The change in enthalpy during a phase change is illustrated in Figure 4.2. The enthalpy change is called the *enthalpy of transformation* or *latent heat*.

There are four main types of enthalpy changes resulting from a phase transformation:

- *Enthalpy of transition ($\Delta_{trs}H$)*. This applies to transformations from one solid phase to another, such as the transformation from δ-Fe to γ-Fe to β-Fe to α-Fe as solid iron cools from its melting point to ambient temperature.
- *Enthalpy of fusion or melting ($\Delta_{fus}H$)*. This applies to the transformation of a solid to a liquid, or *vice versa*.
- *Enthalpy of vapourisation ($\Delta_{vap}H$)*. This applies to the transformation of a liquid to a vapour, or *vice versa*.
- *Enthalpy of sublimation ($\Delta_{sub}H$)*. This applies to the transformation of a solid to a vapour, or *vice versa*.

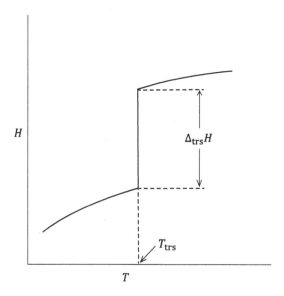

FIGURE 4.2   The change in enthalpy of a substance as it is heated from below a transition temperature to above the transition temperature, or as it is cooled from above the transition temperature to below the transition temperature.

Values of enthalpy change are usually quoted for the temperature at which the phase change occurs at 1 bar pressure and are designated thus: ΔH^0. For example, $\Delta_{fus}H^0(H_2O)=6.01$ kJ mol^{-1} implies a temperature of 273.15 K (0°C) and pressure of 1 bar.

When thermal energy is removed from a pure solid, liquid or gaseous substance (in other words, it cools), the temperature decreases until a phase change temperature is reached. With further cooling, the temperature remains constant while the phase transformation takes place. The amount of substance that transforms is a function of the amount of heat removed. As for heating, the process is reversible, and the temperatures of transformation during heating and cooling have the same value. For example, at 1 bar water freezes at 0°C and ice melts at 0°C. Similarly, the enthalpies of transformation during heating and cooling of a substance have the same values. For example, for the volatilisation of zinc, $\Delta H^0 = 115.3$ kJ mol^{-1} at the boiling point of zinc (907°C), where ΔH^0 is the heat required to convert 1 mole of molten zinc at 907°C to gaseous zinc at 907°C and 1 bar pressure and is also the heat released when 1 mole of gaseous zinc at 907°C and 1 bar is condensed to molten zinc at 907°C.

4.6   HEAT CAPACITY

The *heat capacity*, C, of a substance is the quantity of heat required to raise the temperature of 1 mole of the substance by 1 K (or 1°C). It is the molar counterpart of the more familiar term specific heat s, which is the quantity of heat q required to raise the temperature of a unit mass (usually 1 g or 1 kg) of the substance by 1°C:

$$s = \frac{q}{m \times \Delta T}$$

Since for most substances heat capacity varies with temperature, heat capacity is defined as a limit:

$$C = \lim_{T_1 \to T_2} \frac{q_2 - q_2}{T_2 - T_1} = \frac{\delta q}{dT} \tag{4.16}$$

The units of heat capacity are J mol^{-1} K^{-1}. The heat capacity of a substance has two unique values – one at constant volume C_V, because then $\delta q = dU$, and one at constant pressure C_p, because then $\delta q = dH$. Therefore,

$$C_V = \left(\frac{dU}{dT}\right)_V \tag{4.17}$$

and

$$C_p = \left(\frac{dH}{dT}\right)_p \tag{4.18}$$

Suppose it is desired to produce a certain temperature increase in a system; then if the process is carried out at constant volume all the heat added is used to raise the temperature. However, if the process is carried out at constant pressure then in addition to increasing the temperature the heat added is required to provide the work necessary to expand the system. Hence, the value of C_p for a substance will be greater than the value of C_V. In practice, it is frequently found that the heat capacity at constant pressure is the more useful.

It can be shown* that, for ideal gases, the relation between C_p and C_V is given by

$$C_p - C_V = R \tag{4.19}$$

and for all substances by

$$C_p - C_V = \frac{\alpha^2 V T}{\beta} \tag{4.20}$$

where α is the volumetric coefficient of expansion and β is the volumetric coefficient of compressibility defined, respectively, as:

$$\alpha = \frac{1}{V}\left(\frac{dV}{dT}\right)_p ; \quad \beta = -\frac{1}{V}\left(\frac{dV}{dp}\right)_T$$

Values of α and β can be determined experimentally.

The heat capacity of substances usually increases with increasing temperature as illustrated for some common substances in Figure 4.3. A step change in a curve indicates a phase transformation, for example, changing from one crystal form (or allotrope) to another, melting or boiling. The heat capacities of all common substances have been determined (see Section 4.10), and compilations of these values are widely available and are discussed in Chapter 5.

In 1819, the French physicists P. L. Dulong and A. T. Petit found experimentally that the average value of heat capacity of many elements (except the very light ones) at constant pressure and at intermediate temperatures is approximately the same and equal to about 25 J mol^{-1} K^{-1}. This is known as

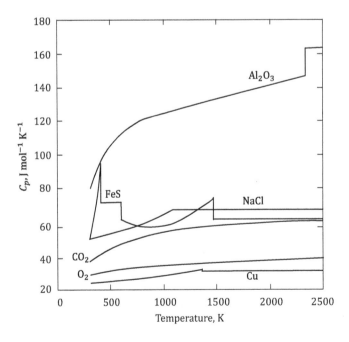

FIGURE 4.3 The variation of heat capacity with temperature of some common substances.

---

* See, for example, Denbigh, K. 1981. *The Principles of Chemical Equilibrium*, 4th ed. Cambridge: Cambridge University Press.

the Dulong–Petit law. Boltzmann,* in 1871, provided a theoretical basis for the law by assuming that a solid consists of a space lattice of independent atoms vibrating about their respective equilibrium positions without interacting with one another. From this, he showed that $C_p = 3R = 3 \times 8.314 = 24.9$ J mol^{-1} K^{-1}. The theoretical determination of heat capacity was further extended by Einstein (1907) and Debye (1912) using quantum theory.

Another generalisation, based on experimental observation, is Kopp's rule (1864) which states that the heat capacity of a solid compound is approximately equal to the sum of the heat capacities of the elements comprising it. The rule has been found to be reasonably accurate for ionic compounds at temperatures at which the constituent elements obey the Dulong–Petit law. Thus, if there are n atoms in the formula of a compound, the heat capacity is approximately $n \times 3R = n \times 24.9$. For example, for sodium chloride (NaCl), $n = 2$ and $C_p(\text{NaCl}) \approx 2 \times 24.9 = 49.8$ J mol^{-1} K^{-1}. This is close to the experimental value, which varies from 49.8 J mol^{-1} K^{-1} at 0°C to 52.0 J mol^{-1} K^{-1} at 100°C.

A convenient way of expressing the effect of temperature on the heat capacity of substances is by means of empirical equations. Several forms are used, but one that has found wide acceptance is

$$C_p = a + bT + cT^2 + \frac{d}{T^2} \tag{4.21}$$

where a, b, c and d are experimentally determined constants which apply over a specified temperature range and T is the temperature in Kelvin. Additional terms in the polynomial are sometimes added for greater accuracy, for example,

$$C_p = a + bT + cT^2 + eT^3 + \frac{d}{T^2} + \frac{f}{T^3}$$

4.7 THE ENTHALPY OF SUBSTANCES

From the definition of C_p, $dH = C_p\, dT$ and integrating between temperatures T_1 and T_2,

$$H(T_2) - H(T_1) = \int_{T_1}^{T_2} C_p\, dT \tag{4.22}$$

Thus, the enthalpy difference $H(T_2) - H(T_1)$ is the area under the curve of C_p *versus* temperature between the temperatures T_2 and T_1. Absolute values of enthalpies of substances are not known since there is no absolute zero for internal energy ($H = U + pV$). However, the following arbitrary convention has been universally adopted:

Convention: The enthalpy of all elements in their standard state is zero at 25°C (298.15 K) at a pressure of 1 bar (10^5 Pa).[†]

As will be shown shortly, this convention enables enthalpy values at 298.15 K to be assigned to compounds based on direct experimental measurement or measurement of their heat capacity. The convention for writing the standard enthalpy of a substance is as follows:

---

* Ludwig Boltzmann (1844–1906) was an Austrian physicist and philosopher. His great achievement was the development of statistical mechanics, which explains and predicts how the properties of atoms, such as mass, charge and structure, determine the physical properties of matter such as viscosity, thermal conductivity and diffusion.

† In older compilations of thermodynamic data, a pressure of 1 atmosphere (101 325 Pa) was used, but this convention is now obsolete.

Convention: H^0 (formula, state of aggregation, temperature)

For example, the enthalpy of 1 mole of the compound FeO in the solid state at 298.15 K and 1 bar is written as H^0(FeO, s, 298.15)=−272.04 kJ. The convention for assigning enthalpy values to substances also enables values to be calculated for the internal energy of a substance since from Equation 4.10, $U = H - pV$.

EXAMPLE 4.2   The relation between U and H for substances

i. Calculate the internal energy of 1 mole of iron at 500 K and 1 bar.

SOLUTION

The standard enthalpy of 1 mole of iron (55.85 g) is 5474 J at 500 K. Its density is approximately 7860 kg m^{-3}. The volume of 1 mole of iron, therefore, is:

$$V_m(\text{Fe}) = \frac{55.85}{7860 \times 10^3} = 7.11 \times 10^{-6} \text{ m}^3$$

At 1 bar pressure,

$$pV = 10^5 \times 7.11 \times 10^{-6} = 0.711 \text{ Joules}$$

Since $U = H - pV$, $U = 5470 - 0.711 = 5473.3$ Joules

ii. Calculate the internal energy of 1 mole of oxygen at 500 K.

The standard enthalpy of 1 mole of oxygen gas at 500 K is 6085 J. Assume ideal behaviour; therefore

$$pV = RT = 8.314 \times 500 = 4157 \text{ J}$$

Since $U = H - pV$, $U = 6085 - 4157 = 1928$ J

COMMENT

Since the value of pV for a mole of an ideal gas is constant at constant temperature, the internal energy of an ideal gas depends only on the quantity of gas (U is an extensive property) and its temperature.

It is clear from the above example that for condensed phases (solids and liquids), for which the term pV is small relative to the enthalpy of the substance, the difference between the enthalpy of the substance and its internal energy is very small at normal pressures. It is only at very high pressures that pV is no longer negligible for solids and liquids.

4.7.1   VARIATION OF ENTHALPY WITH TEMPERATURE

On setting the lower temperature limit to 298.15 K, Equation 4.22 can be written as:

$$H^0(T) = H^0(298.15) + \int_{298}^{T} C_p \, dT \tag{4.23}$$

Since the heat capacity of substances usually increases with increasing temperature, it follows that the value of the enthalpy of substances will also usually increase with temperature. Therefore, the value of the enthalpy of elements at temperatures above 298.15 K will be greater than zero. Substituting an empirical equation for C_p, of the form of Equation 4.21, enables the enthalpy of a substance to be calculated at any temperature if the value at 298.15 K is known from experimental measurement:

$$H^0(T) = H^0(298) + \int_{298}^{T} \left(a + bT + cT^2 + \frac{d}{T^2} \right) dT \tag{4.24}$$

$$= H^0(298) + a(T - 298) + \frac{b}{2}(T^2 - 298^2) + \frac{c}{3}(T^3 - 298^3) - d\left(\frac{1}{T} - \frac{1}{298}\right) \tag{4.25}$$

Equation 4.24 applies over the temperature range for which the substance does not undergo a phase transformation (for example, melting, boiling) and for which the empirical equation for C_p applies. Often, two or more empirical equations are needed to accurately fit the heat capacity data over a range of temperatures, even where there is no phase transformation. In the latter case Equation 4.24 is extended as follows:

$$H^0(T_n) = H^0(298.15) + \int_{298}^{T_1} \left(a_1 + b_1T + c_1T^2 + \frac{d_1}{T^2} \right) dT$$

$$+ \int_{T_1}^{T_2} \left(a_2 + b_2T + c_2T^2 + \frac{d_2}{T^2} \right) dT + \cdots \tag{4.26}$$

When a phase change occurs in the temperature range of interest, the enthalpy of transformation must be taken into account. For example, when a solid substance melts then boils within the temperature range, the enthalpy of the substance at a temperature T above the boiling point is given by:

$$H^0(T) = H^0(298.15) + \int_{T_{298}}^{T_{fus}} C_p(s)\, dT + \Delta_{fus}H^0 + \int_{T_{fus}}^{T_{vap}} C_p(l)\, dT + \Delta_{vap}H^0 + \int_{T_{vap}}^{T} C_p(g)\, dT$$

The calculation procedure is illustrated schematically in Figure 4.4 and illustrated graphically in Figure 4.5 for the example of FeO, which melts at 1650 K. The enthalpies of some common substances, calculated from their C_p values, are shown as a function of temperature in Figure 4.6. As seen previously, the enthalpy of substances increases gradually with increasing temperature, and the enthalpy of elements is zero at 298.15 K (25°C) and positive at temperatures greater than 298.15 K. Note, the enthalpy of compounds is less than that of elements, reflecting the lower internal energy of compounds.

4.7.2 ENTHALPY INCREMENTS

Frequently, the enthalpy difference between two temperatures is of more interest than the actual enthalpy of a substance. This is given by:

$$H^0(T_2) - H^0(T_1) = \int_{T_1}^{T_2} C_p\, dT = \int_{T_1}^{T_2} \left(a + bT + cT^2 + \frac{d}{T^2} \right) dT \tag{4.27}$$

$$= a(T_2 - T_1) + \frac{b}{2}(T_2^2 - T_1^2) + \frac{c}{3}(T_2^3 - T_1^3) - d\left(\frac{1}{T_2} - \frac{1}{T_1}\right)$$

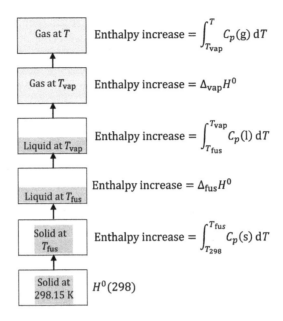

FIGURE 4.4 The increase in enthalpy of a substance as it is heated from 298 K to a temperature above its boiling point. If the substance is an element $H^0(298)=0$; if the substance is a compound $H^0(298)<0$. The enthalpy of the substance at T is the sum of $H^0(298)$ and the enthalpy increases.

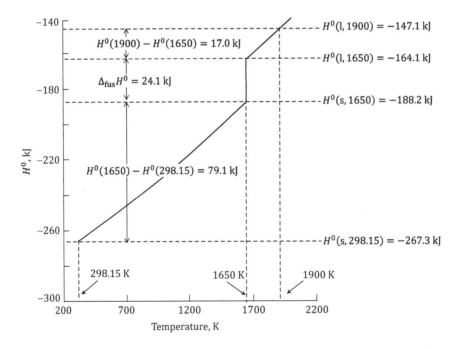

FIGURE 4.5 The increase in enthalpy of FeO as it is brought from 298.15 to 1900 K. Note, since FeO is a compound, its enthalpy at 298.15 K is less than zero.

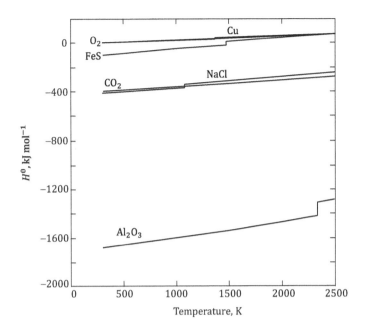

FIGURE 4.6 The variation of enthalpy with temperature for a number of common substances.

The difference in enthalpy of a substance at two temperatures, $H^0(T_2) - H^0(T_1)$, is called the *enthalpy increment*. At constant pressure, $H^0(T_2) - H^0(T_1)$ is the amount of heat required to raise the temperature of 1 mole of a substance from T_1 to T_2 or, alternatively, the amount of heat released when the substance cools from T_2 to T_1. In many applications we are interested in the heat required to heat a substance from room (or ambient) temperature to some other temperature, or the heat to be removed to cool a substance to ambient temperature. *By convention, ambient temperature is chosen arbitrarily to be 25°C (298.15 K)*. Referred to this temperature, the enthalpy increment can be assigned a value. This value is referred to as the *sensible heat* or the *heat content* of a substance. Figure 4.7 shows the variation with temperature of $H^0(T) - H^0(298.15)$ for a number of common substances. Discontinuities in the data represent phase changes. For convenience, we will sometimes write $H - H(298)$ rather than the more correct form $H^0(T) - H^0(298.15)$.

It is rarely necessary to perform calculations using equations based on Equation 4.23 in practice because compilations of enthalpies and enthalpy increments are readily available in various formats, such as hardcopy tables, digitised tables and graphs, or they can be generated as needed using thermodynamic software packages. Sources of enthalpy data are discussed in Chapter 5.

4.8 THE ENTHALPY OF FORMATION

A reaction of fundamental importance in chemical thermodynamics is the reaction to form a compound from its constituent elements. This has the form:

$$k\,A + l\,B = A_k B_l$$

where A and B are elements which react to form the compound $A_k B_l$. The *standard enthalpy of formation* $\Delta_f H^0$ of a compound is defined as the change in enthalpy when 1 mole of the compound

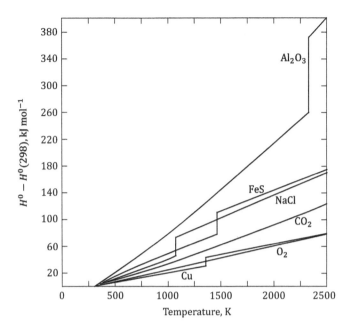

FIGURE 4.7   The variation with temperature of $H^0(T) - H^0(298)$ for a number of common substances.

is formed from its constituent elements under standard state conditions at a specified temperature; that is,

$$\Delta_f H^0(A_k B_l) = H^0(A_k B_l) - k H^0(A) - l H^0(B) \qquad (4.28)$$

For example, the standard enthalpy of formation of water at 298.15 K is −286 kJ mol^{-1}; that is,

$$H_2(g) + 1/2\ O_2(g) = H_2O(l); \quad \Delta_f H^0(298.15) = -286\ \text{kJ}.$$

For convenience, the above information is summarised as follows:

$$\Delta_f H^0(H_2O, l, 298.5) = -286\ \text{kJ}$$

This means that the value refers to the formation of 1 mole of liquid water from its elements under standard state conditions at 298.15 K.

It follows from Equation 4.28 that the enthalpy of a compound at 298.15 K is equal to the enthalpy of formation of the compound at 298.15 K since by convention the enthalpy of the elements forming the compound are zero at 298.15 K; that is, for any substance,

$$H^0(298.15) = \Delta_f H^0(298.15) \qquad (4.29)$$

The standard enthalpy of formation of elements in their standard state is zero at all temperatures, since for the 'formation' reaction (where E is an element):

$$E = E$$

$$\Delta_f H^0(T) = H^0(T) - H^0(T) = 0$$

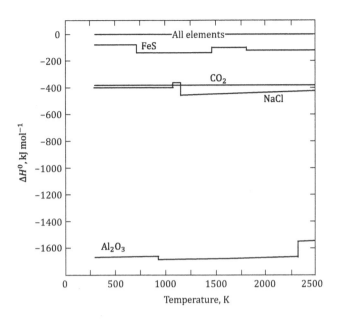

FIGURE 4.8   The variation with temperature of the enthalpy of formation for some common substances.

Values of the enthalpy of formation of substances at 298.15 K are readily available in compilations – in books, as digitised tables and graphs. Values of $\Delta_f H^0$ at temperatures other than 298.15 K can be calculated from the enthalpy values of the individual reactants and product at the relevant temperature by means of the equation:

$$\Delta_f H^0\left(A_k B_l, T\right) = H^0\left(A_k B_l, T\right) - k\, H^0(A,T) - l\, H^0(B,T) \tag{4.30}$$

The variation with temperature of the standard enthalpy of formation of some common substances is shown in Figure 4.8. For many substances the enthalpy of formation is nearly independent of temperature. Sources of enthalpy of formation data are discussed in Chapter 5.

4.9   THE ENTHALPY OF REACTION

The changes in energy of a reacting system are expressed in terms of changes in U (for reactions at constant volume) and H (for reactions at constant pressure). The manner in which this is done is governed by a number of conventions. The following discussion refers specifically to enthalpy since it is more commonly used, but it applies equally to internal energy. For the chemical reaction

$$k\,A + l\,B = m\,C + n\,D \tag{4.31}$$

where k, l, m and n are the number of moles of substances A, B, C and D respectively, the change in enthalpy of the system $\Delta_r H$ when k moles of A react completely with l moles of B to form m moles of C and n moles of D is given by:

$$\Delta_r H = m\, H(C) + n\, H(D) - k\, H(A) - l\, H(B) \tag{4.32}$$

$\Delta_r H$ is called the *enthalpy of reaction*. It follows from this convention that:

- When $\Delta_r H$ is positive (that is, the enthalpy of the products is greater than the enthalpy of the reactants) heat is absorbed at constant temperature and the reaction is *endothermic*; and

- When $\Delta_r H$ is negative (that is, the enthalpy of the products is less than the enthalpy of the reactants) heat is evolved at constant temperature and the reaction is *exothermic*.

When all the substances are in their standard state, the enthalpy of reaction is called the *standard enthalpy of reaction*, and is written as follows: $\Delta_r H^0$.

The enthalpy change due to the complete combustion of a substance at constant pressure is called the *heat of combustion* and has the same value as the enthalpy of reaction. For organic substances, complete combustion means the total conversion of carbon and hydrogen in the fuel to carbon dioxide and water; for example,

Combustion of methane:

$$CH_4(g) + 2\ O_2(g) = CO_2(g) + 2\ H_2O(l); \quad \Delta_r H^0(298.15) = -863\ kJ$$

Combustion of ethanol:

$$C_2H_5OH(l) + 3\ O_2(g) = 2\ CO_2(g) + 3\ H_2O(l); \quad \Delta_r H^0(298.15) = -1365\ kJ$$

The value of $\Delta_r H^0$ depends on the temperature at which a reaction is carried out, and the value at a particular temperature is indicated as follows: $\Delta_r H^0(T)$. The standard state of each reactant and product must also be specified since the enthalpy of a substance, and therefore the value of $\Delta_r H^0$, depends on its state of aggregation. The state of aggregation of the reacting substances is indicated by placing the appropriate symbol from Table 2.1 in parentheses after the chemical symbol or formula for the substance. Thus, for the reaction

$$FeO(c) + H_2(g) = Fe(s) + H_2O(g)$$

the value of $\Delta_r H^0(700)$ is equal to the change in enthalpy when 1 mole of crystalline FeO reacts with 1 mole of gaseous hydrogen to form 1 mole of solid iron and 1 mole of water vapour, all at 1 bar pressure at 700 K.

Phase transformations can also be written as chemical reactions. The enthalpy change of the reaction then equals the enthalpy of the phase change; for example,

$$H_2O(s) = H_2O(l); \quad \Delta_r H^0(273.15) = \Delta_{fus} H^0 = 6.01\ kJ\ mol^{-1}$$

or

$$H_2O(l) = H_2O(s); \quad \Delta_r H^0(273.15) = -\Delta_{fus} H^0 = -6.01\ kJ\ mol^{-1}$$

EXAMPLE 4.3   The relation between enthalpy of reaction and internal energy of reaction

Calculate the change in internal energy for the reaction of Fe_2O_3 with carbon at 1000°C and 1 bar pressure given that the standard enthalpy of the reaction at 1000°C is 467.26 kJ.

SOLUTION

The equation for the reaction is:

$$Fe_2O_3(s) + 3\ C(s) = 2\ Fe(s) + 3\ CO(g)$$

The internal energy of reaction $\Delta_r U^0$, defined in the same manner as for $\Delta_r H^0$, is

$$\Delta_r U^0(1273) = 3 \times \Delta_f U^0(CO, g, 1273) + 2 \times \Delta_f U^0(Fe, s, 1273)$$

$$-3 \times (C, s, 1273) - \Delta_f U^0\left(Fe_2O_3, s, 1273\right)$$

At constant pressure, $\Delta H = \Delta U + p\Delta V$; therefore,

$$\Delta_r U^0 = \Delta_r H^0 - p\Delta V$$

The change in volume ΔV is due to the change in the number of moles of gas since the change in volume of condensed phases (Fe_2O_3, C and Fe) is negligible in comparison. Therefore,

$$\Delta n = \text{moles of gaseous products} - \text{moles of gaseous reactants} = 3 - 0 = 3$$

Since $pV = nRT$,

$$p\Delta V = \Delta n RT = 3 \times 8.314 \times (1000 + 273.15) = 31\ 755 \text{ Joules}$$

and, therefore

$$\Delta_r U^0 = 467.26 - 31.76 = 435.50 \text{ kJ}$$

EXAMPLE 4.4   Calculate the enthalpy change of a reaction

Using readily available data, calculate the standard enthalpy change at 1000°C for the reaction:

$$Fe_2O_3(s) + 3C(s) = 2\,Fe(s) + 3CO(g)$$

SOLUTION

METHOD 1: COMBINING EQUATIONS FOR FORMATION

The reaction can be considered as being the sum of two reactions of formation:

$Fe_2O_3(s) = 2\,Fe(s) + 1.5\,O_2(g)$	$-\Delta_f H^0(Fe_2O_3)$
$3\,C(s) + 1.5\,O_2(g) = 3\,CO(g)$	$3 \times \Delta_f H^0(CO)$
$Fe_2O_3(s) + 3\,C(s) = 2\,Fe(s) + 3\,CO(g)$	

Note that the equation for the formation of Fe_2O_3 is reversed, and hence the negative value of the enthalpy of formation of Fe_2O_3 is used. Also, note that the enthalpy of formation of CO is multiplied by three because there are 3 moles of CO in the equation as written.
   At 1000°C,

$$\Delta_r H^0(1273) = -\Delta_f H^0\left(Fe_2O_3, s, 1273\right) + 3 \times \Delta_f H^0\left(CO, g, 1273\right)$$

$$= -(-808.39) + 3 \times (-113.71) = 467.26 \text{ kJ}$$

The addition of chemical equations in the above manner is justified because chemical equations are also mathematical equations expressing the law of conservation of matter.

METHOD 2: COMBINING ENTHALPIES OF FORMATION

The same result will be obtained if the enthalpies of formation of the reactants and products are combined as follows:

$$\Delta_r H^0(1273) = 3 \times \Delta_f H^0 (CO, g, 1273) + 2 \times \Delta_f H^0 (Fe, s, 1273)$$

$$-3 \times \Delta_f H^0 (C, s, 1273) - \Delta_f H^0 (Fe_2O_3, s, 1273)$$

The enthalpy of formation of elements is zero at all temperatures; therefore,

$$\Delta_r H^0(1273) = 3 \times (-113.71) + 2 \times 0 - 3 \times 0 - (-808.39) = 467.26 \text{ kJ}$$

METHOD 3: COMBINING INDIVIDUAL ENTHALPY VALUES

Yet another approach is to combine the individual enthalpy values of the reactants and products (rather than their enthalpies of formation), at the relevant temperature, according to Equation 4.32. Thus,

$$\Delta_r H^0(1273) = 3 \times H^0 (CO, g, 1273) + 2 \times H^0 (Fe, s, 1273)$$

$$-3 \times H^0 (C, s, 1273) - H^0 (Fe_2O_3, s, 1273)$$

$$= 3 \times (-79.60) + 2 \times 37.64 - 3 \times 17.92 - (-684.54) = 467.26 \text{ kJ}$$

In the above example, calculation Method 1 is an illustration of *Hess's law** which states that the enthalpy change of a reaction is the same whether it takes place in one or in a number of steps. The two equations of formation can be thought of as steps in the overall reaction. Though first discovered experimentally, Hess's law is a logical consequence of the fact that enthalpy is a state function and has a unique value for every state of a system and is independent of the path taken. Note the positive value in this example indicates the reaction is endothermic; that is, heat must be supplied to maintain the reaction temperature of 1000°C.

4.10 EXPERIMENTAL DETERMINATION OF HEAT CAPACITY AND ENTHALPY

The ultimate source of heat capacity and enthalpy data is experimental measurements though there are techniques for estimating values when experimental data are not available. Heat capacities and enthalpies are usually determined by measuring the temperature change in a well-insulated vessel with a known heat capacity. Such a vessel is called a *calorimeter*, and the technique for measuring heat quantities is called *calorimetry*. There are many methods of calorimetry, and only a few examples are given here to illustrate the technique.

The simplest way to determine the heat capacity or enthalpy change for a substance is to heat it to a known temperature (T_2) then drop it into a calorimeter which is at a known initial temperature (T_1). The temperature of the calorimeter will rise as the substance cools, and a new temperature for the calorimeter and substance will be attained (T_3). From the temperature increase of the calorimeter

---

* Germain Henri Hess was born in Geneva in 1802. His family moved to Russia in 1805. He attended high school in Estonia then studied medicine at the University of Tartu and qualified as a physician in 1825. His interests turned to chemistry after meeting Jöns Jakob Berzelius, the famous Swedish chemist. After practicing medicine for a few years, Hess became an adjunct at the Imperial Academy of Sciences in St. Petersburg in 1828 and then professor of chemistry in 1834 at the St Petersburg Technological Institute. His most famous paper, outlining his law on thermochemistry, was published in 1840.

$T_3 - T_1$, and its heat capacity, the heat released by the substance in cooling from T_2 to T_3 is calculated. If the calorimeter is operated at constant pressure this value is equal to $H_{T_2} - H_{T_1}$ for the substance; if it is operated at constant volume the heat released is equal to $U_{T_2} - U_{T_1}$. From a number of measurements at different temperatures the variation of $H_T - H_{T_1}$ with temperature can be found. The heat capacity of the substance can then be calculated from the slope of the curve according to Equation 4.18.

The enthalpy of exothermic reactions (such as heat of combustion and enthalpy of formation of compounds) can be determined in a bomb calorimeter. A steel bomb, containing the reactants, is immersed in a water calorimeter, and the reaction is started by heating the reactants electrically. After the reaction, when the bomb and reactants have cooled and attained thermal equilibrium with the water, the temperature increase of the water is measured. The heat of reaction is thereby obtained, after correcting for the additional heat from the electrical energy required to start the reaction. Since the reaction in a bomb calorimeter occurs at constant volume the heat measured is equal to ΔU for the reaction. The enthalpy of the reaction can be obtained from the relation:

$$\Delta H = \Delta U + \Delta(pV)$$

Since $p\Delta V$ is significant only for gases (Example 4.1, and the following paragraph) a correction to the measured heat of reaction needs to be made only when the reaction involves gaseous substances. Since $pV = nRT$, at constant temperature $\Delta(pV) = \Delta nRT$ where Δn is the increase in the number of moles of gas during the reaction and can be calculated if the composition of the fuel is known. For example, for the reaction $C(s) + O_2(g) = CO_2(g)$, $\Delta n = 0$ and $H = \Delta U$. For fuels that contain hydrogen, water will be formed after cooling of the combustion products, Δn will be negative, and ΔH will be less than ΔU.

Another method for measuring the enthalpy of a reaction is to determine the variation of the equilibrium constant (Section 10.3) of the reaction with temperature. Then, using a form of the van't Hoff Isochore (Equations 10.10 and 10.11) the enthalpy of the reaction is calculated.

PROBLEMS

4.1 Which of the following processes would increase the internal energy of a system by 5 J?
   i.   The system is heated by 20 J and performs 15 J of work.
   ii.   The system is heated by 20 J and has 15 J of work performed on it.
   iii. The system is cooled by 20 J and performs 15 J of work.
   iv.   The system is cooled by 20 J and has 15 J of work performed on it.
4.2 A car engine has six cylinders with a total displacement of 3.00 L and a 10:1 compression ratio (meaning that the volume of each cylinder decreases by a factor of ten when the piston compresses the air–gas mixture inside the cylinder prior to ignition). What is the maximum work that can be done per cylinder when the gas expands at constant temperature against an opposing pressure of 35.0 atm during the engine cycle? Assume that the gas is ideal, the piston is frictionless, and no energy is lost as heat.
4.3 A 70 kg person at rest has a lung volume of 2200 mL. This increased to 2700 mL when the person inhaled. Assuming the lungs maintain a pressure of approximately 1.0 atm, what is the minimum work required to take a single breath? During exercise, the lung volume changed from 2200 to 5200 mL on each breath. How much additional work was done to take a breath while exercising?
4.4 Calculate the reversible work of expansion when 1 kg of water is decomposed to oxygen and hydrogen by electrolysis at constant pressure and at 25°C.
4.5 Calculate the change in internal energy when 1 mole of H_2 reacts with 1/2 mole of O_2 to form 1 mole of water at 1 bar pressure and 298.15 K given that the standard enthalpy of formation of water at 298.15 K is −285.83 kJ mol^{-1}.

4.6   Calculate the change in internal energy for the formation of gaseous ammonia from its elements at 25°C given the standard enthalpy of formation of ammonia at 25°C is –45.94 kJ. The relevant chemical reaction is: $0.5\ N_2(g) + 1.5\ H_2(g) = NH_3(g)$

4.7   Calculate the standard enthalpy (H^0) of CaO at 900 K from its heat capacity equation given that $H^0(CaO, s, 298.15) = 634.92$ kJ and $C_p(CaO,s) = 57.75 - 10.78 \times 10^{-3} \times T - 11.51 \times 10^5 \times T^{-2} + 5.33 \times 10^{-6} \times T^2$ J K^{-1} mol^{-1}.

4.8   Calculate $\Delta_r U^0(298.15)$ for the stoichiometric combustion of 1 mole of propane: $C_3H_8(g) + 5\ O_2(g) = 3\ CO_2(g) + 4\ H_2O(g)$ given $\Delta_r H^0(298.15) = -2043$ kJ.

4.9   Using the data below, calculate $\Delta_r H^0$ for the following chemical reactions at 298 and 1200 K.
   i.   Using the enthalpy values only (columns 2 and 3).
   ii. Using the enthalpies of formation only (columns 4 and 5). Your answers should be the same in both cases.
   iii. Are the reactions endothermic or exothermic?
   Retain the $\Delta_r H^0$ values for use in Problem 8.2.
   Reactions:

$$CH_4(g) + 2\ O_2(g) = CO_2(g) + 2\ H_2O(l,g)$$

$$N_2(g) + 3\ H_2(g) = 2\ NH_3(g)$$

$$Cu_2S(s) + O_2(g) = 2\ Cu(s) + SO_2(g)$$

Data:

	H^0(298.15) kJ	H^0(1200) kJ	$\Delta_f H^0$(298.15) kJ	$\Delta_f H^0$(1200) kJ
CH$_4$(g)	–74.87	–21.61	–74.87	–91.45
CO$_2$(g)	–393.51	–349.03	–393.51	–395.04
Cu	0.00	24.63	0.00	0.00
Cu$_2$S	–81.17	1.62	–81.17	–128.15
H$_2$(g)	0.00	26.80	0.00	0.00
H$_2$O(l)	–285.83		–285.83	
H$_2$O(g)	–241.83	–207.32	–241.83	–249.00
O$_2$(g)	0.00	29.76	0.00	0.00
N$_2$(g)	0.00	28.11	0.00	0.00
NH$_3$(g)	–45.94	–1.53	–45.94	–55.78
SO$_2$(g)	–296.81	–251.37	–296.81	–361.64

4.10 Water was brought to the boil at a pressure of 1 atm. A resistance heating element in the water was then heated by passing a current of 0.5 Amp at 12 V for 7 minutes. The mass of the water decreased by 1.1172 g. Assuming all the heat generated by the element was transferred to the water and that the mass loss was due to water vapourised by the additional heat, calculate the enthalpy of vapourisation of water and the change in internal energy.

4.11 A 1.0 g sample of coal was placed in a bomb calorimeter and ignited using a 150 Watt resistance element through which current was passed for 20 seconds. After the complete combustion of the coal, the temperature of the calorimeter was found to have increased from 25.4 to 27.5°C. The calorimeter was calibrated using the same electrical element, and it was found that its temperature increased by 1°C after 110 seconds. Calculate:
   i.   The heat capacity of the calorimeter.
   ii. The heat of combustion of the coal at constant volume.

iii. If the coal contained 5 mass% hydrogen, calculate the heat of combustion of the coal at constant pressure (assume hydrogen in the coal is in the elemental state).

4.12 Samples of metallic lead each weighing 100 g were heated to various temperatures then dropped into a calorimeter calibrated to measure the heat content of the sample above a reference temperature of 25°C. The average results obtained are shown in the table below. Estimate the latent heat of fusion of lead ($\Delta_{fus}H^0$) at its melting point. The melting point of lead is 327.5°C.

Temperature (K)	400	500	600	700	800
$H(T) - H^0(298)$ (J)	1340	2700	4100	7880	9340

4.13 This exercise is most easily performed by setting up a spreadsheet. *Retain the spreadsheet and results as they will be used again in calculations in Chapters 7 and 8.*

i. Calculate the values of $C_p(O_2)$, $C_p(Al_2O_3)$ and $C_p(Al)$ at 298.15 K and at 100 K intervals from 300 to 1500 K and plot them graphically as a function of temperature. Note, there is a phase change for aluminium within the temperature range.

ii. Calculate the standard enthalpy of O_2, Al_2O_3 and Al at 298.15 K and at 100 K intervals from to 300 to 1500 K and plot the values graphically as a function of temperature.

iii. Calculate values of $H^0(T) - H^0(298.15)$ for O_2, Al_2O_3 and Al at 298.15 K and at 100 K intervals from 3000 to 1500 K and plot the values graphically as a function of temperature.

iv. Using the enthalpy values calculated in part ii, calculate the standard enthalpy of formation of Al_2O_3 at 298.15 K and at 100 K intervals from 300 to 1500 K and plot them graphically as a function of temperature. Is the formation of Al_2O_3 an exothermic or endothermic reaction?

Data:

$$C_p = a + b \times 10^{-3} \times T + c \times 10^{-6} \times T^2 + d \times 10^5 \times T^{-2} \; J\,K^{-1}\,mol^{-1}$$

	a	b	c	d
O_2(g) (298–700)	22.060	20.887	−8.207	1.621
O_2(g) (700–1200 K)	29.793	7.910	−2.204	−6.194
O_2(g) (1200–2500 K)	34.859	1.312	0.163	−14.140
Al(s) (298–933.45 K)	32.974	−20.677	23.753	−4.138
Al(l) (933.45–2790 K)	31.748	0	0	0
Al_2O_3(s) (100–800 K)	9.776	294.725	−198.174	−2.485
Al_2O_3(s) (800–2327 K)	115.977	15.654	−2.358	−44.290

Melting point of Al = 933.45 K
$\Delta_{fus}H$(Al) = 10.711 kJ
H^0(Al_2O_3, s, 298.15) = −1675.79 kJ

5 Sources of thermodynamic data for substances

SCOPE

This chapter describes some important sources of thermodynamic data for substances.

LEARNING OBJECTIVES

1. *Be able to extract thermodynamic data for single substances from online, open access sources.*
2. *Be able to perform simple manipulations of the data such as plotting their variation with temperature.*

5.1 INTRODUCTION

An important step in performing thermodynamic calculations is finding reliable data for the substances involved. Fortunately, thermodynamic data for all common, and many less common, substances have been determined, evaluated, collated and compiled. The first compilations were made more than a hundred years ago, and today there is a range of comprehensive tables of thermochemical data for pure substances. The advent of computer databases facilitated the compilation of data and has made accessing data quite easy. This chapter describes some key sources of thermodynamic data for substances and how some databases can be accessed online.

5.2 COMPILATIONS OF THERMODYNAMIC DATA

The primary sources of thermodynamic data are publications in research journals that report the results of experimental investigations and/or theoretical calculations. Usually, these publications report one or more properties of a substance or a selected range of substances. Complete sets (or compilations) of data for a substance are produced usually by other researchers who critically evaluate the available data and compile them into internally consistent tabular or other forms. There are two main types of compilations:

- Tables of thermodynamic values of substances at 1 bar (or in older compilations at 1 atm) at 298.15 K.
- Tables of thermodynamic values of substances at 1 bar (or 1 atm) in steps of 100 K. The values at 298.15 K and at transition temperatures are also included.

Originally, thermodynamic data were presented as printed tables, but now many compilations are available as computerised databases, often with the facility to download the tabular data into a spreadsheet so calculations can be performed. There are also numerous commercial software packages which, in addition to a database, include programs to calculate specific values at any temperature, perform various types of standard thermodynamic calculations or prepare tables or graphs.

5.2.1 THE REFERENCE STATE

A consistent set of thermodynamic properties requires a reference state be chosen for each element to which thermodynamic properties of all other forms of that element or any compound involving that element are referred. By convention, the usual reference state for an element is the element in its standard state. The standard state for solid and liquid elements is the most stable form of the element at 1 bar pressure and the relevant temperature. For gaseous elements, the standard state is the ideal gas at 1 bar pressure.

If the temperature range of interest is represented by a single phase for a particular element, values of thermodynamic properties for that substance in its reference state will be continuous. In other words, the data can be fitted by a smooth curve. Often however, the temperature range of interest will include more than one phase. For example, for magnesium there will be a solid phase from 0 to 923 K (the melting point of magnesium), a liquid phase from 923 to 1378 K (at which the vapour pressure of magnesium reaches 1 bar) and a gas phase at temperatures > 1378 K. The thermodynamic values in these cases will be discontinuous at these phase change temperatures. The reference states for magnesium will be, respectively, solid magnesium at 1 bar, liquid magnesium at 1 bar and gaseous magnesium as an ideal gas at 1 bar. Thermodynamic properties of magnesium compounds will be referred to the relevant elemental reference state. For magnesium oxide the relevant reactions for formation would be:

$$Mg(s) + 0.5\,O_2(g) = MgO(s) \quad \left(\text{for temperatures} <923\text{ K}\right)$$

$$Mg(l) + 0.5\,O_2(g) = MgO(s) \quad \left(\text{for temperatures } 923 - 1378\text{ K}\right)$$

$$Mg(g) + 0.5\,O_2(g) = MgO(s) \quad \left(\text{for temperatures} >1378\text{ K}\right)$$

Tabulated thermodynamic values for MgO reflect these reference states for magnesium. The data will be smooth within each temperature range, but there will be a discontinuity at 923 and 1378 K.

5.2.2 THE NIST–JANAF, NBS AND US GEOLOGICAL SURVEY TABLES

The NIST–JANAF compilation is a computerised database and can be accessed online at http://kinetics.nist.gov/janaf. The database was developed, and is maintained, by the United States National Institute of Standards and Technology (NIST) and is open access. To access the data, click on the periodic table, then on the element of interest. The compounds of that element are then displayed. Choose the desired compound and click on JANAF table. A table of the form shown in Table 5.1 is produced. If required, the data can be downloaded and the tabulated part copied into a spreadsheet. Column 2 lists heat capacity values, column 5 lists enthalpy increments, and column 6 lists enthalpy of formation values. Columns 3, 4, 7 and 8 contain entropy and Gibbs energy data and will be discussed in later chapters. The units of the properties are listed in Table 5.2.

An older compilation by the National Bureau of Standards (the predecessor of NIST) provides a very comprehensive collection of data at 298.15 K and 1 bar, including data for aqueous solutions:

Wagman, D. D. et al. 1982. The NBS tables of chemical thermodynamic properties. *Journal of Physical and Chemical Reference Data*, 11, Supplement 2.

This is open access, and a scanned copy (in PDF format) can be downloaded from the NIST website: https://srd.nist.gov/JPCRD/jpcrdS2Vol11.pdf

TABLE 5.1

Thermodynamic data for calcium oxide from the NIST–JANAF database. The data were downloaded then copied and pasted into a spreadsheet

Calcium Oxide (CaO)			Ca1O1(cr,l)				
T (K)	C_p	S^0	$-(G^0 - H^0(T))/T$	$H^0 - H^0(298)$	$\Delta_f H^0$	$\Delta_f G^0$	$\log K_f$
0	0	0	INFINITE	−6.749	−631.760	−631.760	INFINITE
100	14.715	6.222	69.166	−6.294	−633.739	−624.099	325.996
200	33.677	23.041	41.825	−3.757	−634.925	−613.887	160.331
298.15	42.120	38.212	38.212	0	−635.089	−603.501	105.731
300	42.242	38.473	38.213	0.078	−635.086	−603.305	105.045
400	46.626	51.300	39.932	4.547	−634.738	−592.755	77.406
500	48.982	61.980	43.305	9.338	−634.242	−582.316	60.834
600	50.480	71.052	47.193	14.315	−633.787	−571.975	49.795
700	51.555	78.917	51.175	19.419	−633.449	−561.701	41.915
800	52.400	85.858	55.085	24.618	−634.112	−551.362	36.000
900	53.112	92.072	58.856	29.894	−634.064	−541.024	31.400
1000	53.735	97.700	62.463	35.237	−634.273	−530.677	27.720
1100	54.300	102.849	65.904	40.639	−634.749	−520.297	24.707
1200	54.831	107.596	69.183	46.096	−643.226	−509.239	22.167
1300	55.329	112.005	72.309	51.604	−643.009	−498.082	20.013
1400	55.810	116.123	75.293	57.161	−642.759	−486.943	18.168
1500	56.275	119.989	78.146	62.766	−642.475	−475.823	16.570
1600	56.727	123.636	80.876	68.416	−642.159	−464.723	15.172
1700	57.174	127.088	83.494	74.111	−641.810	−453.644	13.939
1800	57.609	130.369	86.007	79.850	−790.102	−440.375	12.779
1900	58.045	133.495	88.425	85.633	−788.275	−420.996	11.574
2000	58.471	136.483	90.754	91.459	−786.423	−401.713	10.492
2100	58.894	139.346	93.000	97.327	−784.547	−382.523	9.515
2200	59.317	142.096	95.170	103.238	−782.651	−363.423	8.629
2300	59.735	144.742	97.268	109.190	−780.739	−344.411	7.822
2400	60.149	147.293	99.299	115.184	−778.816	−325.481	7.084
2500	60.563	149.757	101.269	121.220	−776.887	−306.632	6.407
2600	60.978	152.140	103.180	127.297	−774.958	−287.860	5.783
2700	61.388	154.449	105.036	133.415	−773.036	−269.161	5.207
2800	61.798	156.689	106.841	139.575	−771.126	−250.534	4.674
2900	62.208	158.865	108.597	145.775	−769.238	−231.976	4.178
3000	62.618	160.981	110.308	152.016	−767.378	−213.481	3.717
3100	63.024	163.040	111.976	158.298	−765.554	−195.048	3.287
3200	63.434	165.048	113.604	164.621	−763.774	−176.673	2.884
3200.000	63.434	165.048	113.604	164.621	CRYSTAL ↔ LIQUID		
3200.000	62.760	189.890	113.604	244.117	TRANSITION		
3300	62.760	191.821	115.945	250.393	−682.638	−160.836	2.546
3400	62.760	193.695	118.204	256.669	−681.099	−145.047	2.228
3500	62.760	195.514	120.387	262.945	−679.668	−129.302	1.930
3600	62.760	197.282	122.499	269.221	−678.352	−113.596	1.648
3700	62.760	199.002	124.543	275.497	−677.158	−97.925	1.382
3800	62.760	200.676	126.525	281.773	−676.092	−82.285	1.131
3900	62.760	202.306	128.447	288.049	−675.160	−66.670	0.893
4000	62.760	203.895	130.313	294.325	−674.364	−51.079	0.667

(Continued)

TABLE 5.1 (CONTINUED)

Thermodynamic data for calcium oxide from the NIST–JANAF database. The data were downloaded then copied and pasted into a spreadsheet

	Calcium Oxide (CaO)			Ca1O1(cr,l)				
T (K)	C_p	S^0	$-(G^0 - H^0(T))/T$	$H^0 - H^0(298)$	$\Delta_f H^0$	$\Delta_f G^0$	log K_f
4100	62.760	205.444	132.127	300.601	−673.711	−35.505	0.452
4200	62.760	206.957	133.891	306.877	−673.202	−19.946	0.248
4300	62.760	208.434	135.607	313.153	−672.840	−4.396	0.053
4400	62.760	209.876	137.279	319.429	−672.622	11.146	−0.132
4500	62.760	211.287	138.908	325.705	−672.538	26.684	−0.310

TABLE 5.2

The thermodynamic properties (and their units) listed in the NIST–JANAF tables

Column Number	Property	Unit
1	T	K
2	C_p	J K^{-1} mol^{-1}
3	S^0	J K^{-1} mol^{-1}
4	$-(G^0 - H^0(T))/T$	J K^{-1} mol^{-1}
5	$H^0 - H^0(298)$	kJ mol^{-1}
6	$\Delta_f H^0$	kJ mol^{-1}
7	$\Delta_f G^0$	kJ mol^{-1}
8	log K_f	–

The US Geological Survey (USGS) compilation is a scanned copy (in PDF format) of the publication:

Robie, R. A. and B. S. Hemingway. 1995. Thermodynamic properties of minerals and related substances at 298.15 K and 1 Bar (10^5 pascals) pressure and at higher temperatures. *U. S. Geological Survey Bulletin*, no. 2131.

It is open access and can be downloaded from the USGS website: https://pubs.er.usgs.gov/publicatio n/b2131. The table structure is similar to that of the NIST–JANAF tables.

Enthalpy values of substances (as distinct from enthalpies of formation) are not explicitly listed in the NIST–JANAF and USGS tables. However, since (Equation 4.29)

$$H^0(298.15) = \Delta_f H^0(298)$$

the enthalpy of a substance can be calculated if desired from tabulated values of $\Delta_f H^0(298)$ and $H^0(T) - H^0(298)$ as follows:

$$H^0(T) = H^0(298) + \left[H^0(T) - H^0(298) \right]$$

$$= \Delta_f H^0(298) + \left[H^0(T) - H^0(298) \right]$$

For example, using data from Table 5.1 the enthalpy of CaO at 500 K is given by

$$H^0(500) = \Delta H^0(298) + \left[H^0(500) - H^0(298) \right]$$

$$= -635.089 + 9.338 = -625.751 \text{ kJ mol}^{-1}$$

5.2.3 THE FREED SOFTWARE PROGRAM

FREED is a program which combines the thermodynamic databases of the U.S. Bureau of Mines and the U.S. Geological Survey in digital format. Both databases have been combined and configured to run inside the Excel™ spreadsheet program. Data for substances and reactions can be extracted and formatted in graphic or tabular form and exported to documents. This easy-to-use software, developed by Dr. Arthur E Morris, is now in the public domain along with a User's Guide and a set of examples and can be downloaded from the website: www.thermart.net/.

5.2.4 BARIN'S THERMOCHEMICAL TABLES

These tables were originally published in a series of printed volumes:

Barin, I. 1995. *Thermochemical Data of Pure Substances*, 3rd ed. Weinheim, New York: VCH Verlagsgesellschaft mbH.

An electronic version (in PDF format) was published in 2008 (Online ISBN: 9783527619825): http://onlinelibrary.wiley.com/book/10.1002/9783527619825. A wide range of substances is included in this compilation. However, access is normally available only through major research institutes and universities as the database is not open access. The tables have the form shown in Table 5.3 and have a similar structure to that of the NIST–JANAF tables but with two additional columns – the enthalpy values of substance are explicitly listed and the Gibbs energies, in addition to the Gibbs energy of formation, are listed. A comparison of the values of Barin's data for calcium oxide (Table 5.3) with the equivalent NIST–JANAF data shows very close agreement between the two compilations.

5.3 THERMOCHEMICAL SOFTWARE PROGRAMS

There are many commercial thermodynamics software products. Four widely used products are listed below. These have extensive thermodynamic databases, which are periodically updated. The packages can be used to extract thermodynamic data for substances as well as to do thermodynamic calculations. However, to use these to carry out the more advanced types of calculations requires a thorough knowledge of thermodynamics and extensive practice in using the software. Quite misleading conclusions can be made by inexperienced users; this has often occurred in practice. They are powerful packages and, in the hands of experienced users, can perform calculations that would be difficult, extremely time consuming or even impossible to perform manually and greatly expand the power and range of thermodynamics as a tool.

HSC Chemistry™ – developed by Outotec: www.outotec.com/en/Products––services/HSC–Chemistry/

FactSage™ – a fusion of the earlier FACT Win/F*A*C*T and ChemSage/SOLGASMIX thermochemical software: www.factsage.com

MTDATA™ – developed by the UK National Physical Laboratory: http://resource.npl.co.uk/mtdata/mtdatasoftware.htm

Thermo–Calc™ – developed by the Division of Physical Metallurgy at KTH, Stockholm, Sweden: www.thermocalc.com/start

TABLE 5.3

Thermodynamic data for calcium oxide from Barin's tables

CaO   Calcium Oxide

Phase	T	C_p	S	$-(G-H\,298)$	H	$H-H\,298$	G	$\Delta_f H$	$\Delta_f G$	log K
	[K]		[J/K/mol]			56.077	[kJ/mol]			[−]
SOL	298.15	42.122	38.074	38.074	−635.089	0.000	−646.441	−635.089	−603.509	105.732
	300.00	42.239	38.335	38.075	−635.011	0.078	−646.512	−635.085	−603.313	105.046
	400.00	46.628	51.163	39.794	−630.541	4.548	−651.007	−634.675	−592.775	77.409
	500.00	48.981	61.844	43.167	−625.751	9.338	−656.673	−634.107	−582.365	60.839
	600.00	50.479	70.915	47.055	−620.773	14.316	−663.322	−633.567	−572.069	49.803
	700.00	51.555	78.781	51.038	−615.669	19.420	−670.816	−633.157	−561.854	41.926
	800.00	52.401	85.722	54.948	−610.470	24.619	−679.048	−633.668	−551.594	36.015
	900.00	53.111	91.936	58.719	−605.193	29.896	−687.936	−633.565	−541.345	31.419
	1000.00	53−735	97.565	62.326	−599.850	35.239	−697.415	−633.837	−531.087	27.741
	1100.00	54.302	102.714	65.767	−594.448	40.641	−707.433	−634.486	−520.784	24.730
	1200.00	54.830	107.461	69.046	−588.991	46.098	−717.945	−642.459	−509.789	22.191
	1300.00	55.330	111.870	72.173	−583.483	51.606	−728.914	−641.672	−498.764	20.041
	1400.00	55.810	115.988	75.157	−577926	57.163	−740.309	−640.850	−487.802	18.200
	1500.00	56.275	119.854	78.009	−572.321	62.768	−752.103	−639.995	−476.900	16.607
	1600.00	56.729	123.501	80.740	−566.671	68.418	−764.273	−639.107	−466.056	15.215
	1700.00	57.173	126.953	83.358	−560.976	74.113	−776.797	−638.187	−455.268	13.989
	1800.00	57.611	130.234	85.871	−555.237	79.852	−789.658	−790.538	−440.571	12.785
	1900.00	58.043	133.360	88.289	−549.454	85.635	−802.838	−788.712	−421.178	11.579
	2000.00	58.470	136.348	90.618	−543.628	91.461	−816.325	−786.859	−401.882	10.496
	2100.00	58.894	139.211	92.864	−537.760	97.329	−830.104	−784.983	−382.680	9.519
	2200.00	59.315	141.961	95.034	−531.850	103.239	−844.163	−783.085	−363.567	8.632
	2300.00	59.734	144.607	97.132	−525.897	109.192	−858.493	−781.167	−344.541	7.825
	2400.00	60.150	147.158	99.164	−519.903	115.186	−873.082	−779.232	−325.598	7.086
	2500.00	60.564	149.622	101.133	−513.867	121.222	−887.921	−777.283	−306.737	6.409
	2600.00	60.977	152.005	103.044	−507.790	127.299	−903.003	−775.321	−287.954	5.785

(Continued)

TABLE 5.3 (CONTINUED)
Thermodynamic data for calcium oxide from Barin's tables

CaO	Calcium Oxide					56.077					
Phase	T	C_p	S	$-(G-H\,298)$	H	$H-H\,298$	G	$\Delta_f H$	$\Delta_f G$	$\log K$
	[K]		[J/K/mol]				[kJ/mol]			[–]
	2700.00	61.389	154.314	104.900	−501.672	133.417	−918.320	−773.349	−269.246	5.209
	2800.00	61.799	156.554	106.705	−495.512	139.577	−933.864	−771.370	−250.612	4.675
	2900.00	62.208	158.730	108.462	−489.312	145.777	−949.629	−769.385	−232.048	4.180
	3000.00	62.617	160.846	110.173	−483.071	152.018	−965.608	−767.396	−213.554	3.718
	3100.00	63.025	162.906	111.841	−476.789	158.300	−981.796	−765.407	−195.125	3.288
	3200.00	63.432	164.913	113.468	−470.466	164.623	−998.187	−763.419	−176.760	2.885
		24.843			79.496					
LIQ	3200.00	62.760	189.755	113.468	−390.970	244.119	−998.187	−683.923	−176.760	2.885
	3300.00	62.760	191.687	115.809	−384.694	250.395	−1017.260	−682.027	−160.941	2.547
	3400.00	62.760	193.560	118.069	−378.418	256.671	−1036.523	−680.177	−145.178	2.230
	3500.00	62.760	195.379	120.252	−372.142	262.947	−1055.970	−678.376	−129.470	1.932

PROBLEMS

5.1   Calculate the value of H^0 for CaO at 900 K using the data of Table 5.1 and confirm the answer using the data of Table 5.3. Compare it with the value calculated in Problem 4.7. At what temperature does CaO melt? What is the enthalpy of melting of CaO? Is the process endothermic or exothermic?

5.2   Download the data for elemental aluminium from the NIST–JANAF website, and convert it to spreadsheet format.

   i.   Construct plots showing the variation with temperature of C_p, $\Delta_f H^0$, $H^0(T) - H^0(298.15)$ and H^0 over the range 0 to 1500 K. Compare these with the values calculated using empirical equations in Problem 4.13.

   ii.   What is the value of H^0 at 298.15 K? Why does it have this value?

   iii.   Explain the discontinuity in $H^0(T) - H^0(298.15)$ at 933 K. What is the enthalpy of melting of aluminium?

5.3   Download data for the compound Al_2O_3 from the NIST–JANAF website, and convert it to spreadsheet format.

   i.   Construct plots showing the variation with temperature of C_p, $\Delta_f H^0$, $H^0(T) - H^0(298.15)$ and H^0 with temperature over the range 0 to 1500 K. Compare these with the values calculated in Problem 4.13.

   ii.   What is the value of H^0 at 298.15 K? What is the significance of this value?

5.4   Silicon metal has been proposed as a storage material of thermal energy for solar energy systems. The stored energy would be released in periods of demand when solar energy is unavailable (for example, at night) and used to generate electricity.

   i.   How much sensible energy can be stored per kilogram of silicon if the metal is raised in temperature to 1500, 1600 and 1700 K, respectively?

   ii.   How do you account for the very much larger storage capacity at 1700 K compared to 1600 K?

5.5   Using data from the USGS or NBS tables, calculate the standard enthalpy of reaction for the following chemical reactions at 298.15 K. State which are endothermic and which are exothermic.

$$2\,CaO(s) + SiO_2(s) = Ca_2SiO_4(s)$$

$$N_2(g) + 3\,H_2(g) = 2\,NH_3(g)$$

$$CH_4(g) + 2\,S_2(g) = 2\,H_2S(g) + CS_2(g)$$

$$CH_4(g) + NH_3(g) = HCN(g) + 3\,H_2(g)$$

$$ZnSO_4.7H_2O(s) = ZnSO_4.6H_2O(c) + H_2O(l)$$

$$Na_2SiO_3(s) + H_2SO_4(l) = SiO_2(s) + Na_2SO_4(aq) + H_2O(l)$$

6 Some applications of the first law

SCOPE

This chapter shows how the law of conservation of energy and the concept of enthalpy are used to determine the energy requirement of processes.

LEARNING OBJECTIVES

1. *Be able to perform energy balance calculations for simple non-reactive and reactive systems given the input and output conditions.*
2. *Be able to calculate the adiabatic reaction temperature of a simple reactive system.*
3. *Understand the concept of available heat and the limits on heating by combustion imposed by thermodynamics.*

6.1 INTRODUCTION

Modern society depends heavily on the use of energy. Such use requires transformations in the forms of energy and control of energy flows. When coal or natural gas is burned at a power station, part of the chemical potential energy stored in the molecules comprising the coal is converted to heat (to raise steam) which is converted to mechanical (kinetic) energy in turbines which in turn drive generators to convert it to electrical energy. The electrical energy is distributed through a grid and eventually converted back, by customers, into mechanical energy (by means of electric motors in all sorts of applications) or directly into heat.

Apart from their clear positive benefits, energy flows and transformations also cause environmental problems. Waste heat from power stations can result in increased temperatures in rivers used for cooling water; greenhouse gases (particularly CO_2 from fossil fuel combustion) released into the atmosphere alter the energy balance of the earth and, over time, cause increases in global temperatures and climate change; and many of our uses of energy (for example, in manufacturing and mining) themselves cause further emission of pollutants.

One of the uses of thermodynamics is to calculate the efficiency of energy conversion processes so that choices can be made between options in order to minimise the use of resources and limit the losses and environmental impact. The main practical applications of the first law involve determining the energy requirements of processes – natural processes (for example, geological, biological), industrial processes (such as manufacturing, mining, chemical processing, etc.) and domestic and consumer processes. These calculations are performed using only two types of thermodynamic data:

- Enthalpy increments $H^0(T) - H^0(298.15)$, sometimes written as H–H(298)
- Enthalpies of formation at 298.15 K of the substances involved in the process (reactants and products)

We will examine some of these applications in this chapter.

6.2   HEATING AND COOLING OF SUBSTANCES

Probably the simplest application of the first law is to calculate the thermal energy required to heat a substance to a specified temperature or to cool it to a specified temperature. To do this use is made of the enthalpy increment values of the substance. This type of calculation is used in practice to calculate the quantity of fuel or electrical energy required to heat or melt substances and the amount of coolant required to cool or solidify substances.

EXAMPLE 6.1   Melting of aluminium

i. Scrap aluminium is frequently melted in furnaces and recast for recycling. How much energy is required per kilogram of aluminium? Assume the scrap is initially at 25°C and its temperature is to be raised to 1100 K.

SOLUTION

Table 6.1 lists thermodynamic data for aluminium from which it can be seen that aluminium melts at 933 K (660°C). Figure 6.1 shows the enthalpy increment data from the table as a function of temperature. The interval between 298.15 and 933 K represents the heat required to raise the temperature of 1 mole of solid aluminium from 298.15 to 933 K. The vertical section represents the heat required to melt the aluminium at 933 K (the latent heat or enthalpy of fusion). The section above 933 K represents the heat required to raise the temperature of the molten aluminium to 1100 K. The heat required to raise the temperature of 1 mole of aluminium to its melting point is 17.89 kJ. The heat required to fully melt 1 mole of aluminium, if its starting temperature is 298.15 K, is 28.69 kJ. The heat required to melt 1 mole of aluminium and raise its temperature from 298.15 to 1100 K is 33.98 kJ.

TABLE 6.1
Thermodynamic data for aluminium (NIST–JANAF tables)

T	C_p	S	$-(G^0-H(T))/T$	$H-H(298)$	$\Delta_f H^0$	$\Delta_f G^0$	$\log K_f$
K		J/K/mol				J/mol	
0	0	0	INFINITE	−4.539	0	0	0
100	12.997	6.987	47.543	−4.056	0	0	0
200	21.338	19.144	30.413	−2.254	0	0	0
298.15	24.209	28.275	28.275	0	0	0	0
300	24.247	28.425	28.276	0.045	0	0	0
400	25.784	35.63	29.248	2.553	0	0	0
500	26.842	41.501	31.129	5.186	0	0	0
600	27.886	46.485	33.283	7.921	0	0	0
700	29.1	50.872	35.488	10.769	0	0	0
800	30.562	54.85	37.663	13.749	0	0	0
900	32.308	58.548	39.78	16.89	0	0	0
933.45	32.959	59.738	40.474	17.982	CRYSTAL ↔ LIQUID		
933.45	31.751	71.213	40.474	28.693	TRANSITION		
1000	31.751	73.40	42.594	30.806	0	0	0
1100	31.751	76.426	45.534	33.981	0	0	0
1200	31.751	79.189	48.225	37.156	0	0	0
1300	31.751	81.73	50.706	40.331	0	0	0
1400	31.751	84.083	53.007	43.506	0	0	0

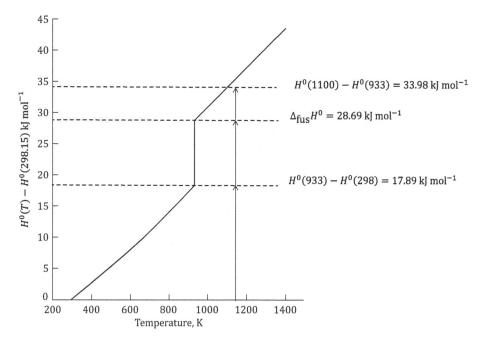

FIGURE 6.1 Enthalpy increments for aluminium.

TABLE 6.2
Conversion of kJ mol⁻¹ Al to kJ kg⁻¹ Al

	kJ mol⁻¹	kJ kg⁻¹ (or MJ tonne⁻¹)
Heat from 298.15 K to 933 K (solid)	17.98	$\frac{17.98}{26.98} \times 1000 = 666$
Heat to melt Al at 933 K (molten)	28.69	$\frac{28.69}{26.98} \times 1000 = 1063$
Heat to 1100 K	33.98	$\frac{33.98}{26.98} \times 1000 = 1259$

Note: The relative atomic mass of aluminium is 26.98.

In practical applications 1 kg (or 1 tonne) would normally be used as the basis rather than 1 mole so the above values need to be converted to kJ/kg (or MJ/tonne). The conversion is summarised in Table 6.2. To melt 1 kg of aluminium initially at 298.15 K and raise its temperature to 1100 K requires 1259 kJ of heat.

COMMENT

The above values also apply if molten aluminium is cooled. In this case they represent the amount of heat that needs to be removed (and some of it potentially recovered) from the aluminium.

  ii. If the heat to melt the aluminium is to be supplied electrically, for example by resistance heating or induction heating, calculate the quantity of electrical energy required to heat 1 kg Al scrap to 1100 K.

SOLUTION

Electrical energy is expressed in units of kilowatt hour (kW h) where 1 kW h is defined as the electrical energy equivalent to 1 kW of power expended for 1 hour:

$$1 \text{ kW h} = 1[\text{kW}] \times 3600[\text{s}] = 3600[\text{s}]\left[\frac{\text{kJ}}{\text{s}}\right] = 3600 \text{ kJ} = 3.6 \text{ MJ}$$

Therefore the electrical energy required to melt 1 kg of aluminium and raise its temperature to 1100 K is 1259/3600 = 0.35 kW h. In practice some heat would be lost through the walls of the furnace (by conduction to the outer wall, then convection from the outer wall to the air), and an additional allowance would need to be made for this.

Alternatively, the heat could be supplied by burning a fuel, for example natural gas, in the space in the furnace around the aluminium charge. The quantity of fuel required per kilogram can also be calculated. This is discussed later.

iii. If the aluminium is placed in the furnace initially at a temperature of 400 K rather than 298.15 K, how much energy is required in this case to heat it to 1100 K?

We would expect the energy required would be less because it already contains some sensible heat. The value is the difference between the value of $H^0(1100) - H^0(298)$ and $H^0(400) - H^0(298)$; that is, $33.98 - 2.553 = 31.427$ kJ per mole of aluminium or $31.427 \times 1000/26.98 = 1165$ kJ per kilogram of aluminium.

6.3   ENERGY BALANCES

The movement of energy and changes in its form during a process can be tracked using energy balances based on the law of conservation of energy. Calculations to do this involve using enthalpies of the input and output substances and enthalpies of formation. In Section 6.2 we saw how to calculate the enthalpy change for the simple process of heating or cooling of a single substance, and in Section 4.9 we saw how to calculate the enthalpy change for a single reaction at constant temperature. Many processes (natural or otherwise) are complex and may involve several phases, phase changes and numerous chemical reactions, and the reactants and products may be at different temperatures. The system in these cases is rarely at one uniform temperature throughout, and the conditions within the system are seldom known accurately. However, the states (that is, the temperature, amounts and compositions of the input substances and of the products, the output substances) can usually be measured, calculated or estimated, and this makes it possible to calculate the heat surplus or deficit for a process.

When complete data on the initial and final states of materials in a process are known, we can envisage an idealised process that will accomplish the same overall change in state in the materials as the actual process but which draws on readily available thermodynamic data for the substances. Since enthalpy is a state function, this ideal process will have the same enthalpy change as the actual process. Such an idealised process is illustrated in Figure 6.2. The full line arrow at the top shows the actual process while the dashed line arrows show the idealised process which achieves the same outputs given the same inputs. In the idealised process, the reactants are heated or cooled from their input temperature to 298.15 K, reacted by means of the appropriate chemical reaction(s) at 298.15 K, then the products are heated or cooled to their output temperature. This idealised process usually bears little resemblance to the actual process beyond the required identity of the initial and final states. Its sole purpose is to enable the change in enthalpy (that is, change in heat content

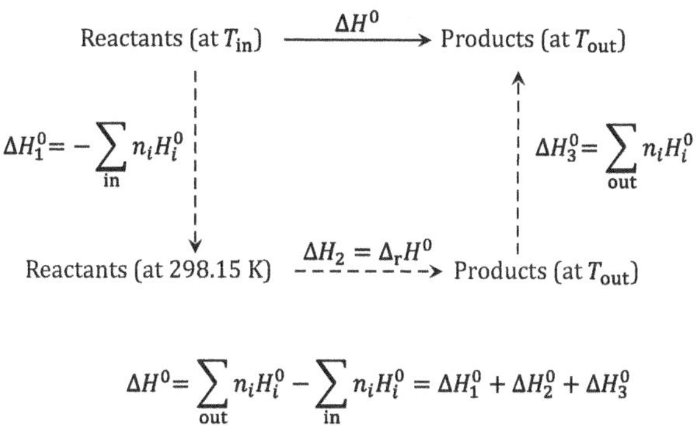

Reactants (at T_{in}) $\xrightarrow{\Delta H^0}$ Products (at T_{out})

$$\Delta H^0_1 = -\sum_{in} n_i H^0_i$$

$$\Delta H^0_3 = \sum_{out} n_i H^0_i$$

Reactants (at 298.15 K) $\xrightarrow[\text{---------}]{\Delta H_2 = \Delta_r H^0}$ Products (at T_{out})

$$\Delta H^0 = \sum_{out} n_i H^0_i - \sum_{in} n_i H^0_i = \Delta H^0_1 + \Delta H^0_2 + \Delta H^0_3$$

FIGURE 6.2   An idealised process to convert input substances to output substances.

if the process operates at constant pressure) of the process to be calculated. The enthalpy change of the system in moving from the input state to the output state is given by

$$\Delta H^0 = \sum_{out} n_i H^0_i - \sum_{in} n_i H^0_i$$

where the symbol Σ (upper-case sigma) indicates 'the sum of'. ΔH^0 is also equal to the algebraic sum of the three steps of the idealised process, that is,

$$\Delta H^0 = \Delta H^0_1 + \Delta H^0_2 + \Delta H^0_3$$

The values of ΔH^0_1, ΔH^0_2 and ΔH^0_3 can be determined from tables of thermodynamic data, thus enabling the enthalpy change ΔH^0 of the process to be calculated.

An alternative idealised process is shown in Figure 6.3 in which the reactants are decomposed to their constituent elements at 298.15 K then recombined into the products at the relevant product temperatures. This approach is equally acceptable and will produce the same overall result and, in fact, is the method used within the software of many thermodynamics packages. However, we will follow the approach illustrated in Figure 6.2.

An energy balance for a system or process is usually presented in tabular form showing the input energy, output energy and the difference (if any). In some ways it resembles a financial balance sheet showing income, expenses and profit or loss. The previous example of melting of aluminium

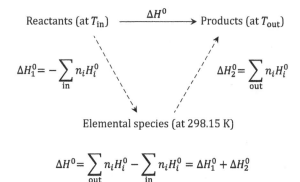

Reactants (at T_{in}) $\xrightarrow{\Delta H^0}$ Products (at T_{out})

$$\Delta H^0_1 = -\sum_{in} n_i H^0_i$$

$$\Delta H^0_2 = \sum_{out} n_i H^0_i$$

Elemental species (at 298.15 K)

$$\Delta H^0 = \sum_{out} n_i H^0_i - \sum_{in} n_i H^0_i = \Delta H^0_1 + \Delta H^0_2$$

FIGURE 6.3   An alternative idealised process to convert input substances to output substances.

TABLE 6.3
Energy balance for the heating of 1 kilogram of aluminium from 400 to 1100 K

INPUT	T (K)	Amount (mol)	Amount (kg)	$-(H-H(298))$ (kJ)
Al(s)	400	−37.06	−1.00	−95
OUTPUT				$H-H(298)$ (kJ)
Al(l)	1100	37.06	1.00	1259
DEFICIT		**0**	**0**	**1164**

is an example of a non-reactive system – the aluminium changes state but does not change chemically; hence there is no reaction. Using the data from Example 6.1(iii), the energy balance for heating aluminium from 400 to 1100 K is shown in Table 6.3. Columns 1 to 4 summarise the known input and output states. Column 5 shows the sensible heats. The surplus/deficit row gives the sum of the sensible heats and shows that 1164 kJ is required per kilogram of Al heated to 1100 K. That heat required is indicated by the sign of the value: A negative value indicates heat is generated (a surplus), a positive value that heat is required (a deficit). This is a trivial example of an energy balance, but the format of the table remains largely the same for more complex systems, as illustrated in subsequent examples.

EXAMPLE 6.2   Combustion of methane

Methane (CH_4) is a commonly used fuel for domestic heating and cooking and for industrial processes. It makes up approximately 95% by volume of natural gas. Calculate the heat generated by combusting methane in air to provide heat to melt aluminium at 1100 K.

SOLUTION

The combustion reaction is

$$CH_4(g) + 2\ O_2(g) = CO_2(g) + 2\ H_2O(g)$$

Air is composed of approximately 21 vol% oxygen and 79 vol% nitrogen. The nitrogen is relatively inert so it can be assumed it doesn't react, although in practice a small amount usually reacts in high temperature combustion processes to form nitrogen oxides – one of the sources of pollution from fossil fuel combustion. Assume a stoichiometric amount of oxygen is used, that is, 2 moles of oxygen per mole of methane. For every mole of oxygen there will be 79/21 = 3.76 moles of nitrogen. Assume also the air and the methane are both initially at 25°C (298.15 K). These are the inputs.

The outputs are 1 mole of CO_2 and 2 moles of water vapour per mole of methane combusted. Let us also assume that the combustion products (CO_2, H_2O and N_2) leave the furnace at 1100 K. The actual system is illustrated in Figure 6.4(a), and the energy balance calculation procedure is summarised in Figure 6.4(b).

The thermodynamic data required are the enthalpy of formation at 298.15 K of all the substances in the reaction (CH_4, O_2, CO_2 and H_2O) and the enthalpy increments $H(T)-H(298)$ for all the input and output substances at the relevant temperature, 298.15 K for the input substances and 1100 K for the output substances. These are all readily available from one of the common databases. The results are shown in Table 6.4 from which it is seen that 517 kJ of thermal energy are produced per mole of methane burned.

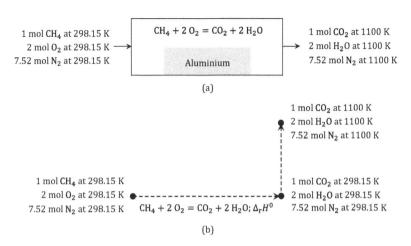

FIGURE 6.4 Idealised process for the combustion of methane: (a) actual process, (b) idealised process.

TABLE 6.4
Energy balance for the combustion of methane in a furnace

INPUT	T (K)	Amount (mol)	Amount (g)	$-\Delta_f H(298)$ (kJ)	$-(H-H(298))$ (kJ)	Total H (kJ)
$CH_4(g)$	298.15	−1.00	−16.04	74.60	0.00	74.60
$O_2(g)$	298.15	−2.00	−64.00	0.00	0.00	0
$N_2(g)$	298.15	−7.52	−210.66	0.00	0.00	0
Total in			−290.70	74.60	0.00	74.6
OUTPUT				$-\Delta_f H(298)$ (kJ)	$H-H(298)$ (kJ)	
$N_2(g)$	1100	7.52	210.66	0.00	186.19	186.19
$CO_2(g)$	1100	1.00	44.01	−393.50	38.90	−354.60
$H_2O(g)$	1100	2.00	36.03	−483.65	60.49	−423.16
Total out			290.70	−877.15	285.58	−591.57
SURPLUS		**0.00**	**0.00**	**−802.55**	**285.58**	**−516.97**

It will be noted that Table 6.4 has two more columns than Table 6.3. This is the form of table used for energy balances for reactive systems. Columns 1 to 4 summarise the known input and output states. It is important at this stage to confirm that the total mass of input substances equals the total mass of output substances – in other words that the law of conservation of matter has been obeyed. The enthalpy values are then entered into columns 5 and 6 (taking into account the relevant number of moles). Negative values are entered for the input substances. Column 7 lists the sums of the two enthalpy components for each input and output substance and gives their overall sum at the bottom. The values in the surplus/deficit row are the sums of the values for the inputs and outputs for each column. The value −802.55 kJ is the enthalpy of reaction at 298.15 K for the combustion of methane, and the value 285.58 kJ is the shortfall in sensible heat. Their sum is the surplus or deficit heat, namely, −516.97 kJ. That this is a surplus is indicated by the negative sign. This is called the *available heat* since this is the energy from the combustion of methane that is available for heating of scrap aluminium.

The heat balance in Example 6.2 assumes that the metal and combustion gases leave the furnace at 1100 K. In a batch process, scrap is added to the furnace, then heated to the required temperature and removed. In this case, the energy balance in Table 6.4 applies only at the end of the heating process since before then the temperature of the metal will be less than 1100 K, and the combustion gases leaving the furnace will also be less than 1100 K because they are giving up heat to the cooler metal. In a continuous melting process, scrap is continuously fed to the furnace (at one end) and molten metal at 1100 K is continuously removed from the furnace (at the other end). In this case, the energy balance is a snapshot at any time during steady-state operation.

EXAMPLE 6.3   The effect of pre-heating air

If in Example 6.2, the combustion air was pre-heated to 400 K (using some of the heat from the flue gases), calculate the thermal energy available to heat the aluminium scrap assuming the combustion products leave the furnace at 1100 K.

SOLUTION

The calculation procedure is as shown in Figure 6.5, and the energy balance is shown in Table 6.5. In this case more heat is available, 545 kJ compared with 517 kJ, for heating the contents of the furnace. Thus, pre-heating the air reduces the amount of fuel required per unit of available heat.

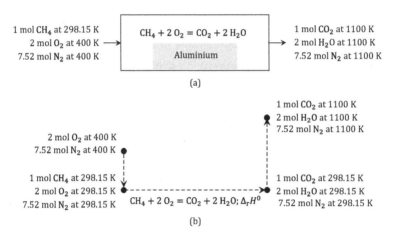

FIGURE 6.5   Idealised process for the combustion of methane with pre-heated combustion air: (a) actual process, (b) idealised process.

TABLE 6.5

Energy balance for the combustion of methane in a furnace with combustion air pre-heated

INPUT	T (K)	Amount (mol)	Amount (g)	$-\Delta_f H(298)$ (kJ)	$-(H-H(298))$ (kJ)	Total H kJ
$CH_4(g)$	298	−1.00	−16.04	74.60	0.00	74.60
$O_2(g)$	400	−2.00	−64.00	0.00	−6.05	−6.05
$N_2(g)$	400	−7.52	−210.66	0.00	−22.34	−22.34
Total in			−290.70	74.60	−28.39	
OUTPUT				$-\Delta_f H(298)$ (kJ)	$H-H(298)$ (kJ)	
$O_2(g)$	1100	0.00	0.00	0.00	0.00	0.00
$N_2(g)$	1100	7.52	210.66	0.00	186.19	186.19
$CO_2(g)$	1100	1.00	44.01	−393.50	38.90	−354.60
$H_2O(g)$	1100	2.00	36.03	−483.65	60.49	−423.16
Total out			290.70	−877.15	285.58	
SURPLUS		**0.00**	**0.00**	**−802.55**	**257.19**	**−545.36**

6.4   ADIABATIC TEMPERATURE OF REACTION

In an adiabatic process no heat is exchanged between the system and the surroundings, and in an exothermic chemical reaction the heat produced in an adiabatic system will raise the temperature of the system. The temperature attained by a system in undergoing an adiabatic reaction is called the adiabatic reaction temperature, and it is the theoretically highest temperature attainable by the system by utilising the heat of reaction. It is a particularly useful concept in combustion of fuels, the adiabatic flame temperature (AFT) of a fuel being the highest temperature attainable with that fuel. If the composition of the fuel and the products of combustion and their relative amounts are known it is possible to calculate the adiabatic temperature by means of an energy balance. In this case, the output temperature is unknown and the heat surplus/deficit will be zero. In other words, all the heat is used to raise the temperature of the output gases. The object is to find the temperature at which the sum of the sensible heat of the output products makes the balance equal to zero. When performed manually from tabulated data this can most easily be done by interpolating values calculated for the surplus/deficit over a range of temperatures to find the value which makes the surplus/deficit equal to zero. This is illustrated in the following example. Many computer-based thermodynamic packages perform these calculations routinely.

EXAMPLE 6.4   The adiabatic flame temperature of methane

Calculate the AFT of methane if it is burned stoichiometrically in air, assuming the air and methane are both initially at 25°C.

SOLUTION

Using the input data from the previous example, the AFT can be determined by substituting values of $H - H(298)$ of the output gases at various temperatures until the temperature is found at which the overall energy change of the system is zero. The exact temperature can be determined by calculating the energy balance over a range of temperatures which span the expected value of the AFT. This is illustrated in Table 6.6 and in Figure 6.6 which shows the relation is nearly linear; hence a linear interpolation can be used to determine the temperature at which the surplus/deficit has a value of zero. This is found to be 2325 K (2052°C). The energy balance is summarised in Table 6.7.

TABLE 6.6
Energy balance for the combustion of methane at temperatures from 2000 to 2600 K

T (K)	INPUT Total H (kJ)	OUTPUT Total H (kJ)	SURPLUS/DEFICIT (kJ)
2000	−74.6	−217.31	−142.71
2200	−74.6	−129.89	−55.29
2400	−74.6	−41.46	33.14
2600	−74.6	47.84	122.44

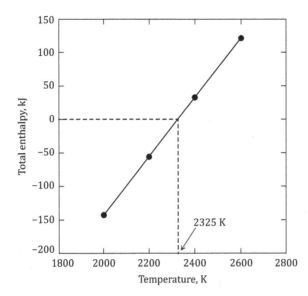

FIGURE 6.6   The total enthalpy change (surplus/deficit) for the combustion of methane at temperatures from 2000 to 2600 K.

TABLE 6.7
Energy balance to calculate the adiabatic flame temperature of methane combusted with a stoichiometric amount of air at 25°C

INPUT	T (K)	Amount (mol)	Amount (g)	$-\Delta_f H(298)$ (kJ)	$-(H - H(298))$ (kJ)	Total H (kJ)
CH4(g)	298	−1.00	−16.04	74.60	0.00	74.60
O2(g)	298	−2.00	−64.00	0.00	0.00	0.00
N2(g)	298	−7.52	−210.66	0.00	0.00	0.00
Total in			−290.70	74.60	0.00	74.60
OUTPUT				$-\Delta_f H(298)$ (kJ)	$H - H(298)$ (kJ)	
O_2(g)	2325	0.00	0.00	0.00	0.00	0.00
N_2(g)	2325	7.52	210.66	0.00	510.72	510.72
CO_2(g)	2325	1.00	44.01	−393.50	111.23	−282.27
H_2O(g)	2325	2.00	36.03	−483.66	180.61	−303.05
Total out			290.70	−877.16	802.56	−74.60
SURPLUS/ DEFICIT		**0.00**	**0.00**	**−802.56**	**802.56**	**0.00**

AFT is not a unique property of a fuel but depends on the conditions of combustion, particularly the inlet temperature of the fuel and air, the air-to-fuel ratio and the degree of oxygen enrichment. Therefore, the AFT of different fuels must be compared under the same conditions of combustion. Values of AFT for some common fuels are listed in Table 6.8. The AFT is never achieved in real flames for two reasons. Complete combustion, for example of CO to CO_2, does not occur at high

---

TABLE 6.8
Approximate values of adiabatic flame temperatures of some common fuels for complete combustion with stoichiometric dry air with air and fuel initially at 25°C

Hydrogen gas	2250°C
Bituminous coal	2100–2250°C
Fuel oil	2100°C
Kerosene	2090°C
Methane gas	2050°C
Wood	1980°C

---

temperatures because of chemical equilibrium considerations. This will become clearer when the equilibrium of reactions is examined in Chapter 10. It is possible to correct for incomplete combustion by calculating the equilibrium composition of the combustion gases. The second, more important reason, is that although combustion is rapid it is not instantaneous, and during combustion some heat is lost to the surroundings from a flame; hence, combustion only approximates an adiabatic process. Typically, the temperature of flames is between 50 and 75% of the AFT.

6.5 HEAT UTILISATION IN FURNACES

In an energy balance the total input and output of heat are of most interest, and little emphasis is placed on the question of the supply of heat at a temperature appropriate for the process. An input of a certain amount of heat to a process tells us nothing about the temperature at which this quantity of heat must be supplied. In the example of melting aluminium and heating it to 1100 K, thermal energy is transferred from the flame and hot combustion products to the aluminium (by radiation and convection). Heat can be transferred spontaneously only from a hotter to a cooler body. We know this from everyday experience, but it is also embodied in the second law of thermodynamics (Chapter 7). Therefore, in a process in which scrap is continually being added to the furnace and molten aluminium at 1100 K is being removed at the same rate, the maximum thermal energy that can be given up by the combustion gases to the aluminium scrap (to heat it) is the energy released when the combustion gases cool from the flame temperature to 1100 K.

EXAMPLE 6.5   Available heat

Calculate the thermal energy available to heat aluminium scrap to 1100 K by the stoichiometric combustion of methane gas.

SOLUTION

This involves calculating the heat released when the combustion gases cool from the flame temperature to 1100 K since it is this energy that is available to be transferred to the scrap. No chemical reaction with the aluminium occurs so the energy balance has the form shown in Table 6.9 using the AFT as the flame temperature. The value of 516.98 kJ of heat transferred from the combustion gases to heat the aluminium is, of course, the same as the value calculated in Example 6.2 as the heat transferred to the furnace charge.

TABLE 6.9

Energy balance for cooling methane combustion gases from the AFT to 1100 K

INPUT	T (K)	Amount (mol)	Amount (g)	$-(H-H(298))$ (kJ)
$N_2(g)$	2325	7.52	210.66	−510.72
$CO_2(g)$	2325	1.00	44.01	−111.23
$H_2O(g)$	2325	2.00	36.03	−180.61
Total in			290.70	−802.56
OUTPUT				$H-H(298)$ (kJ)
$N_2(g)$	1100	7.52	210.66	186.19
$CO_2(g)$	1100	1.00	44.01	38.90
$H_2O(g)$	1100	2.00	36.03	60.49
Total out			290.70	285.58
SURPLUS		**0.00**	**0.00**	**−516.98**

The above calculation is of interest because it shows that of the total sensible heat in the combustion gases, 802.59 kJ, only 516.98 kJ, the available heat, is transferred to the aluminium. This is an efficiency of 64.4%. The remaining 285.58 kJ is sensible heat in the flue gases, which leave the furnace at 1100 K. In practice, the efficiency would be much lower for two reasons. Firstly, the AFT is never actually achieved in practice, and, secondly, the flue gas will actually leave the furnace at a temperature greater than 1100 K because complete transfer of all the available heat to the charge is never attained.

Clearly, the available heat of the process depends on the flame temperature of the fuel; the higher this is the greater will be the fraction of the sensible heat available to the process. The way to increase the flame temperature of a fuel is to minimise the thermal load by, for example, minimising the amount of excess air for combustion and the moisture content of the fuel, increasing the heat input to the combustion process by pre-heating the air and/or fuel and reducing the amount of nitrogen to be heated by enriching the air with oxygen. The amount of available heat actually transferred to the charge depends on the furnace design and operating practice.

If the amount of gas leaving the reaction zone can be reduced then the amount of sensible heat leaving the reactor will also be reduced. Electrical heating has the advantage of not generating combustion gases at the point of utilisation, and the only flue gases from electrically heated processes are those formed by chemical reactions between the charge materials. This volume is relatively small compared with the volume produced from burning fuels for heat. Furthermore, the 'flame temperature' of electric arcs and plasmas is very high. Therefore, the available heat for processes using electrical heating is very high (in most cases probably greater than 90%). Electrical heating, which is expensive per unit of heat produced compared with combustion of fuels, becomes economical when the temperature required by the particular process is high because then the fraction of available heat from fuels is small.

PROBLEMS

Use data from the USGS, NBS or NIST–JANAF tables where required in answering these questions.

6.1   Calculate the quantity of heat required to be removed to cool a 1 tonne billet of steel from 1700 to 800 K. For the purpose of the calculation assume the steel is pure iron (actually it is typically around 98 mass% Fe). If recycled water at 100°C is used to cool the steel, calculate the consumption of water (that is, what mass of water is lost through vapourisation?).

6.2   Calculate the quantity of heat required to raise the temperature of 1 m^3 of air at 298.15 K to 900 K at a constant pressure of 1 atm.

6.3   Zinc dust and powdered sulfur are mixed in stoichiometric amounts according to the reaction: $Zn(s) + S(s) = ZnS(s)$. The mixture is ignited at 25°C, and the reaction goes to completion. Calculate the adiabatic reaction temperature.

6.4   Calculate the enthalpy of combustion of propane at 298.15 K according to the reaction: $C_3H_8(g) + 5\ O_2(g) = 3\ CO_2(g) + 4\ H_2O(l)$. Now calculate the adiabatic temperature for combustion of propane in air with a stoichiometric amount of oxygen for the following cases:

   i.   The air and propane are both initially at 298.15 K.
   ii.   The air is pre-heated to 500 K, and the propane is initially at 298.15 K.

   If an excess amount of air was used in the combustion, discuss the effect this would have on the value of the adiabatic temperature.

6.5   Zinc sulfide is roasted in air according to the reaction: $ZnS + 1.5\ O_2 = ZnO + SO_2$

   i.   Calculate the enthalpy of the reaction at 298.15 K.
   ii.   Assume that in practice 20% excess air is used, that the zinc sulfide and the air are initially at 25°C and the reaction products are withdrawn at 1100 K. Make an energy balance for the process, and calculate the heat surplus or deficit.

7 The second and third laws

All forms of energy are equal, but some forms are more equal than others.*

Anon

SCOPE

This chapter explains the second and third laws and their implications for chemical thermodynamics.

LEARNING OBJECTIVES

1. *Understand the meaning of the second and third laws, both qualitatively and quantitatively.*
2. *Understand the concept of entropy, its physical interpretation and its significance for predicting the spontaneity of processes.*
3. *Be able to calculate the entropy change of a system due to mixing, the entropy of a substance at any temperature and the entropy change of reactions.*
4. *Understand how the entropy of substances and of reactions can be determined experimentally.*

7.1 INTRODUCTION

The second law of thermodynamics finds its greatest application in predicting whether a particular process, often a chemical reaction, can occur and to what extent, and in indicating ways in which the extent of a particular process or reaction can be increased or decreased. Around the middle of the 19th century, some thermodynamicists suggested that the heat evolved by a chemical reaction (its enthalpy of reaction) is a measure of its tendency to occur since it had been observed that spontaneous reactions are usually accompanied by the release of heat and that the larger the heat of reaction the more stable the product seemed to be. However, there are many common exceptions to this. The melting of ice and evaporation of water are endothermic processes which proceed spontaneously. When table salt (NaCl), and most other inorganic salts, are dissolved in water (a spontaneous process) heat is absorbed (the heat of solution) from the water and its temperature falls. When solid barium hydroxide is mixed with solid ammonium nitrate a spontaneous endothermic reaction occurs,

$$Ba(OH)_2.8 H_2O(s) + 2 NH_4NO_3(s) = Ba(NO_3)_2(s) + 2 NH_3(g) + 10 H_2O(l)$$

for which $\Delta_r H^0 = 131.1$ kJ. Many other examples could be given. Clearly, the sign of the enthalpy change is not a criterion of the spontaneity of a process.

Experience has shown that spontaneous processes can never be entirely reversed without leaving a change of some kind outside the system in which the process was carried out. This is one way of stating the second law of thermodynamics and it provides the basis for establishing the criterion for whether a process will occur and to what extent. But, as we will see, there is no single, simple statement of the second law that conveys its full significance. In order to understand the second law, we

---

* This statement sums up the first and second laws of thermodynamics in a parody of the sentence from George Orwell's novel *Animal Farm*: 'All animals are equal, but some animals are more equal than others'.

first introduce the concept of entropy in this chapter. We then introduce the concepts of Gibbs and Helmholtz energy in Chapter 8 which provide the desired criteria for determining the spontaneity of a process.

7.2 ENTROPY AND THE SECOND LAW

We introduce the concept of entropy initially by simply defining it, then follow this with a discussion of its nature and significance. The *entropy** of a system is defined by the relation

$$dS = \frac{\delta q_{rev}}{T} \tag{7.1}$$

where dS is the change in entropy due to an amount δq of heat flowing into the system reversibly at temperature T. Entropy is an extensive property of a system with units J K^{-1}. It is a state function, that is,

$$S = S(T, P, n_1, n_2, \ldots)$$

It follows, therefore, from Equation 2.11 that the entropy change due to a system moving from state 1 to state 2 is given by:

$$\Delta S = S_2 - S_1 = \int_{S_1}^{S_2} dS \tag{7.2}$$

The significance of entropy is summarised in the *second law of thermodynamics*:

Second law: The entropy of an adiabatic system can never decrease. dS is zero for reversible changes and positive for irreversible changes.

The second law is an empirical law. There is no mathematical proof just as there is no proof of the first law of thermodynamics. It is based purely on experimental observations made over many decades. Mathematically, the second law is expressed by the following two equations:

$$\text{For reversible processes:} \quad \delta q = T\, dS \tag{7.3}$$

$$\text{For irreversible processes:} \quad \delta q < T\, dS \tag{7.4}$$

The implications of Equations 7.3 and 7.4 are wide ranging, and their implications for chemical systems in particular are the subject of much of the remainder of this book.

7.2.1 THE NATURE OF ENTROPY

Classical thermodynamics is not concerned with the atomic or molecular nature of substances, only with their macroscopic nature, and to be able to make use of the concept of entropy it is not

---

* The German scientist Rudolf Clausius (1822–1888) was the first to use the name entropy (in 1856), though he had introduced the concept earlier. He named it after the Greek word τροπη meaning transformation.

necessary to even enquire as to its physical nature. Notwithstanding this, when thermodynamics and atomistics are considered together a greater insight into the nature of matter and its changes is achieved than if the two are studied in isolation, and it is worthwhile to examine the physical interpretation of entropy.

At the atomic and molecular (microscopic) level, entropy is the measure of the randomness of a system. The more random a system, the higher is its entropy. One mole of a gas at 1 bar pressure is in a more random state than 1 mole of a crystalline solid at the same temperature and pressure because the molecules of a gas are free to move independently of one another whereas the movement of the atoms or molecules in a solid is much more restricted. Gases, therefore, have much higher entropies than do solids. Liquids have higher entropies than crystalline solids but lower entropies than gases because although the molecules of liquids can move relative to one another, they are not independent – liquids have a fixed volume at a particular temperature, whereas gases occupy whatever space is available. Similarly, the entropy of a substance (solid, liquid or gas) increases as its temperature is increased because the kinetic energy of its atoms or molecules increases, thereby increasing its randomness.

If a match is ignited, which is a spontaneous process, the reverse process of taking the combustion products plus heat and reconstituting a match and air never occurs naturally. A book resting on a table is never seen to absorb energy from the table and jump back onto the bookshelf. Gases always expand to fill available space; they never contract spontaneously. If a hot body is placed in contact with a cold one, heat flows down the temperature gradient from the hot to the cold – never the other way. In all these processes the entropy increases. In all of the reverse processes, energy is conserved and the first law is obeyed, yet the processes never occur naturally. In all naturally occurring (that is, spontaneous) processes, energy is conserved but entropy is increased.

While the first law states that in all processes energy is conserved, the second law qualifies this by stating that the only energy conserving processes that actually occur spontaneously are those which result in an increase in total entropy.

At the microscopic level, spontaneous processes can be thought of as processes of mixing, of which there are two distinct types:* The spreading of atoms or molecules over positions in space (physical mixing) and the sharing or spreading of the available energy of a system between the atoms or molecules themselves (heat transfer and chemical reactions). For a physical mixture consisting of different atoms and/or molecules, because the number of particles is so large (1 mole of a substance contains 6.0234×10^{23} atoms or molecules) the probability of it returning to the unmixed state (the reverse of a spontaneous process) is infinitesimally small, and the probability of the system being found in the well-mixed state is virtually certain. A spontaneous mixing process, therefore, is one in which a system changes from a state of low probability (unmixed) to one of higher probability (mixed). In the case of gaseous diffusion, the spontaneous mixing tendency is simply the intermingling of the constituent particles in space. The same is true for the inter-diffusion of solids and liquids. The spontaneous expansion of a gas may be thought of as a process in which the particles become more completely mixed over the available space.

In other processes, it is not spatial mixing but mixing or sharing of total energy. The spontaneous process of temperature equalisation which occurs when a hot body is in contact with a cooler body can be thought of as the mixing of available energy through the mechanism of atomic vibrations at the interface of the bodies. There is a greater range of quantised energy levels available when the hot body contacts the cold body, and, when the available energy redistributes itself, the most probable mixed energy state is that corresponding to an increased occupation of the mid-range energy levels. In the case of spontaneous chemical reactions, the total energy of a system

---

* Denbigh, K. 1981. *The Principles of Chemical Equilibrium*, 4th ed., Cambridge: Cambridge University Press.

becomes spread over the whole range of quantised energy levels of the reagents and products. When a reaction occurs, a larger number of quantum states become available; namely, those corresponding to the products. The final equilibrium composition of an isolated reaction system is the composition at which the available energy is distributed over the various quantum states in the most random manner.

When atomic mixing occurs spontaneously in an isolated system, the change is accompanied by an increase in entropy of the system (the second law). It is also accompanied by an increase in the number of arrangements for the system. A relationship between these, therefore, is to be expected, and this is given by the Boltzmann equation,

$$S = k \ln W \tag{7.5}$$

where W is the number of possible arrangements (or number of microstates) and k is the Boltzmann constant (1.38065×10^{-23} J K^{-1}) given by the relation:

$$k = \frac{R}{L} \tag{7.6}$$

This statistical definition of entropy was developed by Ludwig Boltzmann* in the 1870s, and he showed that it is equivalent to the thermodynamic definition. To calculate S using Equation 7.5, the total number and types of atoms in a system and the number of distinguishable positions in space which they can occupy must be known. This calculation is difficult, and its solution involves the averaging techniques of statistical mechanics and, except for the simple case of mixing of inert particles (Section 7.3.2), is beyond the scope of this book.

SUMMARY

- *The entropy of a system is a measure of the degree of disorder of the system. An increase in entropy corresponds to a spreading of the system over a larger number of spatial and/ or energy configurations.*
- *High entropy states are those which have a high probability of existing.*
- *When a spontaneous process occurs in an isolated system, the system moves from a state of low probability to a state of high probability, that is, from a state of low entropy towards a state of maximum entropy.*

7.2.2 BROAD IMPLICATIONS OF THE FIRST AND SECOND LAWS

The first and second laws of thermodynamics were formulated on the basis of scientific observation and everyday experience. The first law gained acceptance almost immediately, probably because it appeared reasonable and in accord with human perception of the permanence of nature. However, it took much longer for the second law to become universally accepted, probably because of its disturbing implications. Since every system plus its surroundings makes up the universe, the universe is (by this definition) an isolated system. It follows that every time a spontaneous event occurs the entropy of the universe increases, and the universe is permanently changed. This has some interesting implications:

- The universe is changing from a low entropy (more ordered) state to a more high entropy (low ordered) state.

---

* Ludwig Eduard Boltzmann (1844–1906) was an Austrian physicist and philosopher whose great contribution was the development of statistical mechanics.

- Entropy is the only quantity in the physical sciences that requires a direction for time, sometimes called the arrow of time.* As times goes forward, the entropy of an isolated system can increase but never decrease. Hence, from one perspective, entropy change is a way of distinguishing the past from the future.
- Since the entropy of the universe is increasing with time, eventually there must come a time when the entropy has reached an overall maximum, equivalent to the final equilibrium state, with all matter and energy uniformly dispersed.

Clausius referred to this latter state as the 'heat death of the universe'. In that state the same total amount of energy would be present in the universe as at its beginning (the first law), but none of it would be available for work.

7.2.3 ALTERNATIVE STATEMENTS OF THE SECOND LAW

While Equations 7.3 and 7.4 are the usually accepted statements of the second law, the law has been stated in other ways which reflect the historical thermo-mechanical origins of thermodynamics as distinct from the chemical approach adopted in this text. One statement is:

Heat will not flow spontaneously from a substance at lower temperature to a substance at higher temperature.

This statement was first made by Clausius in 1854. Put another way, it means heat doesn't flow from cold to hot without external energy being added. This is known from everyday experience. In a system where heat flows from cold to hot, entropy has to be decreasing. This can occur only if more entropy is created somewhere else, so that the total entropy of the system and its surroundings increases. For example, in a refrigerator (the system), heat flows from cold to hot (across the boundary between the system and its surroundings) but only under the action of a compressor, which is driven by a motor. The electrical energy consumed by the motor is converted to heat and results in an overall net increase in entropy of the universe.

That this statement follows from Equation 7.4 can be demonstrated as follows. Consider the transfer of energy as heat from one large reservoir at temperature T_h (the source) to another (the sink) at a lower temperature T_c. When an amount of heat Δq leaves the hotter source the entropy of the source changes by $-\Delta q / T_h$. This is a decrease in entropy. When Δq enters the colder sink, its entropy changes by $+\Delta q / T_c$. This is an increase in entropy. The total change in entropy is the sum of these,

$$\Delta S = \frac{\Delta q}{T_c} - \frac{\Delta q}{T_h} = \Delta q \left(\frac{1}{T_c} - \frac{1}{T_h} \right) \tag{7.7}$$

which is positive because $T_h > T_c$. Since ΔS is positive for the process, cooling is a spontaneous process.

Another statement is:

It is not possible to convert a given quantity of heat completely into work.

This means that, for any device that converts heat into work (such as a steam engine, internal combustion engine or gas turbine), there will always be a quantity of heat that cannot be converted into work. For example, in a car engine or gas turbine, hot gases are expelled which still contain thermal energy. Some of this energy is not converted to work due to inefficiencies in the engine, but

---

* The phrase 'time's arrow' was coined by the British astronomer, physicist and mathematician Arthur Eddington to describe the apparent one-way direction, or asymmetry, of time (Eddington, A.S. 1928. *The Nature of the Physical World*, Cambridge: Cambridge University Press).

there is a theoretical limit beyond which efficiency cannot be increased. This limit was established by Carnot* in 1824 by calculating the maximum amount of work that, theoretically, could be performed by a gas cooling between two known temperatures. This led to the development of *Carnot's theorem*, which can be considered as yet another statement of the second law:

> *No heat engine can be more efficient than a reversible engine operating between the same temperature limits, and all reversible engines operating between the same temperature limits have the same efficiency.*

A heat engine is any device for converting heat into work. It must do this spontaneously to be useful. Therefore, the flow of heat from the hot source (for example, the combusted fuel) to the colder sink (for example, the exhaust) must be accompanied by an overall increase in entropy. We saw above that the change in entropy when heat Δq leaves a hot source and enters a colder sink is given by Equation 7.7. Now, suppose than an amount of heat less than Δq is transferred to the sink and the difference is converted to mechanical work (the whole purpose of the heat engine). Then

$$\Delta S = \frac{\Delta q_c}{T_c} - \frac{\Delta q_h}{T_h}$$

where $\Delta q_h > \Delta q_c$. In this case ΔS will be positive if

$$\frac{\Delta q_c}{T_c} - \frac{\Delta q_h}{T_h} \geq 0$$

that is, if

$$\Delta q_c \geq \Delta q_h \times \frac{T_c}{T_h}$$

Within this limitation we are able to utilise the difference $\Delta q_h - \Delta q_c$ as work (by means of a suitable device) and still have a spontaneous process. The maximum work the engine can do is:

$$w_{max} = \Delta q_h - \Delta q_{c,min}$$

where $\Delta q_{c,min}$ is the minimum value Δq_c can have. This is when $\Delta q_c = \Delta q_h \times T_c / T_h$. Therefore,

$$w_{max} = \Delta q_h - \Delta q_h \times \frac{T_c}{T_h} = \Delta q_h \left(1 - \frac{T_c}{T_h}\right)$$

This is the reversible case since then ΔS will be zero. The efficiency of the engine ε is defined as:

$$\varepsilon = \frac{\text{Maximum work performed by the engine}}{\text{Thermal energy absorbed by the engine}} = \frac{w_{max}}{\Delta q_h}$$

Therefore, the maximum efficiency the engine can have is:

$$\varepsilon = 1 - \frac{T_c}{T_h} = \frac{T_h - T_c}{T_h} \tag{7.8}$$

This efficiency limit can never be achieved in practice, since it was derived for the reversible case, but it sets the theoretical upper limit for the efficiency of any process for converting thermal energy into work.

---

* Nicolas Léonard Sadi Carnot (1796–1832) was a French military engineer. He wrote the book *Reflections on the Motive Power of Fire*, published in 1824, in which he presented a generalised theory of heat engines. Carnot is often called the 'father of thermodynamics'.

EXAMPLE 7.1   The efficiency of a coal-fired power station

A typical 1000 MW coal-fired power station for generating electricity consumes about 400 tonnes of bituminous coal per hour. Assume the boiler operates at a steam temperature of 550°C and the heat exchanger cools the condensed steam to 50°C. A typical bituminous coal has a heat of combustion of around 28 500 kJ kg^{-1}. Calculate the theoretical, maximum efficiency of the process. How does this compare with the efficiency actually achieved?

SOLUTION

The theoretical efficiency is obtained from Equation 7.8. $T_h = 540 + 273.15 = 823.15$ K and $T_c = 50 + 273.15 = 323.15$ K. Therefore the efficiency is:

$$\varepsilon = 1 - \frac{323.15}{823.15} = 0.61$$

The theoretical maximum amount of heat that can be converted into mechanical energy (in the turbine) to generate electricity is 61% of the heat of combustion of the coal. The remaining 39% of the heat is lost to the environment in the combustion gases that are discharged at a temperature of 50°C.

The actual efficiency can be calculated knowing the actual power output of the station: 1000 MW (or 1000 MJ s^{-1}). Four hundred tonnes per hour of coal is equivalent to 111 kg s^{-1}. Therefore, the heat generated per second by combustion of the coal is:

$$q = 111 \times 28\ 500 = 3\ 163\ 000 \text{ kJ s}^{-1} = 3163 \text{ MW}$$

Therefore, the actual efficiency is 1000/3163 = 0.32; that is, only 32% of the energy released by burning the coal is converted into mechanical, then electrical energy.

The above example demonstrates the theoretical difficulty of converting heat into work efficiently and illustrates the inefficiency of an actual (spontaneous) process compared with the theoretical (reversible) process.

7.3   THE ENTROPY OF MIXING

7.3.1   THE MIXING OF IDEAL GASES

Imagine a container of volume V_2 with a partition. Now assume one side of the container has a volume V_1 and contains n moles of an ideal gas and the other side is empty (a vacuum), as shown in Figure 7.1(a). If the partition is removed, the gas will expand spontaneously to fill the entire container (Figure 7.1(b)) and its volume will change from V_1 to V_2 and its pressure will change from p_1 to p_2. At constant temperature, the internal energy of the gas will not change ($dU = 0$) and, therefore, from Equation 4.7, $\delta q = pdV$. Since $dS = q/T$ and $pV = nRT$, the entropy change of the gas due to expansion will be:

$$\Delta S = \int dS = \int_{V_1}^{V_2} \frac{p}{T} dV = \int_{V_1}^{V_2} \frac{nR}{V} dV$$

$$= nR \ln \frac{V_2}{V_1} = nR \ln \frac{p_1}{p_2} \tag{7.9}$$

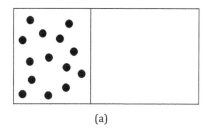 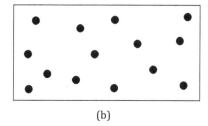

FIGURE 7.1   Isothermal expansion of a gas.

This is the equation for the entropy change of a gas expanding freely (spontaneously) from pressure p_1 to pressure p_2 at constant temperature.

Now consider the system in Figure 7.2(a). This time there is a different gas in each of the sections: n_A moles of gas A of volume V_A and n_B moles of gas B of volume V_B. We assume that both gases are at the same temperature and pressure and that they do not react chemically. If the partition is removed the two gases will randomly diffuse and form a homogenous mixture at the same temperature and pressure, as shown in Figure 7.2(b). Each gas will now occupy the volume $V_A + V_B$

To calculate the entropy change due to the mixing of the gases we can treat the mixing as two separate gas expansions, one for gas A and the other for gas B. From Equation 7.9 the respective entropy changes are:

$$\Delta S_A = n_A R \ln \frac{V_A + V_B}{V_A}$$

and

$$\Delta S_B = n_B R \ln \frac{V_A + V_B}{V_B}$$

Since entropy is an extensive property, the entropy of mixing of the gases is the sum of the individual entropies of expansion:

$$\Delta_{mix} S = n_A R \ln \frac{V_A + V_B}{V_A} + n_B R \ln \frac{V_A + V_B}{V_B}$$

Since $pV = nRT$, volume is directly proportional to the number of moles (at constant temperature and pressure), and since we know the number of moles we can substitute these for volume:

$$\Delta_{mix} S = n_A R \ln \frac{n_A + n_B}{n_A} + n_B R \ln \frac{n_A + n_B}{n_B}$$

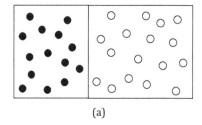 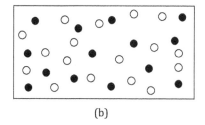

FIGURE 7.2   The spontaneous mixing of two gases.

and, from the definition of mole fraction (Equation 2.2),

$$\Delta_{\mathrm{mix}}S = n_{\mathrm{A}}R\ln\frac{1}{x_{\mathrm{A}}} + n_{\mathrm{B}}R\ln\frac{1}{x_{\mathrm{B}}} = -n_{\mathrm{A}}R\ln x_{\mathrm{A}} - n_{\mathrm{B}}R\ln x_{\mathrm{B}}$$

Therefore,

$$\Delta_{\mathrm{mix}}S = -R(n_{\mathrm{A}}\ln x_{\mathrm{A}} + n_{\mathrm{B}}\ln x_{\mathrm{B}}) \tag{7.10}$$

Alternatively, since $n = n_{\mathrm{A}} + n_{\mathrm{B}}$,

$$\Delta_{\mathrm{mix}}S = -n\,R(x_{\mathrm{A}}\ln x_{\mathrm{A}} + x_{\mathrm{B}}\ln x_{\mathrm{B}})\ \text{Joules K}^{-1} \tag{7.11}$$

where n is the total number of moles of gas. Generalising for a mixture of A, B, … , for 1 mole of gas mixture:

$$\Delta_{\mathrm{mix}}S_m = -R(x_{\mathrm{A}}\ln x_{\mathrm{A}} + x_{\mathrm{B}}\ln x_{\mathrm{B}} + \cdots)\ \text{Joules mol}^{-1}\ \text{K}^{-1}$$

$\Delta_{\mathrm{mix}}S_m$ is the *integral molar entropy of mixing* or simply the *entropy of mixing* of ideal gases at constant temperature and pressure. It can be understood as the difference between the entropy of 1 mole of gas mixture (at a particular pressure and temperature) S_m and sum of the entropies of the individual gas components prior to mixing S_m^0 at the same pressure and temperature as the mixture. Thus,

$$\Delta_{\mathrm{mix}}S_m = S_m - S_m^0 = -R(x_{\mathrm{A}}\ln x_{\mathrm{A}} + x_{\mathrm{B}}\ln x_{\mathrm{B}} + \cdots)\ \text{Joules mol}^{-1}\ \text{K}^{-1} \tag{7.12}$$

When the two gases mix, their mole fractions will be less than 1, making the term inside the parentheses in Equation 7.12 negative. Thus the entropy of mixing will always be positive. This is consistent with the statistical understanding of entropy as a measure of randomness: Randomness has increased from the state of the system in Figure 7.2(a) to that in Figure 7.2(b). Note also that the value of $\Delta_{\mathrm{mix}}S$ for a gas mixture is independent of temperature and pressure.

7.3.2 THE GENERAL EQUATION FOR MIXING

A mixture is a physical combination of two or more substances, without chemical bonding or other chemical change, in which the identities of the substances are retained. The mixing of substances, whether solid, liquid or gas, is an irreversible process and results in an increase in entropy (increased randomness). Mixing may be, and often is, constrained to occur under particular conditions. For example, the different substances may or may not be at the same temperature and pressure, and the final volume need not necessarily be the sum of the initially separate volumes. So work may be done on or by the new system during the process of mixing. However, if the final volume is the sum of the initial separate volumes and if there is no heat transfer, then no work is done. In that case, the entropy of mixing is entirely accounted for by the movement of each substance into a final volume not initially available to it.

It is possible to derive a general equation for the mixing of inert particles using statistical mechanics. The particles may be macroscopic (for example, grains of sand, droplets) or microscopic (atoms or molecules); however the assumption is they do not interact in any way. The following is a simplified approach, and a more complete account can be found in advanced texts. Assume there are N_{A} particles of substance A and N_B particles of substance B, where N is a very large number. We assume the particles of A and B are approximately the same size and consider the space they occupy as being subdivided into a lattice with cells the size of the particles. The entropy of mixing of the particles according to Boltzmann's equation is

$$\Delta_{mix}S = S_m - S_m^0 = k \ln W$$

where W is the number of different ways the particles of substances A and B can be arranged to give the same macroscopic properties. For ordered arrangements of the particles, W is small, but for a completely random arrangement W has its maximum value. It can be shown in this case that

$$W = \frac{N!}{N_A! N_B!}$$

Applying Stirling's approximation for the factorial of a large integer and substituting into the Boltzmann equation yields

$$\Delta_{mix}S = -k \left(N_A \ln \frac{N_A}{N} + N_B \ln \frac{N_B}{N} \right)$$

where N, the number of lattice sites, is given by $N = N_A + N_B$. The fractions N_A/N and N_B/N can be replaced by mole fractions and N by $n \times L$, where L is Avogadro's constant and n is the total number of moles:

$$\Delta_{mix}S = -k n L \left(x_A \ln x_A + x_B \ln x_B \right)$$

Since $kL = R$ (Equation 7.6),

$$\Delta_{mix}S = -n R \left(x_A \ln x_A + x_B \ln x_B \right) \tag{7.13}$$

or, for 1 mole of a mixture of A, B, C …

$$\Delta_{mix}S_m = -R \left(x_A \ln x_A + x_B \ln x_B + \cdots \right) \tag{7.14}$$

This is the same as Equation 7.12 for the mixing of ideal gases but is more general and applies to the mixing of solids, liquids or gases provided there is no interaction between the particles. This type of mixing is referred to as *ideal mixing* and is considered further in Chapter 9. The variation of $\Delta_{mix}S_m$ with composition for a binary ideal mixture, calculated using Equation 7.14, is shown in Figure 7.3. Note the values are always positive and equal to zero at pure A and pure B.

7.4 THE ENTROPY OF PHASE CHANGES

Just as for enthalpy, there are four main types of entropy changes resulting from a phase transformation:

Entropy of transition ($\Delta_{trs}S$). This applies to transformations from one solid phase to another.
Entropy of fusion or melting ($\Delta_{fus}S$). This applies to the transformation of a solid to a liquid.
Entropy of vapourisation ($\Delta_{vap}S$). This applies to the transformation of a liquid to a vapour.
Entropy of sublimation ($\Delta_{sub}S$). This applies to the transformation of a solid to a vapour.

Reversible phase changes occur at constant temperature and pressure. From Equation 7.1, the entropy change will be the reversible heat transferred during the phase change, namely, the enthalpy change (Equation 4.14), divided by the temperature. For fusion (melting) of a solid to a liquid at the melting point (T_{fus}), the entropy of fusion is:

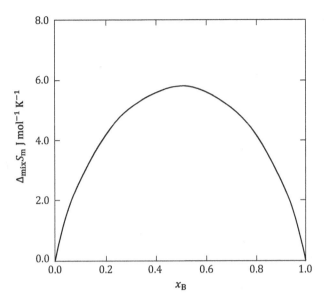

FIGURE 7.3 The variation of entropy of mixing for a binary mixture of A and B, assuming ideal mixing.

$$\Delta_{fus}S = \frac{\Delta_{fus}H}{T_{fus}}$$

Similarly, for vapourisation of a liquid to a gas at the boiling point (T_{vap}), the entropy of vapourisation is:

$$\Delta_{vap}S = \frac{\Delta_{vap}H}{T_{vap}}$$

Values of ΔS for phase changes are usually quoted for the temperature at which the phase change occurs at 1 bar pressure and are designated thus: ΔS^0. Phase transformations can be written as chemical reactions and the entropy change of the reaction then equals the entropy of the phase change, for example,

$$H_2O(s) = H_2O(l); \quad \Delta_r S^0(273.15) = \Delta_{fus}S^0(273.15) = 22.00 \text{ J mol}^{-1} \text{ K}^{-1}$$

**EXAMPLE 7.2   Calculate the entropy of transformation of
a substance from its enthalpy of transformation**

What is the entropy of fusion of ice at its melting point at 1 bar pressure?

SOLUTION

$\Delta_{fus}H^0(H_2O) = 6.01$ kJ mol^{-1} at 273.15 K and 1 bar.

Therefore, $\Delta_{fus}S^0(H_2O) = 6010 / 273.15 = 22.00$ J mol^{-1} K^{-1}

COMMENT

This is an increase in entropy, as expected, since the ice transforms from the solid to the liquid state.

7.5   THE THIRD LAW AND THE ENTROPY OF SUBSTANCES

Since $\delta q = C_P \, dT$ (Equation 4.16) it follows from the definition of entropy that for a reversible process

$$dS = \frac{C_p}{T} \, dT \tag{7.15}$$

which on integrating gives

$$\Delta S = S(T_1) - S(T_2) = \int_{T_1}^{T_2} \frac{C_p}{T} \, dT \tag{7.16}$$

Equation 7.16 enables calculation of the entropy of a substance if its heat capacity is known and if a value of the entropy at one temperature is known. In the case of the enthalpy of substances (Equation 4.22), an arbitrary value of zero for elements at 298.15 K and 1 bar pressure was chosen. However, for entropy an arbitrary assumption is not necessary since the *third law of thermodynamics** provides an absolute zero for the entropy of a substance:

Third law: The entropy of all substances in internal equilibrium is zero at 0 K.

The entropy of a substance has two types of atomic contributions: The configuration contribution and the vibration contribution. The entropy of a substance is their sum, that is,

$$S = S_{conf} + S_{vib} = k \ln W_{conf} + k \ln W_{vib}$$

where W_{conf} is the maximum number of ways the atoms or molecules comprising the substance can be arranged over the available lattice positions, and W_{vib} is the maximum number of ways they can vibrate about these positions. For a pure substance, the individual atoms or molecules are all the same and are indistinguishable and, for a perfectly ordered structure, the atoms or molecules can be arranged in one way only. In this case

$$S_{conf} = k \ln 1 = 0$$

For a pure substance with a perfectly ordered atomic or molecular structure, therefore, the entropy is determined by the vibrational contribution alone. The vibrational probability W_{vib} is one at absolute zero, that is, $\ln W_{vib} = 0$, and, therefore, $S = 0$ for a perfectly ordered substance at 0 K.

Absolute values for the entropy of a substance can be calculated at temperatures above 0 K using Equation 7.16 if the heat capacity of the substance is known:

$$S(T) = \int_0^T \frac{C_p}{T} \, dT$$

This requires knowledge of heat capacity values down to 0 K. Values at very low temperatures cannot be determined experimentally and can be obtained only by extrapolating from the lowest

---

* The third law was developed by the German chemist Walther Nernst during the years 1906–1912. It is also sometimes referred to as Nernst's heat theorem.

experimentally measured values. To do this the Debye model is used. Debye (1912) derived the following relation for solids from statistical mechanics considerations:

$$C_V = 1943 \left(\frac{T}{\theta} \right)^3 \tag{7.17}$$

where θ is a constant characteristic of each substance and is derived from the slope of the C_p curve above ~50 K. Since the relation between C_V and C_p is known (Equation 4.20), values of C_p down to 0 K can be obtained. The value of entropy at any higher temperature can then be calculated. In this manner, values of the entropies of most substances at 298.15 K have been calculated and are readily available in compilations. These values may then be used to calculate the entropy at other temperatures:

$$S(T) - S(298.15) = \int_{298.15}^{T} \frac{C_p}{T} dT \tag{7.18}$$

For calculation purposes, empirical equations for heat capacity can be substituted. For example, substituting Equation 4.21 into Equation 7.18 yields

$$S^0(T) = S^0(298.15) + \int_{298.15}^{T} \frac{a + bT + cT^2 + \dfrac{d}{T^2}}{T} dT$$

$$= S^0(298.15) + \int_{T298.15}^{T} \left(\frac{a}{T} + b + cT + \frac{d}{T^3} \right) dT$$

$$= S^0(298.15) + a \ln T + bT + \frac{cT^2}{2} - \frac{d}{2T^2} \tag{7.19}$$

$$= S^0(298) + a \left(\ln T - \ln 298 \right) + b \left(T - 298 \right)$$

$$+ \frac{c}{2} \left(T^2 - 298^2 \right) - \frac{d}{2} \left(\frac{1}{T^2} - \frac{1}{298^2} \right)$$

When a phase transformation occurs in the temperature range of interest, the entropy of transformation must be included, for example,

$$S^0(T) = S^0(298.15) + \int_{298.15}^{T_{fus}} \frac{C_p(s)}{T} dT + \frac{\Delta_{fus} H^0}{T_{fus}} + \int_{T_{fus}}^{T} \frac{C_p(l)}{T} dT + \cdots \tag{7.20}$$

where $298.15 < T_{fus} < T \ldots$ and the substance melts at T_{fus} for which the enthalpy of fusion is $\Delta_{fus} H^0$. The calculation procedure is illustrated schematically in Figure 7.4, and the results of such a calculation for ferrous oxide (FeO), which melts at 1650 K, over the temperature range 298.15 to 1900 K are shown graphically in Figure 7.5.

As for enthalpy, the *molar entropy* $S_m(B)$ is the entropy of 1 mole of substance B; it has the unit J mol^{-1} K^{-1}. Values of the molar entropy of substances at 298.15 K and at 100 K intervals are tabulated in the NIST–JANAF, US Geological Survey and Barin compilations (Section 5.2). The variation of entropy with temperature for some common substances is shown in Figure 7.6. Discontinuities in the trends indicate a phase change. Note the entropy of substances decreases with

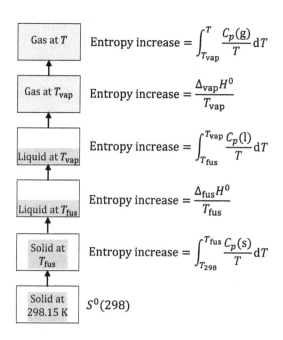

FIGURE 7.4   The increase in entropy of a substance as it is heated from 298 K to a temperature above its boiling point. The entropy of the substance at T is the sum of $S^0(298)$ and the entropy increases.

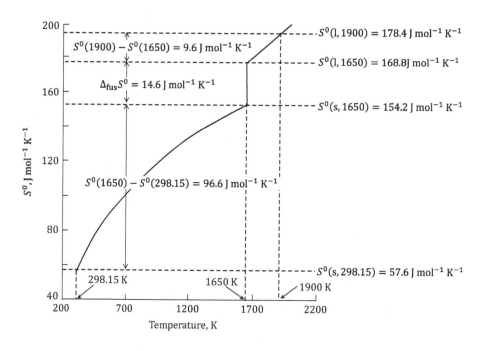

FIGURE 7.5   The increase in entropy of FeO as it is heated from 298.15 to 1900 K.

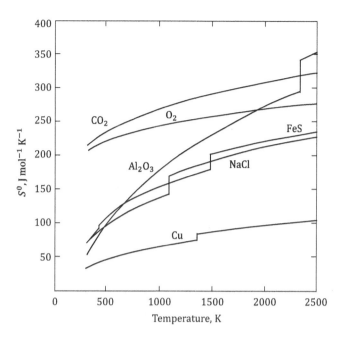

FIGURE 7.6   The variation of entropy with temperature of some common substances.

decreasing temperature and becomes zero at 0 K (as required by the third law). Note also that the entropy of gaseous substances is greater than that of liquids which in turn is greater, though not very much greater, than that of solids. This is as expected from our understanding of the physical nature of entropy.

7.6   THE ENTROPY OF FORMATION AND ENTROPY OF REACTION

The entropy of formation of a compound is defined in the same manner as for the enthalpy of formation. Thus, the *standard entropy of formation* $\Delta_f S^0$ of a compound is the change in entropy when 1 mole of the compound is formed from its constituent elements under standard state conditions at a specified temperature. For the reaction

$$k\,A + l\,B = A_k B_l$$

where A and B are elements which react to form the compound $A_k B_l$, the standard entropy of formation of $A_k B_l$ is given by:

$$\Delta_f S^0(A_k B_l) = S^0(A_k B_l) - k\,S^0(A) - l\,S^0(B) \tag{7.21}$$

The standard entropy of formation of elements in their standard state is zero at all temperatures, since for the reaction (where E is an element):

$$E = E$$

$$\Delta S^0(T) = S^0(T) - S^0(T) = 0$$

Values of entropy of formation of substances are not tabulated in the NIST–JANAF, US Geological Survey and Barin compilations of data (Section 5.2) but may be calculated readily at any temperature by means of Equation 7.21 from the entropy values of the reactants and product. The variation with temperature of the standard entropy of formation of some common substances is shown in Figure 7.7. Note that, as is also the case for enthalpy of formation, $\Delta_f S^0$ values are nearly independent of temperature but change in value at each temperature of transformation.

The *entropy of reaction*, $\Delta_r S$, is defined in the same manner as for the enthalpy of reaction. For the reaction

$$k\,A + l\,B = m\,C + n\,D \qquad (4.31)$$

the *standard entropy of reaction* is:

$$\Delta_r S^0 = mS_C^0 + nS_D^0 - kS_A^0 - lS_B^0 \qquad (7.22)$$

The value of $\Delta_r S^0$ can be calculated by means of Equation 7.22 from the values of the standard entropy of the individual substances taking part in the reaction at the appropriate temperature.

It follows from this definition that:

- When ΔS^0 is positive the entropy has increased as a result of the reaction, for example, from solid to gas such as in the reaction:

$$CaCO_3(s) = CaO(s) + CO_2(g); \quad \Delta S^0(298.15) = 160.2 \text{ J K}^{-1}$$

- When ΔS^0 is negative the entropy has decreased as a result of the reaction, for example, from gas to solid such as the reaction:

$$Na(s) + 0.5\,Cl_2(g) = NaCl(s); \quad \Delta S^0(298.15) = -90.71 \text{ J K}^{-1}$$

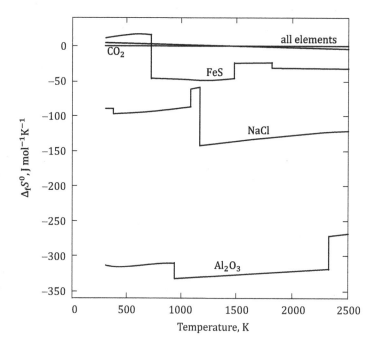

FIGURE 7.7   The variation with temperature of the entropy of formation of some common substances.

- When ΔS^0 is small there has been little change in entropy, for example solids reacting to form solids or no net increase or decrease in gas volume:

$$Si(s) + C(s) = SiC(s); \quad \Delta S^0(298.15) = -8.07 \text{ J K}^{-1}$$

$$C(s) + O_2(g) = CO_2(g); \quad \Delta S^0(298.15) = 2.88 \text{ J K}^{-1}$$

EXAMPLE 7.3   Calculate the entropy change of a reaction

Using readily available data, calculate the standard entropy change at 1000°C for the reaction:

$$Fe_2O_3(s) + 3 \text{ C}(s) = 2 \text{ Fe}(s) + 3 \text{ CO}(g)$$

SOLUTION

The approach is the same as in Example 4.4 for calculating the standard enthalpy change of a reaction. Again, three methods are available.

METHOD 1: COMBINING EQUATIONS FOR FORMATION

$Fe_2O_3(s) = 2 \text{ Fe}(s) + 1.5 \text{ O}_2(g)$	$-\Delta_f S^0(Fe_2O_3)$
$3 \text{ C}(s) + 1.5 \text{ (g)} = 3 \text{ CO}(g)$	$3 \times \Delta_f S^0(CO)$
$Fe_2O_3(s) + 3 \text{ C}(s) = 2 \text{ Fe}(s) + 3 \text{ CO}(g)$	

Substituting values at 1273.15 K and summing the equations:

$$\Delta_r S^0(1273) = -\Delta_f S^0(Fe_2O_3, s, 1273) + 3 \times \Delta_f S^0(CO, g, 1273)$$

$$= -(-249.04) + 3 \times (86.78) = 509.38 \text{ J K}^{-1}$$

METHOD 2: COMBINING ENTROPIES OF FORMATION

$$\Delta_r S^0(1273) = 3 \times \Delta_f S^0(CO, g, 1273) + 2 \times \Delta_f S^0(Fe, s, 1273)$$

$$- 3 \times \Delta_f S^0(C, s, 1273) - \Delta_f S^0(Fe_2O_3, s, 1273)$$

$$= 3 \times (86.78) + 2 \times 0 - 3 \times 0 - (-249.04) = 509.38 \text{ J K}^{-1}$$

METHOD 3: COMBINING INDIVIDUAL ENTHALPY VALUES

$$\Delta_r S^0(1273) = 3 \times S^0(CO, g, 1273) + 2 \times S^0(Fe, s, 1273)$$

$$- 3 \times S^0(C, s, 1273) - S^0(Fe_2O_3, s, 1273)$$

$$= 3 \times (242.72) + 2 \times 78.50 - 3 \times 29.86 - 286.17 = 509.38 \text{ J K}^{-1}$$

Note the positive value in the above example indicates an increase in entropy in changing the state from that of the reactants to that of the products. This is expected since there is a volume increase due to formation of a gaseous product.

7.7   ENTROPY AS A CRITERION OF SPONTANEITY

Entropy can be used as a criterion for whether a reaction (or other process) will occur spontaneously. Consider an isolated system consisting of a mixture of reactive substances inside a reaction vessel. According to the second law, during a spontaneous change the entropy of the system increases ($dS > 0$) while for a reversible process (or one at equilibrium) there is no change in entropy ($dS = 0$). Calculation of the change in entropy between the initial and a hypothesised final state of the isolated system, therefore, provides a means of determining, without the need to perform the actual experiment, whether the change from initial to final state will occur spontaneously; if the entropy change is positive the change will occur, otherwise it will not. The usefulness of being able to do this type of calculation is readily apparent.

The practicality of using entropy as a criterion of whether a process will occur or not is limited, however, by the difficulty in applying the test to actual systems which are usually non-adiabatic and open. In these cases, it is necessary to consider the entropy change not only of the reaction mixture but also of the surroundings, which together constitute an isolated system. Thus,

$$\Delta S_{\text{system}} + \Delta S_{\text{surroundings}} > 0 \quad \text{process is spontaneous} \tag{7.23}$$

$$\Delta S_{\text{system}} + \Delta S_{\text{surroundings}} = 0 \quad \text{process is reversible/at equilibrium} \tag{7.24}$$

Determining $\Delta S_{\text{surroundings}}$ is not only tedious but liable to error since it is necessary to include in the surroundings all substances outside the system which will undergo change during the process under consideration. This difficulty is overcome by introducing two new state variables, Gibbs and Helmholtz energy, which arise from combining the first and second laws. This is discussed in Chapter 8.

7.8   EXPERIMENTAL DETERMINATION OF ENTROPY

The entropy of substances can be determined from their heat capacity values using Equation 7.18 if values of heat capacity are known. These can be determined experimentally. At lower temperatures, where it is not possible to determine experimental values, heat capacities are calculated using the Debye model (Section 7.5). The entropy change of reactions can be determined using Equation 7.22 from the entropies of the reactants and products. The entropy of reaction can also be determined indirectly for reactions that are electrochemical in nature from the variation of the cell voltage with temperature as discussed in Chapter 16: Since $\Delta G = \Delta H - T\Delta S$, $(d\Delta G/dT) = -\Delta S$ and, since $\Delta G = -zFE$, then $\Delta S = z\,F(dE/dT)$.

PROBLEMS

7.1   Calculate the entropy change for the following phase transformations:
   i.   The melting of gold at 1 bar pressure given the melting point of gold is 1337.6 K and the standard enthalpy of fusion is 12.552 kJ mol^{-1}
   ii. The melting of lithium hydride at 1 bar pressure given $\Delta_{\text{fus}}H^0(\text{LiH}) = 22.594$ kJ and $T_{\text{fus}}(\text{LiH}) = 961.8$ K
   iii. The boiling of toluene (C_7H_8) at 1 bar pressure given $\Delta_{\text{vap}}H^0(C_7H_8) = 33.5$ kJ and $T_{\text{vap}}(C_7H_8) = 383.27$ K
   Why is the answer to part iii much larger than the answers to parts i and ii?
7.2   Calculate the change in entropy when 1 mole of an ideal gas at atmospheric pressure is compressed isothermally to 1000 atm. What is the significance of the sign of the answer?
7.3   A rigid vessel is divided into two sections of equal volume by a partition. One section contains 1 mole of ideal gas A at 1 atm, and the other contains 1 mole of ideal gas B at 1 atm.

Calculate the entropy change when the partition is removed, assuming the temperature is constant. If in another rigid vessel, there were initially 3 moles of gas A and 1 mole of gas B, both at the same pressure, what would be the entropy change due to the gases mixing when the partition is removed, assuming constant temperature? What would be the entropy change in both cases if both sections of the vessel had contained gas A?

7.4 Calculate the molar entropy of mixing of two ideal gases, A and B, at 0, 0.2, 0.4, 0.6, 0.8 and 1.0 mole fraction of gas A, and show the results graphically as a function of mole fraction of gas A. At what value of mole fraction of gas A is the entropy a maximum?

7.5 Calculate the maximum efficiency of a piston-driven steam engine operating on steam at 100°C and discharging it at 60°C, and compare it with the maximum efficiency of a modern steam turbine operating at 350°C and discharging at 70°C.

7.6 When molten iron and nickel are mixed and cooled they form a metallic solid solution which is almost ideal. Calculate the approximate entropy increase when 50 g nickel and 100 g of iron are melted to form a solution.

7.7 Calculate the change in entropy when 5 moles of silicon tetrachloride ($SiCl_4$) are heated from 350 to 380 K given that the average value of $C_p(SiCl_4)$ over the temperature range is 95.3 J mol^{-1} K^{-1}. What is the significance of the sign of ΔS?

7.8 Use entropy data from the NIST–JANAF tables for the following calculations. In each case assume no heat is lost to the surroundings.
   i. Calculate the entropy change when a 1 kg block of metallic copper cools from 500 to 400 K. Explain the significance of the sign.
   ii. Calculate the value of S when a 1 kg block of metallic copper is heated from 300 to 500 K. Again, explain the significance of the sign.
   iii. Explain what will happen when two 1 kg blocks of copper are contacted if one block is initially at 500 K and the other at 300 K. Assume no heat is lost to the surroundings. Calculate the entropy change between the initial state and the equilibrium state. Explain the significance of the sign and how it is consistent with the second law of thermodynamics.

7.9 Using data from the table below calculate the entropy of reaction of the following chemical reactions at 298.15 and 1200 K as follows:
   i. Using the entropy values only (columns 2 and 3).
   ii. Using the entropy of formation values only (columns 3 and 4). Your answers should be the same in both cases.
   iii. Comment on the significance and magnitude of the sign of $\Delta_r S^0$ in each case.
   Retain the $\Delta_r S^0$ values for use in Problem 8.2.
   Reactions:

$$CH_4(g) + 2\, O_2(g) = CO_2(g) + 2\, H_2O(l,g)$$

$$N_2(g) + 3\, H_2(g) = 2\, NH_3(g)$$

$$Cu_2S(s) + 2\, O_2(g) = 2\, Cu(s) + SO_2(g)$$

Data:

	$S^0(298.15)$ J K^{-1}	$S^0(1200)$ J K^{-1}	$\Delta S^0(298.15)$ J K^{-1}	$\Delta S^0(1200)$ J K^{-1}
$CH_4(g)$	186.21	261.24	−80.89	−110.90
$CO_2(g)$	213.77	279.37	2.88	0.80
Cu	33.16	70.36	0.00	0.00
Cu_2S	116.15	249.45	17.77	−30.06
$H_2(g)$	130.68	171.79	0.00	0.00
$H_2O(l)$	69.95		−162.40	0.00
$H_2O(g)$		240.61		−56.19
$O_2(g)$	205.15	250.01	0.00	0.00
$N_2(g)$	191.61	234.23	0.00	0.00
$NH_3(g)$	192.78	257.21	−99.05	−117.59
$SO_2(g)$	248.22	315.78	11.02	−73.02

7.10 What sign and magnitude would you expect for the entropy of reaction for each of the following reactions? Briefly explain your reason.
Select three of the equations, and calculate their entropy of reaction at 25°C using entropy data from the NBS, USGS or NIST–JANAF tables.

$$Si + C = SiC$$

$$2\ CaO(s) + SiO_2(s) = Ca_2SiO_4(s)$$

$$N_2(g) + 3\ H_2(g) = 2\ NH_3(g)$$

$$CH_4(g) + 2\ S_2(g) = 2\ H_2S(g) + CS_2(g)$$

$$CH_4(g) + NH_3(g) = HCN(g) + 3\ H_2O(g)$$

$$ZnSO_4.7H_2O(s) = ZnSO_4.6H_2O(c) + H_2O(l)$$

$$H_2(g) + 0.5\ O_2(g) = H_2O(g)$$

$$Zn(s) + H_2SO_4(aq) = ZnSO_4(s) + H_2(g)$$

$$CaCO_3(s) = CaO(s) + CO_2(g)$$

$$SiO_2(s) + 3\ C(s) = SiC(s) + 2\ CO(g)$$

$$Na_2SiO_3(s) + H_2SO_4(l) = SiO_2(s) + Na_2SO_4(s) + H_2O(l)$$

7.11 This chapter was introduced with the quote '*All forms of energy are equal, but some forms are more equal than others*'. Do you think this is a succinct summary of the first and second laws of thermodynamics? Briefly explain why.

7.12 This is a continuation of Problem 4.13. It utilises equations of the form of Equations 7.19 and 7.20 to generate data from empirical heat capacity and other data. The calculations are best performed by extending the spreadsheet used in Problem 4.13. *You should retain the spreadsheet and results as they will be used again in Chapter 8.*
  i.   Calculate the standard entropy of O_2, Al_2O_3 and Al at 298.15 K and at 100 K intervals from 300 to 1500 K, and plot the values graphically as a function of temperature. Note, there is a phase change for aluminium within the temperature range.

ii. Using the entropy values, calculate the standard entropy of formation of Al_2O_3 at 298.15 K and at 100 K intervals from 300 to 1500 K, and plot them graphically as a function of temperature.

Data:

$$C_p = a + b \times 10^{-3} \times T + c \times 10^{-6} \times T^2 + d \times 10^5 \times T^{-2} \text{ J K}^{-1} \text{ mol}^{-1}$$

	a	b	c	d
O_2(g) (298–700)	22.060	20.887	−8.207	1.621
O_2(g) (700–1200 K)	29.793	7.910	−2.204	−6.194
O_2(g) (1200–2500 K)	34.859	1.312	0.163	−14.140
Al(s) (298–933.45 K)	32.974	−20.677	23.753	−4.138
Al(l) (933.45–2790 K)	31.748	0	0	0
Al_2O_3(s) (100–800 K)	9.776	294.725	−198.174	−2.485
Al_2O_3(s) (800–2327 K)	115.977	15.654	−2.358	−44.290

Melting point of Al = 933.45 K

$\Delta_{fus}H(Al) = 10.711 \text{ J K}^{-1}$

$S^0(O_2, g, 298.15) = 205.15 \text{ J K}^{-1}$

$S^0(Al, s, 298.15) = 28.27 \text{ J K}^{-1}$

$S^0(Al_2O_3, s, 298.15) = 50.95 \text{ J K}^{-1}$

8 Gibbs and Helmholtz energies

Curiouser and curiouser!

Lewis Carroll*

SCOPE

This chapter introduces the state functions Gibbs energy and Helmholtz energy and discusses their implications for chemical thermodynamics.

LEARNING OBJECTIVES

1. *Understand qualitatively and quantitatively the concepts of Gibbs energy and Helmholtz energy.*
2. *Understand the use of Gibbs and Helmholtz energies to determine the spontaneity of processes.*
3. *Be able to calculate the Gibbs energy change of a system due to mixing, the Gibbs energy of a substance at any temperature and the Gibbs energy of reactions.*
4. *Understand how the Gibbs energy of substances and of reactions can be determined experimentally.*

8.1 INTRODUCTION

We saw in Section 7.7 that the usefulness of entropy as a criterion of whether a process occurs spontaneously or not is limited by the difficulty in applying the test to actual systems since it is necessary to consider the entropy changes not only of the process itself but of the surroundings which together constitute an isolated system. Usually in practice interest is limited to the process – whether it will occur or not or to what extent it will occur – and not the surroundings. This difficulty is overcome by introducing two new state variables, Gibbs energy and Helmholtz energy. The numerical value of the change in these, whether it is negative or positive, determines whether a process will occur spontaneously under constant pressure or constant volume conditions, respectively.

To illustrate the approach, consider a reaction taking place at constant pressure. The reaction in this case is the system of interest. From Equation 7.23, the reaction will occur spontaneously if

$$\Delta_r S + \Delta S_{surroundings} > 0$$

At constant pressure, $\Delta_r H = \Delta q$ (Equation 4.13) and from the definition of entropy (Equation 7.1)

$$\Delta S_{surroundings} = \frac{-\Delta_r H}{T}$$

where T is the temperature of the system. Note, by convention (Section 4.2.2) a negative sign is placed in front of $\Delta_r H$ because heat is transferred to the surroundings.[†] The reaction will be spontaneous if

---

* Lewis Carroll (1865). *Alice's Adventures in Wonderland*, London: Macmillan.
† $\Delta_r H$ may itself be positive or negative depending on whether the reaction is endothermic or exothermic.

$$\Delta_r S - \frac{\Delta_r H}{T} > 0$$

Multiplying throughout by $-T$ and rearranging gives the expression

$$\Delta_r H - T \Delta_r S < 0$$

This is the criterion for spontaneity of the reaction at constant pressure. Its usefulness is that it requires only information about the reaction ($\Delta_r H$ and $\Delta_r S$) and nothing about the surroundings. In general, a process will occur spontaneously at constant temperature and pressure if,

$$\Delta H - T \Delta S < 0$$

The term $H - TS$ is called Gibbs energy. It, and the corresponding term for constant volume processes, Helmholtz energy, are discussed in the following sections.

8.2   COMBINED STATEMENT OF THE FIRST AND SECOND LAWS

Combining the first law statement (Equation 4.1)

$$dU = \delta q - \delta w$$

with the second law statements (Equations 7.3 and 7.4), namely

$$\text{For reversible processes:} \quad \delta q = TdS$$

$$\text{For irreversible processes:} \quad \delta q < TdS$$

yields:

$$\text{For reversible processes:} \quad dU = TdS - \delta w \tag{8.1}$$

$$\text{For irreversible processes:} \quad dU > T\,dS - \delta w \tag{8.2}$$

where δw is the reversible work done *by* the system and is the sum of all the various forms of work done by the system. Equations 8.1 and 8.2 are forms of the *combined statement of the first and second laws*.

If the external pressure is constant

$$\delta w = pdV + \delta w'$$

where $\delta w'$ is the sum of all forms of energy other than the work of expansion. In chemical applications $\delta w'$ is usually electrical energy generated by chemical reactions in batteries or used in electrolysis. Substituting into Equations 8.1 and 8.2

$$\delta w' \leq TdS - pdV - dU \tag{8.3}$$

where the equality sign refers to a reversible, and the inequality sign to a spontaneous, process. Further, since (Equation 4.11)

$$dH = dU + pdV - Vdp$$

substituting for dU yields

$$\delta w' \leq TdS + Vdp + dH \qquad (8.4)$$

If a particular process is carried out reversibly, $\delta w'$ will have its maximum value. This is called the *available work** of the process. For any spontaneous process, the actual work done will be less than the available work.

8.3 HELMHOLTZ AND GIBBS ENERGIES

We now introduce two new thermodynamic functions, Helmholtz and Gibbs energies:[†]

$$\text{Helmholtz energy:} \quad A = U - TS \qquad (8.5)$$

$$\text{Gibbs energy:} \ G = U + pV - TS = H - TS \qquad (8.6)$$

Since Helmholtz and Gibbs energies are both defined in terms of state properties they are themselves state properties. They are also extensive properties with the unit of energy (Joules).

Equation 8.6 can be understood as follows. The internal energy U is the energy required to create a system at constant temperature and pressure. An additional amount of work pV must be done to create space for the system against an external pressure. The system surroundings will contribute[‡] an amount of thermal energy TS to the system (since from the definition of entropy, $TdS = \delta q$) in order to maintain the temperature of the system at T. The total energy required to create the system is the sum of these three components and is the Gibbs energy of the system. If no energy is required to create the space for the system (that is, it already exists, as in a constant volume enclosure) then $pV = 0$ and Equation 8.5 gives the energy required to create the system at temperature T.

Differentiating Equations 8.5 and 8.6:

$$dA = dU - TdS - SdT \qquad (8.7)$$

$$dG = dH - TdS - SdT \qquad (8.8)$$

Substituting dU and dH from Equations 8.3 and 8.4 in Equations 8.7 and 8.8, respectively, yields

$$dA \leq -pdV - SdT - \delta w' \qquad (8.9)$$

and

$$dG \leq Vdp - SdT - \delta w' \qquad (8.10)$$

It follows from Equation 8.10 that for a reversible process which performs no work other than work of expansion

$$dG = Vdp - SdT \qquad (8.11)$$

---

* $\delta w'$ will be positive for a spontaneous process and will be negative for a non-spontaneous process; that is, in the latter case work must be done on the system for the process to occur.

[†] In older texts, Gibbs energy is referred to as *free energy* or *Gibbs free energy*, and Helmholtz energy is referred to as the *work function* or *Helmholtz free energy*. These terms are now obsolete.

[‡] The contribution will be positive for an endothermic reaction and negative for an exothermic reaction.

Equation 8.10 is another form of the *combined statement of the first and second laws.* At constant temperature ($dT = 0$),

$$\left(\frac{dG}{dp} \right)_T = V \qquad (8.12)$$

and, at constant pressure ($dp = 0$),

$$\left(\frac{dG}{dT} \right)_p = -S \qquad (8.13)$$

For a process occurring at constant volume and temperature, it follows from Equation 8.9 that

$$dA \leq -\delta w' \quad \text{(at constant } V \text{ and } T) \qquad (8.14)$$

and for a process occurring at constant pressure and temperature, it follows from Equation 8.10 that

$$dG \leq -\delta w' \quad \text{(at constant } p \text{ and } T) \qquad (8.15)$$

If the process is reversible,

$$dA = -\delta w' \quad \text{(at constant } V \text{ and } T) \qquad (8.16)$$

and

$$dG = -\delta w' \quad \text{(at constant } p \text{ and } T) \qquad (8.17)$$

where $\delta w'$ will be the maximum work obtainable from the process, other than work of expansion. If $\delta w' > 0$ work is done by the process; if $\delta w' < 0$ work must be done on the process for the change to take place.

SUMMARY

- *The change in Gibbs energy of a process occurring at constant temperature and pressure is the minimum amount of work (other than against an external pressure) required to make the process occur (for a non-spontaneous reaction) or the maximum amount of work that can be obtained from the process if the process is spontaneous.*
- *The change in Helmholtz energy of a process occurring at constant temperature and volume is the minimum amount of work (other than against an external pressure) required to make the process occur (for a non-spontaneous reaction) or the maximum amount of work that can be obtained from the process if the process is spontaneous.*
- *For processes occurring at constant pressure or constant volume, thermal energy equal to $T\Delta S$ must be added to the system for endothermic processes or removed for exothermic processes to maintain a constant temperature.*

8.3.1 THE CRITERIA FOR SPONTANEITY

In the case where the process performs no work other than the work due to volume expansion, $\delta w' = 0$, and it follows from Equations 8.14 and 8.15

$dA < 0$ for a spontaneous process occurring at constant volume and temperature   (8.18)

$dG < 0$ for a spontaneous process occurring at constant pressure and temperature   (8.19)

and

$dA = 0$ for a reversible process occurring at constant volume and temperature   (8.20)

$dG = 0$ for a reversible process occurring at constant pressure and temperature   (8.21)

The change in the Helmholtz or Gibbs energy between two states of a system, therefore, is a criterion for whether the system will change spontaneously from one state to the other, that is, whether a process will occur. Since this criterion applies to a system alone (and does not require any knowledge of the surroundings, except the temperature and pressure), the Helmholtz and Gibbs energy changes of a system are much more useful criteria of spontaneity of a process than is the entropy change of the isolated system.

Changes in Helmholtz and Gibbs energies are defined in the same manner as for U, H and S. For a system changing from state 1 to state 2, $\Delta A = A_2 - A_1$ and $\Delta G = G_2 - G_1$. Thus, from Equations 8.5 and 8.6, at constant temperature

$$\Delta A = \Delta U - T\Delta S \tag{8.22}$$

and

$$\Delta G = \Delta H - T\Delta S \tag{8.23}$$

Most chemical reactions, and many other processes, occur at constant pressure rather than constant volume, and, accordingly, Gibbs energy is generally more useful than Helmholtz energy. For a process occurring at constant temperature and pressure:

When $\Delta G < 0$ (that is, has a negative value) the process will proceed.
When $\Delta G > 0$ (that is, has a positive value) the reverse process will proceed.
When $\Delta G = 0$ the system is at equilibrium.

8.3.2 THE GIBBS–HELMHOLTZ EQUATION

At constant pressure and for a reversible reaction when no work other than the work of expansion is done (Equation 8.13):

$$dG = -SdT$$

Substituting for S from Equation 8.6

$$dG = -\left(\frac{H-G}{T}\right)dT$$

Multiplying throughout by T and rearranging,

$$TdG - GdT = -HdT$$

By further manipulation it can be shown that

$$\frac{d(G/T)}{dT} = \frac{-H}{T^2} \tag{8.24}$$

or

$$\frac{d(G/T)}{d(1/T)} = H \tag{8.25}$$

Equations 8.24 and 8.25 are forms of the *Gibbs–Helmholtz equation*. The corresponding relation for Helmholtz energy is:

$$\frac{d(A/T)}{d(1/T)} = U \tag{8.26}$$

For any change of state of a system,

$$\frac{d(\Delta G/T)}{dT} = \frac{-\Delta H}{T^2} \tag{8.27}$$

or

$$\frac{d(\Delta G/T)}{d(1/T)} = \Delta H \tag{8.28}$$

These equations are particularly useful because they enable the enthalpy change ΔH of a process to be determined from experimentally measured variation of ΔG with temperature, or, conversely, they enable ΔG to be determined from experimentally measured values of ΔH as a function of temperature.

8.4   THE GIBBS ENERGY OF PHASE CHANGES

Since phase changes at constant temperature and pressure occur at equilibrium of the two phases involved (Section 4.5), the Gibbs energy change of phase transformations is zero. For example, for melting,

$$\Delta_{\text{fus}}G^0 = \Delta_{\text{fus}}H^0 - T_{\text{fus}}\Delta_{\text{fus}}S^0 = \Delta_{\text{fus}}H^0 - T_{\text{fus}} \times \frac{\Delta_{\text{fus}}H^0}{T_{\text{fus}}} = 0$$

8.5   THE GIBBS ENERGY OF MIXING

It follows from Equation 8.23 that

$$\Delta_{\text{mix}}G_m = \Delta_{\text{mix}}H_m - T\Delta_{\text{mix}}S_m \tag{8.29}$$

For physical mixing, without any interaction between the components (that is, ideal mixing), $\Delta_{\text{mix}}H_m = 0$. Substituting $\Delta_{\text{mix}}S_m$ from Equation 7.14 into Equation 8.29 gives for the mixing of components A, B, ...

$$\Delta_{\text{mix}}G_m = RT\left(x_A \ln x_A + x_B \ln x_B + \cdots\right)$$

$\Delta_{\text{mix}}G_m$ is the *molar Gibbs energy of mixing* at constant temperature and pressure. The Gibbs energy of mixing can be understood as the difference between the Gibbs energy of 1 mole of the mixture (at a particular composition, pressure and temperature) G_m and the sum of the Gibbs energies of the individual components comprising the mixture prior to mixing (and at the same pressure and temperature as the mixture) G_m^0, where

$$G_m^0 = x_A G_A^0 + x_B G_B^0 + \cdots$$

Therefore,

$$\Delta_{mix} G_m = G_m - G_m^0 = RT\left(x_A \ln x_A + x_B \ln x_B + \cdots\right) \tag{8.30}$$

When components are mixed, their mole fractions are less than 1, making the term inside the parentheses in Equation 8.30 is negative. Thus the Gibbs energy of mixing will always be negative, indicating it is a spontaneous process. The variation of $\Delta_{mix} G_m$ with composition for a binary ideal mixture at three temperatures calculated using Equation 8.30 is shown in Figure 8.1. Note the $\Delta_{mix} G_m$ curves are related to the $\Delta_{mix} S_m$ curve in Figure 7.3 by the factor $(-T)$.

8.6 THE GIBBS ENERGY OF SUBSTANCES

The standard Gibbs energy of a substance at any temperature is related to the standard enthalpy and entropy of the substance at the same temperature by the relation:

$$G^0(T) = H^0(T) - TS^0(T) \tag{8.31}$$

Since H does not have absolute values, it is not possible to assign absolute values to the Gibbs energies of substances. A value can be assigned, however, by combining the enthalpy value of a substance (based on the arbitrary convention of a value of zero for enthalpy at 298.15 K at 1 bar pressure for all elements) with its absolute value of entropy, based on the third law, both at the relevant temperature. This is illustrated in Example 8.1.

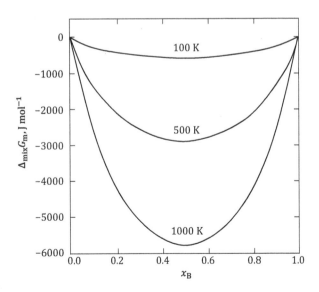

FIGURE 8.1 The variation of Gibbs energy of mixing for a binary mixture of A and B at 100, 500 and 1000 K, assuming ideal behaviour.

EXAMPLE 8.1   Calculate the Gibbs energy of a substance using Equation 8.31

Calculate the Gibbs energy of Fe_2O_3 at 1200 K from its enthalpy and entropy values.

SOLUTION

$$H^0\left(Fe_2O_3, s, 1200\right) = -694.88 \text{ kJ mol}^{-1}$$

$$S^0\left(Fe_2O_3, s, 1200\right) = 277.80 \text{ J mol}^{-1}$$

Therefore, from Equation 8.31

$$G^0\left(Fe_2O_3, s, 1200\right) = -694.88 - \frac{1200 \times 277.80}{1000} = -1028 \text{ kJ mol}^{-1}$$

It follows from Equation 8.31, and the fact that entropies of pure substances are always positive, that Gibbs energies of pure substances decrease with an increase in temperature. As for enthalpy and entropy, the *molar Gibbs energy* $G_m(B)$ is the Gibbs energy of 1 mole of substance B; it has the unit J mol^{-1}. The variation with temperature of the Gibbs energy of some common substances is shown in Figure 8.2. Values of the molar Gibbs energy for substances are tabulated at 298.15 K and at 100 K intervals in the compilation of thermodynamic data by Barin (as illustrated in Table 5.3, column 7). They are not tabulated explicitly in the NIST–JANAF and USGS tables but can be calculated. All three compilations list values of the *Gibbs energy function*

$$Gef = \frac{G^0(T) - H^0(298.15)}{T} \tag{8.32}$$

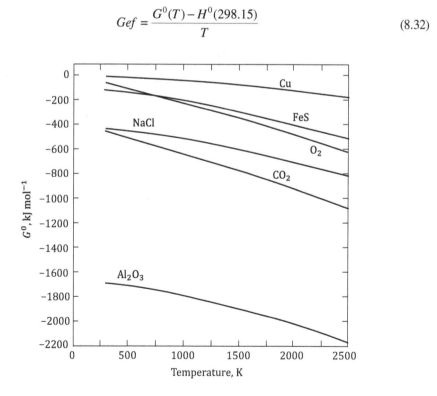

FIGURE 8.2   The variation of Gibbs energy with temperature of some common substances.

for substances at 100 K intervals. The value of G^0 of a substance can be found at any particular temperature by substituting the corresponding value of H^0 at 298.15 K. This form is useful for interpolation between temperature intervals since the variation of *Gef* with temperature is quite gradual.

EXAMPLE 8.2   Calculate the Gibbs energy of a substance from its Gibbs energy function

Calculate the Gibbs energy of Fe_2O_3 at 1200 K from its Gibbs energy function.

SOLUTION

$$Gef\left(Fe_2O_3, s, 1200\right) = 171.39 \text{ J K}^{-1} \text{ mol}^{-1}$$

$$H^0\left(Fe_2O_3, s, 298.15\right) = -825.50 \text{ kJ mol}^{-1}$$

Therefore,

$$G^0\left(Fe_2O_3, s, 1200\right) = \frac{-171.39 \times 1200}{1000} + \left(-825.50\right) = -1031 \text{ kJ mol}^{-1}$$

8.7   THE GIBBS ENERGY OF FORMATION

The *standard Gibbs energy of formation* of a compound is defined (in the same manner as for enthalpy and entropy of formation) as the change in Gibbs energy when 1 mole of the compound is formed from its constituent elements under standard state conditions at a specified temperature. Thus, for the reaction

$$k\,A + l\,B = A_k B_l$$

$$\Delta_f G^0\left(A_k B_l\right) = G^0\left(A_k B_l\right) - kG^0(A) - lG^0(B) \tag{8.33}$$

where A and B are elements which react to form the compound $A_k B_l$. The standard Gibbs energy of formation of elements in their standard state is zero at all temperatures and 1 bar, since for the reaction (where E is an element):

$$E = E$$

$$\Delta_f G^0(T) = G^0(T) - G^0(T) = 0$$

Some texts do not clearly distinguish between the Gibbs energy of a substance (G^0) and its Gibbs energy of formation ($\Delta_f G^0$), and care needs to be exercised when using data from different sources in calculations. For example,

$$G^0\left(H_2O, 298.15\right) = -306.7 \text{ kJ mol}^{-1}$$

but

$$\Delta_f G^0\left(H_2O, 298.15\right) = -237.1 \text{ kJ mol}^{-1}$$

**EXAMPLE 8.3 Calculation of the Gibbs energy of formation
from the Gibbs energies of the reactants and products**

Calculate the standard Gibbs energy of formation of water at 298.15 K.

SOLUTION

$$H_2(g) + 1/2\, O_2(g) = H_2O(l)$$

$$\Delta_f G^0\left(H_2O,l,298.15\right) = G^0\left(H_2O,l,298.15\right) - G^0\left(H_2,g,298.15\right) - 0.5 \times G^0\left(O_2,g,298.15\right)$$

$$= -306.69 - \left(-38.96\right) - 0.5 \times \left(-61.16\right) = -237.1 \text{ kJ mol}^{-1}$$

Values of the Gibbs energy of formation of substances are tabulated in the NIST–JANAF, US Geological Survey and Barin compilations of data (Section 5.2) at 298.15 K and at 100 K intervals. The variation with temperature of the standard Gibbs energy of formation of some common compounds is shown in Figure 8.3. Note that, unlike for $\Delta_f H^0$ and $\Delta_f S^0$ which are approximately constant over large temperature ranges, $\Delta_f G^0$ can vary considerably with temperature and that the relations are very nearly linear. Assuming a linear relationship then, since, $\Delta G^0 = \Delta H^0 - T\Delta S^0$, it follows that the slope of the lines corresponds to the average value of $-\Delta_f S^0$, and the intercept at 0 K corresponds to the average value of $\Delta_f H^0$ over the relevant temperature range.

In some compilations* of thermodynamic data $\Delta_f G^0$ is expressed as a function of temperature using empirical equations of the form:

$$\Delta_f G^0 = p + q\,T \ln T + r\,T^2 + s\,T + t\,T^{-1}$$

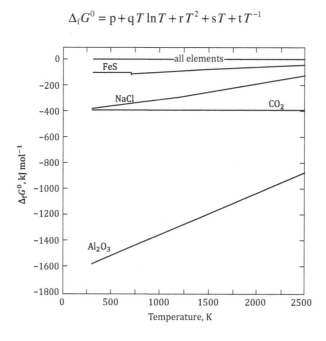

FIGURE 8.3 The variation with temperature of the Gibbs energy of formation of some common substances.

---

* For example: Kubaschewski, O. and C. B. Alcock. 1979. *Metallurgical Thermochemistry*, 5th ed., Oxford: Pergamon Press. 378–384.

Values of p, q, r, s and t are tabulated for common substances over wide temperature ranges. Often, the terms containing r and t are not needed to accurately represent the data, and the equation then has the form:

$$\Delta_f G^0 = p + q\,T \ln T + s\,T \tag{8.34}$$

Sometimes, the q term is also omitted, and the relationship becomes linear:

$$\Delta_f G^0 = p + s\,T \tag{8.35}$$

In this case, p is the average value of $\Delta_f H^0$ and $-s$ is the average value of $\Delta_f S^0$ over the temperature range for which the relation applies.

8.8 THE GIBBS ENERGY OF REACTION

The *Gibbs energy of reaction* is defined in the same manner as enthalpy and entropy of reaction. For the reaction

$$k\,A + l\,B = m\,C + n\,D \tag{4.31}$$

the Gibbs energy change is:

$$\Delta G = m\,G(C) + n\,G(D) - k\,G(A) - l\,G(B) \tag{8.36}$$

When all the reactants and products are in their standard state, the *standard Gibbs energy change of the reaction* is given by

$$\Delta G^0 = m\,G^0(C) + n\,G^0(D) - k\,G^0(A) - l\,G^0(B) \tag{8.37}$$

As we saw previously,

When $\Delta G > 0$ (that is, has a positive value) the reaction will proceed to the left.
When $\Delta G < 0$ (that is, has a negative value) the reaction will proceed to the right.
When $\Delta G = 0$ the reaction is at equilibrium.

Considering the relation $\Delta G = \Delta H - T\Delta S$, we can distinguish four situations by examining the signs of the terms on the right side of the equation.

- When ΔS is positive and ΔH is negative, ΔG is negative and a process is spontaneous at all temperatures.
- When ΔS is positive and ΔH is positive, the relative magnitudes of ΔS and ΔH determine the magnitude and sign of ΔG and whether the reaction is spontaneous. High temperatures make the reaction more favourable, because the ΔH term becomes small relative to the $T\Delta S$ term.
- When ΔS is negative and ΔH is negative, the relative magnitudes of ΔS and ΔH determine the magnitude and sign of ΔG and whether the reaction is spontaneous. Low temperatures make the reaction more favourable because the ΔH term becomes dominant.
- When ΔS is negative and ΔH is positive, ΔG is positive, and a process is not spontaneous at any temperature, but the reverse process is spontaneous.

EXAMPLE 8.4   Calculate the Gibbs energy change of a reaction using tabulated values

Using available data, calculate the standard Gibbs energy change at 1000°C for the reaction:

$$Fe_2O_3(s) + 3\ C(s) = 2\ Fe(s) + 3\ CO(g)$$

SOLUTION

The approach is the same as in Example 4.4 for calculating the standard enthalpy change of a reaction. Again, the three methods are available, and an additional one.

METHOD 1: COMBINING EQUATIONS FOR FORMATION

$$
\begin{array}{ll}
Fe_2O_3(s) = 2\ Fe(s) + 1.5\ O_2(g) & -\Delta_f G^0(Fe_2O_3) \\
3\ C(s) + 1.5\ (g) = 3\ CO(g) & 3 \times \Delta_f G^0(CO) \\
\hline
Fe_2O_3(s) + 3\ C(s) = 2\ Fe(s) + 3\ CO(g) &
\end{array}
$$

At 1000°C (1273.15 K),

$$\Delta_r G^0(1273) = -\Delta_f G^0\left(Fe_2O_3, s, 1273\right) + 3 \times \Delta_f G^0\left(CO, g, 1273\right)$$

$$= -\left(-491.33\right) + 3 \times \left(-224.20\right) = -181.26\ kJ$$

The negative value indicates that the reduction of Fe_2O_3 by carbon to form metallic iron and gaseous CO will occur spontaneously at 1000°C under standard state conditions.

METHOD 2: COMBINING GIBBS ENERGY OF FORMATION VALUES

$$\Delta_r G^0(1273) = 3 \times \Delta_f G^0\left(CO, g, 1273\right) + 2 \times \Delta_f G^0\left(Fe, s, 1273\right)$$

$$- 3 \times \Delta_f G^0\left(C, s, 1273\right) - \Delta_f G^0\left(Fe_2O_3, s, 1273\right)$$

$$= 3 \times \left(-224.20\right) + 2 \times 0 - 3 \times 0 - \left(-491.33\right) = -181.27\ kJ$$

METHOD 3: COMBINING INDIVIDUAL GIBBS ENERGY VALUES

$$\Delta_r G^0(1273) = 3 \times G^0\left(CO, g, 1273\right) + 2 \times G^0\left(Fe, s, 1273\right)$$

$$- 3 \times G^0\left(C, s, 1273\right) - G^0\left(Fe_2O_3, s, 1273\right)$$

$$= 3 \times \left(-388.61\right) + 2 \times \left(-62.31\right) - 3 \times \left(-20.10\right) - \left(-1048.88\right) = -181.27\ kJ$$

METHOD 4: COMBINING $\Delta_r H^0$ AND $\Delta_r S^0$ VALUES

$$\Delta_r G^0 = \Delta_r H^0 - T\Delta_r S^0$$

Taking the values of $\Delta_r H^0 = 467.26\ kJ\ mol^{-1}$ and $\Delta_r S^0 = 509.38\ J\ mol^{-1}$ calculated in Examples 4.4 and 7.3, respectively

$$\Delta_r G^0 = 467.26 - \frac{1273.15 \times 509.38}{1000} = -181.26\ kJ$$

EXAMPLE 8.5   Calculate the Gibbs energy change of a reaction from empirical Gibbs energy equations

Using equations for the Gibbs energy of formation of Fe_2O_3 and CO, calculate the standard Gibbs change at 1000°C for the reaction:

$$Fe_2O_3(s) + 3\,C(s) = 2\,Fe(s) + 3\,CO(g)$$

SOLUTION

$Fe_2O_3(s) = 2\,Fe(s) + 1.5\,O_2(g)$	$-\Delta_f G^0(Fe_2O_3)$
$3\,C(s) + 1.5\,O_2(g) = 3\,CO(g)$	$3 \times \Delta_f G^0(CO)$
$Fe_2O_3(s) + 3\,C(s) = 2\,Fe(s) + 3\,CO(g)$	$\Delta_r G^0 = 3 \times \Delta_f G^0(CO) - \Delta_f G^0(Fe_2O_3)$

From the literature:*

$$\Delta_f G^0(Fe_2O_3) = -814\,100 + 250.7\,T \text{ J}$$

$$\Delta_f G^0(CO) = -114\,390 - 85.8\,T \text{ J}$$

$Fe_2O_3(s) = 2\,Fe(s) + 1.5\,O_2(g)$	$-\Delta_f G^0(Fe_2O_3) = 814\,100 - 250.7\,T \text{ J}$
$3\,C(s) + 1.5\,O_2(g) = 3\,CO(g)$	$3 \times \Delta_f G^0(CO) = 3 \times (-114\,390 - 85.8\,T) \text{ J}$
$Fe_2O_3(s) + 3\,C(s) = 2\,Fe(s) + 3\,CO(g)$	$\Delta_r G^0 = 470\,930 - 508.1\,T$

Therefore,

$$\Delta_r G^0(1273) = 470\,930 - 508.1 \times 1273 = 175.9 \text{ kJ}$$

EXAMPLE 8.6   The electrolysis of water

How much energy must be supplied electrically to decompose water by electrolysis at 25°C and 1 bar?
Data:

	$H_2O(l)$	$H_2(g)$	$O_2(g)$
$H^0(298.15)$ kJ mol^{-1}	-285.83	0	0
$S^0(298.15)$ kJ mol^{-1}	69.95	130.68	205.15

SOLUTION

The decomposition reaction is: $H_2O(l) = H_2(g) + 0.5\,O_2(g)$

$$\Delta_r H^0 = H^0(H_2) + 0.5 \times H^0(O_2) - H^0(H_2O) = 0 + 0 - (-285.15) = 285.15 \text{ kJ}$$

$$\Delta_r S^0 = S^0(H_2) + 0.5 \times S^0(O_2) - S^0(H_2O) = 130.68 + 0.5 \times 205.15 - 69.95 = 163.31 \text{ J}$$

---

* Turkgogan, E. T. 1980. *Physical Chemistry of High Temperature Technology*, New York: Academic Press.

$\Delta_r G^0$ is the minimum energy required to make the process occur. For each mole of water decomposed,

$$\Delta_r G^0 = \Delta_r H^0 - T\Delta_r S^0 = 285.83 - \left(298.15 \times \frac{163.31}{1000}\right) = 285.83 - 48.69 = 237.14 \text{ kJ} = 0.066 \text{ kW h}$$

Since $dG = -\delta w'$ (at constant p and T), work must be done on the sytem for the process to occur.

COMMENT

Enthalpy is the total energy of a thermodynamic system (Section 4.3.1). Therefore, the enthalpy change, $\Delta H^0 = \Delta G^0 + T\Delta S^0$, is the energy necessary to accomplish the decomposition. However, it is not necessary to supply the whole amount in the form of electrical energy. Since entropy increases in the process of dissociation of water (liquid being converted to gas), $T\Delta S^0$ is positive, and the amount of energy $T\Delta S^0$ (298.15 × 163.31 = 48.7 kJ), which is thermal energy (since $\Delta S = \Delta q/T$), can be provided from the surroundings at temperature T (298.15 K in this case). The energy which must be supplied electrically is the difference, that is, the change in the Gibbs energy $\Delta_r G^0$.

The calculated value is the electrical energy for reversible electrolysis (zero current). For the process to occur at a finite rate the energy must be greater. There are also other barriers (discussed in Chapter 16). Depending on the composition of the liquid and the electrodes, and their configuration, the energy required for practical electrolysis of water is closer to 0.1 kW h mol^{-1}.

EXAMPLE 8.7   The hydrogen fuel cell

How much electrical energy can be generated by a hydrogen fuel cell operating at 25°C and 1 bar, and what is the efficiency of conversion?

SOLUTION

Hydrogen and oxygen can be combined in a fuel cell (Section 17.5) to produce electrical energy by the reverse of the reaction for electrolysis of water:

$$H_2(g) + 0.5\, O_2(g) = H_2O(l)$$

Fuel cells use a chemical reaction to generate a potential difference, as do batteries, but differ from batteries in that the fuel is continuously supplied, in this case in the form of gaseous hydrogen. Fuel cells can produce electrical energy at a higher efficiency than by burning the fuel to produce heat to drive a generator because they are not subject to the limits imposed by Carnot's theorem (Section 7.2.3).

The relevant data are the same as in Example 8.6. The maximum energy (per mole of hydrogen) which can be provided as electrical energy is the change in the Gibbs energy:

$$\Delta_r G^0 = \Delta_r H^0 - T\Delta_r S^0 = -285.83 + 48.69 = -237.14 \text{ kJ} = -0.066 \text{ kWh}$$

Since $dG = -\delta w'$ (at constant p and T), work will be done by the sytem.

The chemical energy in the fuel (equal to the enthalpy of combustion of hydrogen) is converted to electrical energy at a maximum efficiency of:

$$237.1/285.83 \times 100\% = 83\%.$$

COMMENT

In this case, the entropy of the gases decreases by 163.31 J in the process of combination of hydrogen and oxygen. Since the total entropy ($S_{system} + S_{surrondings}$) cannot decrease as a result of the reaction, entropy must be created by an amount of $T\Delta S$ thermal energy (48.7 kJ) being transferred to the surroundings (exothermic reaction) to maintain the temperature at 25°C.

The calculated efficiency is for a reversible process and can never be achieved in practice – typical efficiency values are 50–60%. However, the fuel cell efficiency is much greater than the reversible efficiency of a power station which burned the hydrogen and used the heat to drive a generator.

In summary, the Gibbs energy change of a process is the minimum energy which has to be supplied to drive a non-spontaneous process or the maximum energy that can be obtained from a spontaneous process at constant temperature and pressure. In the electrolysis of water, for which the enthalpy change (endothermic) is 285.8 kJ, an amount of 237.1 kJ of electrical energy is required to drive electrolysis, and the heat from the environment contributes the balance (48.7 kJ) to maintain a constant temperature. In the fuel cell, 237.1 kJ of electrical energy is produced, and 48.7 kJ of heat is lost to the environment to maintain a constant temperature.

8.9   THE USE OF GIBBS ENERGY TO STUDY REACTIONS

One of the main applications of Gibbs energies is to predict whether a reaction will occur under certain conditions and, if so, to what extent. In the simple case of pure substances reacting to form pure solid or liquid products or gaseous products at 1 bar pressure, the sign (positive or negative) of the standard Gibbs energy of the reaction tells in which direction the reaction will proceed. This is illustrated in Example 8.8. Where reactants and products are not in their standard state (such as when they are present in solutions or gas mixtures) reactions can never go entirely to completion; that is, none of the reactants can be completely consumed. This is because as the concentration of a reactant becomes smaller due to it being consumed it becomes increasingly less available for reaction. This is explored more fully in Chapter 10 after the concept of the activity of a species in solution is introduced in Chapter 9. Furthermore, in many reacting systems more than one chemical reaction is possible between the elements present. For example, the elements O, H and Cl can form several compounds – H_2O, H_2O_2, HCl, HOCl. Under a given set of conditions which of these are formed, and what are their relative amounts? This problem can also be addressed using Gibbs energy, but it involves considering competing reactions. This is discussed in Chapter 15. Yet another related application of Gibbs energy is to determine which state of a substance (solid, liquid or gas) is stable under specified conditions. This is examined in Chapter 13.

EXAMPLE 8.8   Determining if a reaction will occur under standard state conditions

 i. At what temperature can silicon carbide be made by reacting silicon metal with silica? The relevant reaction is:

$$Si(s) + C(s) = SiC(s)$$

SOLUTION

The value of $\Delta_f G^0$ obtained from published data is negative at all practical temperatures (for example it varies from −69.49 kJ at 298.15 K to −46.14 kJ at 2000 K). This means that the reaction will proceed at all temperatures, and solid silicon metal will react with solid carbon to form solid silicon carbide until either all the silicon or all the carbon is consumed. However, there is a kinetic constraint that also needs to be taken into account; we consider this in Section 8.9.1.

ii. At what temperature will limestone begin to decompose spontaneously under standard state conditions? The relevant reaction is:

$$CaCO_3(s) = CaO(s) + CO_2(g)$$

SOLUTION

$\Delta_r G^0$ calculated from $\Delta_f G^0(CaCO_3)$, $\Delta_f G^0(CaO)$ and $\Delta_f G^0(CO_2)$ values over a range of temperatures are listed in the table below, and the variation with temperature is shown in Figure 8.4.

T	$\Delta_f G^0(CO_2)$	$\Delta_f G^0(CaCO_3)$	$\Delta_f G^0(CaO)$	$\Delta_r G^0$
K	kJ mol^{-1}	kJ mol^{-1}	kJ mol^{-1}	kJ mol^{-1}
100	−393.686	−1178.31	−623.939	160.683
200	−394.064	−1153.9	−613.696	146.140
300	−394.369	−1127.6	−603.101	130.127
400	−394.648	−1101.4	−592.545	114.203
500	−394.912	−1075.51	−582.11	98.487
600	−395.154	−1049.92	−571.775	82.988
700	−395.369	−1024.57	−561.501	67.696
800	−395.556	−999.309	−551.146	52.607
900	−395.716	−974.203	−540.771	37.716
1000	−395.852	−949.233	−530.363	23.018
1100	−395.967	−924.372	−519.897	8.508
1200	−396.064	−898.977	−508.726	−5.813
1300	−396.143	−873.627	−497.429	−19.945
1400	−396.206	−848.442	−486.126	−33.890
1500	−396.254	−823.424	−474.818	−47.648
1600	−396.289	−798.576	−463.508	−61.221
1700	−396.31	−776.129	−452.198	−72.379
1800	−396.318	−754.013	−440.891	−83.196
1900	−396.313	−732.144	−429.589	−93.758
2000	−396.295	−710.505	−418.296	−104.086

$\Delta_r G^0$ changes from being positive to negative at approximately 1160 K (887°C). This means that the reaction will proceed to the right, and solid limestone will decompose to form pure lime (CaO) and carbon dioxide at 1 bar at temperatures greater than 1160 K. The reaction will proceed until all the limestone has decomposed.

COMMENT

Conversely, if pure CaO and CO_2 at 1 bar pressure are contacted at temperatures below 1160 K, they will react to form $CaCO_3$. This follows because ΔG^0 for the reverse reaction

$$CaO(s) + CO_2(g) = CaCO_3(s)$$

is the negative of the value for the decomposition of $CaCO_3$.

The thermal decomposition of limestone is more complicated than implied by this example and is explored more fully in Example 10.4.

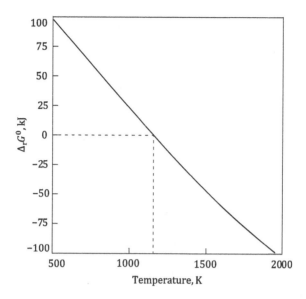

FIGURE 8.4   The variation of Gibbs energy for the decomposition of calcium carbonate.

8.9.1   THE IMPORTANCE OF KINETICS

In Example 8.8i, thermodynamics predicts that the reaction of Si and C will proceed at all temperatures, but in practice it will not actually proceed at low temperatures. A certain temperature needs to be achieved, determined experimentally, before Si and C will start to react. In commercial production of SiC, temperatures in excess of 1600°C are used. The reason for this is that, for a reaction to proceed at a finite rate, besides needing to meet the thermodynamic criterion of a negative change in Gibbs energy, a kinetic energy criterion has also to be met. If molecules have too little kinetic energy, or collide with incorrect orientation, they will not react, but if they are moving fast enough and with the correct orientation such that the kinetic energy upon collision is greater than the minimum energy barrier, then interatomic bonds can be broken, and a reaction will occur if the Gibbs energy change is negative. The amount of energy needed to achieve bond disruption is the *activation energy*, E_a, and this is clearly temperature dependent.

Activation energy can be thought of as the height of the kinetic energy barrier separating two minima of Gibbs energy, namely those of the reactants and products of a reaction. The concept is illustrated in Figure 8.5 which shows the energy of a reacting system as a function of the reaction coordinate (which represents the progress along a reaction pathway as changes occur in one or more molecular entity). In the case illustrated, although $\Delta_r G^0$ is negative, the reaction cannot proceed until the energy of the system is raised sufficiently, by increasing the temperature. For many reactions, although they are thermodynamically feasible at low temperatures, they either will not occur (for all practical purposes) or will occur only very slowly unless the temperature is raised sufficiently.

When the rate of a reaction is changed by the presence of a substance, which is not altered chemically as a result of the reaction, this substance is known as a *catalyst*. The function of a catalyst is to take part in the reaction to form an activated complex which then decomposes, leaving the catalyst unchanged, but lowers the activation energy required to form that particular activated complex. The catalyst can be homogeneous (being of the same phase as the reactants) or heterogeneous (being a different phase from that of the reactants), such as a solid surface on which the overall reaction proceeds.

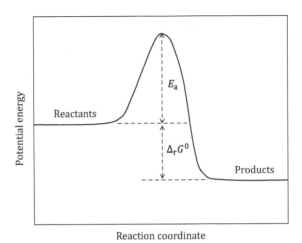

FIGURE 8.5   The relationship between activation energy and Gibbs energy of reaction.

SUMMARY

- *For reactants and products in their standard state, a reaction is thermodynamically feasible if the value of the standard Gibbs energy of the reaction is negative, and the reaction can proceed until one of the reactants is fully consumed. If the value of the standard Gibbs energy is positive, the reverse reaction is thermodynamically feasible and can proceed until one of the reactants is fully consumed.*
- *While a reaction may be thermodynamically feasible, it will proceed in practice only if the kinetic barrier of activation energy is exceeded.*

8.10   EXPERIMENTAL DETERMINATION OF GIBBS ENERGY

The Gibbs energy of substances and Gibbs energy of reactions can be calculated from measured values of the enthalpy and entropy of formation and heat capacities of the reactants and products; from experimentally measured values of ΔH as a function of temperature using the Gibbs–Helmholtz equation (Equations 8.24 and 8.25); and from measured values of the equilibrium constant using Equation 10.6. Gibbs energies of reaction can be also obtained from electrical potential measurements using Equation 16.7 for reactions that can be performed in an electrochemical cell.

PROBLEMS

8.1   Calculate the molar entropy and molar Gibbs energy of mixing at 1200 K of two ideal gases, A and B, at 0, 0.2, 0.4, 0.6, 0.8 and 1.0 mole fraction of gas A, and show the results graphically as a function of mole fraction of gas A. At what value of mole fraction of gas A is the entropy a maximum and Gibbs energy a minimum? Calculate the molar enthalpy of mixing of A and B at each composition. Comment on the significance of the value.

8.2   Using data from the table below calculate the Gibbs energy of reaction of the following chemical reactions at 298.15 and 1200 K:

  i. Using the Gibbs energy values only (columns 2 and 3).

  ii. Using the Gibbs energy of formation values only (columns 3 and 4).

  iii. Using the values of $\Delta_r H^0$ and $\Delta_r S^0$ calculated previously in Problems 4.9 and 7.9, respectively.

    (Your answers in all cases should be the same irrespective of the method of calculation.)

iv. Comment on the significance and magnitude of the sign of $\Delta_r G$ in each case.

$$CH_4(g) + 2\,O_2(g) = CO_2(g) + 2\,H_2O(l,g)$$

$$N_2(g) + 3\,H_2(g) = 2\,NH_3(g)$$

$$Cu_2S(s) + 2\,O_2(g) = 2\,Cu(s) + SO_2(g)$$

Data (in kJ mol^{-1}):

	G^0 (298.15)	G^0 (1200)	ΔG^0 (298.15)	ΔG^0 (1200)
$CH_4(g)$	−130.39	−335.10	−50.76	41.62
$CO_2(g)$	−457.24	−684.27	−394.36	−396.01
Cu	−9.89	−59.80	0.00	0.00
Cu_2S	−115.80	−297.72	−86.47	−92.07
$H_2(g)$	−38.96	−179.35	0.00	0.00
$H_2O(l)$	−306.67	–	−237.14	–
$H_2O(g)$	−298.16	−496.05	−228.62	−181.57
$O_2(g)$	−61.17	−270.25	0.00	0.00
$N_2(g)$	−57.13	−252.96	0.00	0.00
$NH_3(g)$	−103.42	−310.18	−16.41	85.33
$SO_2(g)$	−370.82	−630.31	−300.10	−274.01

8.3 Using data from the NBS, USGS and NIST–JANAF tables, calculate the standard Gibbs energy of reaction for the following chemical reactions at 298.15 K. Indicate which are spontaneous under standard state conditions.

$$2\,CaO(s) + SiO_2(s) = Ca_2SiO_4(s)$$

$$N_2(g) + 3\,H_2(g) = 2\,NH_3(g)$$

$$CH_4(g) + 2\,S_2(g) = 2\,H_2S(g) + CS_2(g)$$

$$CH_4(g) + NH_3(g) = HCN(g) + 3\,H_2(g)$$

$$ZnSO_4.7H_2O(s) = ZnSO_4.6H_2O(c) + H_2O(l)$$

$$Na_2SiO_3(s) + H_2SO_4(l) = SiO_2(s) + Na_2SO_4(s) + H_2O(l)$$

8.4 Calculate the maximum electrical energy that may be obtained from a fuel cell operating on methane (CH_4) at 25°C. What is the efficiency of the energy conversion? Use 1 mole of methane as the basis for calculation.

8.5 To overcome the high cost and difficulty in transporting hydrogen, it has been suggested that aluminium could be used as an intermediate product. Aluminium would be produced from alumina (Al_2O_3) using the conventional Hall–Heroult process (see Section 17.1.1), then transported as a solid to where hydrogen is required where it would be reacted with steam to generate hydrogen: $2\,Al(s) + 3\,H_2O(g) = 3\,H_2(g) + Al_2O_3(s)$. The Al_2O_3 produced in the reaction would be recycled back to the smelter to regenerate aluminium.

i. What is the thermodynamic feasibility of this process?

ii. How energy efficient is the concept? For a very preliminary assessment assume the following: The reaction of Al with steam is carried out at 500°C; all heat generated by the reaction is captured and utilised; the hydrogen produced is used in combustion processes in which all the heat of reaction is fully utilised; the energy required to produce Al from Al_2O_3 (in the Hall–Heroult process) is 180 000 kJ kg^{-1} Al. In practice this is supplied electrically.

    iii. Explain why the second and third assumptions above could never be realised in practice.

8.6   This is a continuation of Problems 4.13 and 7.12. It utilises the enthalpy and entropy values previously calculated. The calculations are most easily performed by extending the spreadsheet used previously.

    i.   Calculate the standard Gibbs energy of O_2, Al_2O_3 and Al at 298.15 K and at 100 K intervals by means of Equation 8.31 from 100 to 1500 K, and plot the values graphically as a function of temperature. Note, there is a phase change for aluminium within the temperature range.

    ii. Using the Gibbs energy values calculated in part i, calculate the standard Gibbs energy of formation of Al_2O_3 at 298.15 K and at 100 K intervals from 300 to 1500 K, and plot them graphically as a function of temperature.

9 Solutions

This chapter describes common types of solutions and examines the thermodynamic properties of solutions, starting first with gas mixtures then extending the concepts to liquid and solid solutions, both ideal and non-ideal.

LEARNING OBJECTIVES

1. *Understand the difference between mixtures and solutions and the common types of solutions.*
2. *Understand the concepts of partial and integral quantities and how they are related.*
3. *Be able to calculate the partial molar Gibbs energy of a component in ideal and non-ideal gas mixtures.*
4. *Understand how the concept of partial molar Gibbs energy of a component in a gas mixture can be extended to liquid and solid solutions leading to the concept of thermodynamic activity.*
5. *Understand the standard states commonly used for the activity of components of solutions and how they are related.*
6. *Be able to calculate the partial and integral properties of ideal and non-ideal solutions given their compositions and other relevant data.*
7. *Understand the principles of the main experimental methods for determining the activity of components of solutions.*

9.1   INTRODUCTION

Solutions are single-phase systems consisting of two or more components. The key point of this definition is the phrase 'single phase'. Water containing dissolved sugar is a solution since it is a single phase, but water containing suspended silt is not a solution because the water and silt are distinct phases. The latter is a mixture which, as defined in Chapter 2, is a physical combination of two or more substances, without chemical bonding or other chemical change, in which the separate identities of the substances are retained. Solutions may be gaseous, liquid or solid. Air is a solution of nitrogen, oxygen, water vapour, carbon dioxide and minor quantities of other gases; sea water is a dilute solution in water of NaCl, $MgCl_2$, $CaCl_2$, carbon dioxide and many other elements and compounds; molten steel is a dilute solution of carbon, silicon, manganese, oxygen and other elements in iron.

Substances in solution can react with other substances added to the solution in much the same way as the pure substances can react. For example, sodium hydroxide will react with HCl by the same overall reaction

$$NaOH + HCl = NaCl + H_2O$$

whether the sodium hydroxide is in the solid form or dissolved in water or whether the HCl is in the gaseous state or in solution as hydrochloric acid. However, the extent to which a reaction can occur in solutions depends on the concentrations of the reactants and products and how they interact with other components in the solution. Generally, the components of a solution interact with one another at the atomic, molecular or ionic level either to enhance or reduce the 'availability' of each component, and the 'availability' of a component is a function of the concentration of the component. As will be seen later the thermodynamic term for 'availability' is activity.

The composition of solutions is expressed in terms of the concentrations of the components forming the solution. When the concentration of one component is very large compared to that of the others, the major component is referred to as the *solvent* and minor components as *solutes*. The concentration of the solvent, if required, is usually calculated by subtracting the concentrations of all the solutes from the total concentration (that is, it is calculated by 'difference'). In some systems, there is a limit to how much solute can dissolve into the solvent. Once this concentration has been reached any further solute component added will remain undissolved and form a separate phase. The solution in this case is said to be *saturated* with respect to the solute. In other systems, there can be complete *miscibility** across the entire composition range. For example, ethanol and water are miscible at all concentrations. In these cases, it is not really necessary, or even possible, to say which component is the solute and which is the solvent.

9.2 TYPES OF SOLUTIONS

Several types of solutions are frequently encountered – gas mixtures, aqueous solutions, organic solutions, molten solutions and solid solutions. Some of these are ionic in nature (many aqueous and molten salt solutions), and others are elemental (solutions consisting only of elements, for example, molten alloys) or molecular in nature (such as many organic solutions) or a combination. We have discussed gas mixtures already in some detail in Chapter 3 and will discuss their thermodynamic properties in Section 9.4. However, we have not yet discussed the other types of solutions, and the following is a brief introduction to their nature as background information.

9.2.1 AQUEOUS SOLUTIONS

In aqueous solutions the solvent is water. These are the most familiar type of solution. For aqueous solutions of soluble organic compounds (for example sucrose, glucose, ethanol) and simple inorganic gases such as oxygen and nitrogen, the solute is present in the solvent as molecules. The large majority of inorganic substances, however, dissociate to varying extents into ions when dissolved in water. These substances are referred to as *electrolytes* when they are dissolved in water. The thermodynamics of electrolytes and electrolyte solutions is discussed in detail in Chapter 12. For thermodynamic purposes, the concentration of solutes and ions in dilute aqueous solutions is usually expressed in terms of molality (Equation 2.6).

9.2.2 ORGANIC SOLUTIONS

The overall capacity of substances to form solutions depends primarily on their polarity, and the general rule is like dissolves like. For example, a very polar substance such as urea is very soluble in highly polar water, less soluble in fairly polar methanol and practically insoluble in non-polar benzene. In contrast, a non-polar substance such as naphthalene is insoluble in water, fairly soluble in methanol and highly soluble in benzene. The composition of organic solutions for thermodynamic purposes is expressed in terms of mole fraction of the components (Equation 2.2).

9.2.3 MOLTEN SOLUTIONS

Important molten solutions are naturally occurring magma (from which igneous rocks form) and various human-made solutions such as molten glass, molten metals and alloys, molten salt mixtures and molten oxide mixtures (or slags), the latter two being used in the smelting and refining of metals. Molten solutions are often called *melts*. The composition of melts is usually reported in terms

---

* Miscibility is the property of substances to form a solution. A substance is said to be miscible if it dissolves in a nominated substance or immiscible if it does not.

of mass percent or, for components which are low in concentration, parts per million. For thermo-dynamic calculations, however, composition is expressed in terms of atom fraction (for elements) and mole fraction (for compounds) as defined by Equation 2.2. An exception, discussed in Section 9.5.3, is for very dilute alloy solutions. In these cases, the concentration is expressed in mass percent or mole percent.

Selection of the components comprising a molten solution is not always straightforward. While the composition of an alloy can be readily expressed in terms of the elements comprising it, this is not so simple for molten magmas, glasses and slags. Though these are formed predominantly from metal oxides the oxides are not present in molecular form in the molten state but in a dissociated ionic form. A particular element may be present in several ionic forms, and representation of melt composition in terms of the ionic components present is difficult and usually impossible. The composition, therefore, is usually expressed in terms of the oxides and other simple compounds which would form the melt were they mixed in the appropriate proportion and melted. This is illustrated in Table 9.1 for the case of a common type of magma, a typical window glass and a typical slag produced in a blast furnace in which iron ore is smelted to make iron. The oxidation state of the element in the melt ideally should be known for those elements which have more than one oxidation state. Iron is represented in slag by FeO for divalent iron and as Fe_2O_3 or Fe_3O_4 for trivalent iron. Frequently only the total iron content is determined analytically, and in that case the composition may be expressed as Fe_TO which means the total iron content is expressed as FeO. This approach is acceptable for some purposes.

9.2.4 SOLID SOLUTIONS

A *solid solution* is a solid-state solution of two or more substances. A solid solution is differentiated from a compound in that the crystal structure of one of the components (which can be thought of as the solvent) remains fundamentally unchanged by addition of the other components (the solutes) and is differentiated from a mixture in that it is a single homogeneous phase. Solid solutions often can form when two elements (generally metals) are close together in the periodic table whereas chemical compounds generally are formed from elements not in close proximity in the periodic table. The solute may be incorporated within the solvent crystal lattice by replacing a solvent atom, molecule or ion (called substitution) or by fitting into the space between solvent atoms, molecules or ions (called interstitial) (Figure 9.1). Both types of solid solution affect the physical properties of the substance by distorting the crystal lattice of the solvent.

TABLE 9.1
Typical compositions of some common melts

	Mafic (or basaltic) magma (Mass%)	Window glass (Mass%)	Iron blast furnace slag (Mass%)
SiO_2	57.0	73.0	35.0
Al_2O_3	16.0	1.4	10.0
FeO	7.5		2.7
MnO			5.3
MgO	4.5	3.8	5.0
CaO	7.5	8.2	42.0
Na_2O	3.5	12.8	
K_2O	2.0	0.8	
TiO_2	1.5		
Dissolved gas	0.5		
Total	100.0	100.0	100.0

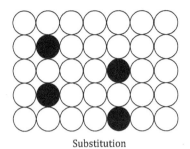 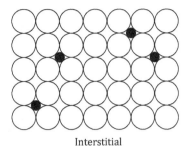

Substitution   Interstitial

FIGURE 9.1   The two types of solid solutions.

Some mixtures readily form solid solutions over a wide range of concentrations while others will not form solid solutions at all. The propensity for any two substances to form a solid solution is a complicated matter involving the chemical and crystallographic properties of the substances as well as temperature and pressure. Substitutional solid solutions may form if the solute and solvent have:

- Similar atomic radii (typically <15% difference)
- The same crystal structure
- Similar electronegativity
- The same valency

High temperature favours the formation of solid solutions. At higher temperatures substances have greater atomic vibration and more open structures which are easier to distort locally to accommodate differently sized cations. More importantly, solid solutions have a higher entropy than the end-members due to the increased disorder associated with the distributed cations, and at high temperatures the $-TS$ term in the Gibbs energy equation ($G = H - TS$) is larger, and the solid solution becomes more stable.

Solid solutions may be metallic in nature (consisting of two or more elements dissolved in a metal), or they may be compounds with some elements substituting for others within the crystal lattice. For example, silver and gold and copper and nickel form solid solutions over the entire composition range. The mineral olivine $(Mg, Fe)_2SiO_4$ can be thought of as a solid solution of forsterite Mg_2SiO_4 and fayalite Fe_2SiO_4. An olivine solid solution that has half of the Mg^{2+} ions in forsterite replaced by Fe^{2+} ions would have the chemical formula $MgFeSiO_4$.

The composition of solid solutions is usually reported in terms of mass percent or, for components which are low in concentration, parts per million (ppm). As for other solutions, for thermodynamic calculations composition is expressed in terms of atom fraction (for elements) and mole fraction (for compounds).

9.3   INTEGRAL AND PARTIAL QUANTITIES

It is an experimental observation that, unlike in mixtures, extensive properties of components of a solution do not necessarily have the same value as that of the same quantity of the pure components. This is illustrated by the following example. The molar volumes of water and ethanol at 20°C are 18.0 and 58.0 mL mol^{-1}, respectively. If the molar volumes of ethanol and water did not change when they are mixed, we would expect a solution made from 1 mole of each to have a volume of 18.0 + 58.0 = 76.0 mL. However, the volume of a solution comprised of 1 mole of each, determined experimentally, is actually 74.3 mL; that is, the volume of the solution is 1.7 mL less than the volume of the separate components. Clearly, ethanol and water occupy a smaller volume when mixed compared with their pure forms. This type of observation has led to the concept of *partial quantity* for describing the thermodynamic properties of components of solutions.

Consider a solution at temperature T and pressure p consisting of components A, B, C, … which we will consider to be a system. Extensive thermodynamic properties of a system can be expressed by Equation 2.10,

$$X = X(T, P, n_A, n_B, n_C, \ldots)$$

where X is any extensive state function (such as V, U, H, S or G) and n_A, n_B, n_C … are the number of moles of components A, B, C … . X is the *total* value of the property for the system. The corresponding value for 1 mole of the system X_m, the *integral molar quantity*, is given by Equation 2.8:

$$X_m = \frac{X}{n_A + n_B + n_C + \cdots}$$

Applying the fundamental theorem of partial differentiation to Equation 2.10,

$$dX = \left(\frac{\partial X}{\partial T}\right)_{p, n_A, n_B, \ldots} dT + \left(\frac{\partial X}{\partial p}\right)_{T, n_A, n_B, \ldots} dp + \left(\frac{\partial X}{\partial n_A}\right)_{p, T, n_B, n_C, \ldots} dn_A$$
$$+ \left(\frac{\partial X}{\partial n_B}\right)_{p, T, n_A, n_C, \ldots} dn_B + \cdots \tag{9.1}$$

If the temperature and pressure of the solution are constant the first two terms are zero, and

$$dX = \left(\frac{\partial X}{\partial n_A}\right)_{p, T, n_B, n_C, \ldots} dn_A + \left(\frac{\partial X}{\partial n_B}\right)_{p, T, n_A, n_C, \ldots} dn_B + \left(\frac{\partial X}{\partial n_C}\right)_{p, T, n_A, n_B, n_D, \ldots} dn_2 + \cdots \tag{9.2}$$

We now define a new quantity X_B, called the *partial molar quantity*, as:

$$X_B = \left(\frac{\partial X}{\partial n_B}\right)_{p, T, n_A, n_C, \ldots} \tag{9.3}$$

where n_A, n_C, … denotes all the components of the system other than component B. Thus X_B is the rate of change of the value of the solution property X with respect to the amount of component B, all other species concentrations remaining constant.

The significance of X_B can be understood as follows. If 1 mole of component B is added to a very large quantity of solution so its composition doesn't change, X_B is the magnitude of the corresponding change in the value of X for the solution; that is:

X_B *is the value of property X that 1 mole of component* B *has in the solution at a particular composition. It will be greater than, equal to or less than* $X_m(B)$*, the value of X for 1 mole of pure* B*.*

In the example of a solution of water and ethanol, the partial molar volumes of water and ethanol, (V_{H_2O} and $V_{C_2H_5OH}$) are less than the molar volumes of pure water and pure ethanol, $V_m(H_2O)$ and $V_m(C_2H_5OH)$.

Now, substituting Equation 9.3 back into Equation 9.2 we have:

$$dX = X_A\, dn_A + X_B\, dn_B + X_C\, dn_C + \cdots \tag{9.4}$$

If Equation 9.4 is integrated term-by-term from zero to 1 total mole at constant composition we obtain (at constant temperature and pressure)

$$(n_A + n_B + n_C + \cdots)X = X_A n_A + X_B n_B + X_C n_C + \cdots \tag{9.5}$$

Dividing throughout by $(n_A + n_B + n_C + \cdots)$ yields

Solution: $$X_m = x_A X_A + x_B X_B + x_C X_C + \cdots \qquad (9.6)$$

X_m is the *integral molar quantity*. This is the value of the state function X for 1 mole of solution of composition x_A, x_B, x_C ... The term integral is used because X_m is the sum of the individual contributions of components A, B, C, ... weighted according to their mole fraction in the solution. In the above example, the integral molar volume of the water–ethanol solution at $x_{H_2O} = 0.5$ is $74.3/2 = 37.15$ mL mol^{-1}.

9.3.1 RELATIVE PARTIAL AND INTEGRAL QUANTITIES

Partial molar quantities defined by Equation 9.3 refer to absolute values. Unlike the state properties volume and entropy, the state properties internal energy, enthalpy and Gibbs energy do not have absolute values. Therefore, for these quantities, it is common to select a particular state as a reference and express the partial molar quantity relative to that state. The *relative partial molar quantity* of component B is defined as:

$$\Delta X_B = X_B - X_B^0 \qquad (9.7)$$

where X_B^0 is the molar quantity for component B in a chosen reference state, the most common being the pure substance standard state.

Similarly, relative integral molar quantities can be defined. Consider the formation of a solution consisting of n_A moles of component A, n_B moles of component B and so on. For the system consisting of the components in their reference states prior to their forming a solution, the value of the extensive property is the sum of the values of X for each of the components, weighted for their relative amounts:

Unmixed components: $$X_m^0 = x_A X_A^0 + x_B X_B^0 + \cdots \qquad (9.8)$$

For the solution containing n_A moles of component A, n_B moles of component B and so on,

$$X_m = x_A X_A + x_B X_B + \cdots \qquad (9.6)$$

Subtracting the two equations,

$$\Delta_{mix} X_m = X_m - X_m^0 = x_A \left(X_A - X_A^0 \right) + x_B \left(X_B - X_B^0 \right) + \cdots \qquad (9.9)$$

that is,

$$\Delta_{mix} X_m = X_m - X_m^0 = x_A \, \Delta X_A + x_B \, \Delta X_B + \cdots \qquad (9.10)$$

where $\Delta_{mix} X_m$ is the *relative integral molar quantity* or, more simply, the *molar quantity of mixing* (for example, molar enthalpy of mixing, etc.). It is the difference in the value of X_m for the solution and X_m of the components in their standard state prior to forming the solution.

EXAMPLE 9.1   Integral and relative integral molar volume

The compounds ZnS and FeS when powdered, mixed, heated to a high temperature (~700°C) and held for sufficiently long (several days) form a solid solution (Zn,Fe)S in which the zinc is partially replaced by iron within the ZnS crystal structure. Calculate the integral molar volume of a solid solution formed from a mixture of 0.2 moles of ZnS and 0.1 mole of FeS. Calculate the difference in volume between the solid solution and the starting ZnS and FeS.

Data: $V_{ZnS}^0 = 23.830$ mL mol^{-1} and $V_{FeS}^0 = 18.187$ mL mol^{-1}; at the given composition, $V_{ZnS} = 24.094$ mL mol^{-1} and $V_{FeS} = 24.305$ mL mol^{-1}.

SOLUTION

V_{ZnS}^0 and V_{FeS}^0 are the molar volumes of pure ZnS and FeS, respectively, and V_{ZnS} and V_{FeS} are the partial molar volumes of ZnS and FeS in the solid solution, respectively. The composition in terms of mole fraction is:

$$x_{ZnS} = \frac{0.2}{0.2 + 0.1} = 0.667$$

and

$$x_{FeS} = \frac{0.1}{0.2 + 0.1} = 0.333$$

The integral molar volume prior to mixing of ZnS and FeS (Equation 9.8) is:

$$V_m^0 = x_{ZnS} V_{ZnS}^0 + x_{FeS} V_{FeS}^0 = 0.667 \times 23.830 + 0.333 \times 18.187 = 21.951 \text{ mL mol}^{-1}$$

The integral molar volume of the solid solution (Equation 9.6) is:

$$V_m = x_{ZnS} V_{ZnS} + x_{FeS} V_{FeS} = 0.667 \times 24.094 + 0.333 \times 24.305 = 24.164 \text{ mL mol}^{-1}$$

The change in volume, or the relative integral molar volume (Equation 9.9), is:

$$\Delta V = V_m - V_m^0 = 24.164 - 21.951 = 2.193 \text{ mL mol}^{-1}$$

Thus, the same mass of solid solution has a larger volume than the corresponding unreacted components.

9.3.2   CALCULATING PARTIAL QUANTITIES FROM INTEGRAL QUANTITIES

Differentiating Equation 9.6 for a binary system yields

$$dX_m = x_A dX_A + X_A dx_A + x_B dX_B + X_B dx_B \tag{9.11}$$

Combining Equations 9.21 and 9.11,

$$dX_m = X_A dx_A + X_B dx_B \tag{9.12}$$

Multiplying throughout by x_A/dx_B, and noting that $dx_A = -dx_B$ (since $x_A + x_B = 1$),

$$x_A \frac{dX}{dx_B} = -x_A X_A + x_A X_B \tag{9.13}$$

Adding Equation 9.13 to Equation 9.6 and rearranging,

$$X_B = X_m + x_A \frac{dX_m}{dx_B} = X_m + (1 - x_B) \frac{dX_m}{dx_B} \tag{9.14}$$

The corresponding equation for X_A is:

$$X_A = X_m + (1 - x_A) \frac{dX_m}{dx_A} \tag{9.15}$$

Analogous relationships apply to relative quantities:

$$\Delta X_A = \Delta X_m + (1 - x_A) \frac{d\Delta X_m}{dx_A} \tag{9.16}$$

$$\Delta X_B = \Delta X_m + (1 - x_B) \frac{d\Delta X_m}{dx_B} \tag{9.17}$$

Equations 9.14 to 9.17 provide a method, the *method of intercepts*, for calculating partial molar quantities from measured integral molar quantities. This is illustrated graphically in Figure 9.2(a) for the case of integral molar Gibbs energy of the solution and in Figure 9.2(b) for the case of relative integral molar Gibbs energy of mixing (which has a similar form to Figure 8.1). This can be understood as follows by reference to Figure 9.2(a) and making the relevant substitutions in Equation 9.14:

$$G_B = CD + (1 - x_B') \frac{BC}{(1 - x_B')} = CD + BC = BD$$

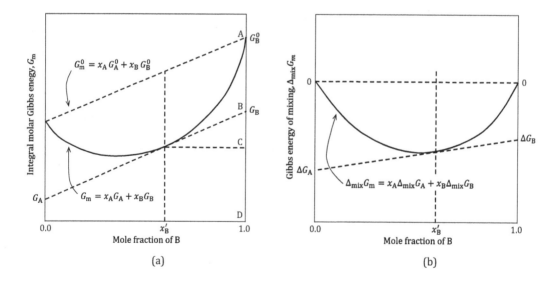

FIGURE 9.2 The graphical relationship between partial and integral molar quantities in a binary solution of A and B. (a) The intercepts at $x_B = 0$ and $x_B = 1$ of the tangent to the integral molar Gibbs energy curve at x_B' give the partial molar Gibbs energy of A and B, respectively, at x_B'. The dashed line joining the integral molar Gibbs energy of pure A and pure B is the integral molar volume assuming A and B do not form a solution. (b) The corresponding relation for relative integral molar Gibbs energy of mixing.

9.3.3 The Gibbs–Duhem equation

If Equation 9.6 is multiplied throughout by $(n_A + n_B + n_C + \cdots)$ we obtain the relation

$$X = n_A X_A + n_B X_B + n_C X_C + \cdots \tag{9.18}$$

which on complete differentiation gives

$$dX = n_A \, dX_A + n_B \, dX_B + \cdots + X_A \, dn_A + X_B \, dn_B + \cdots \tag{9.19}$$

Subtracting Equation 9.4 from Equation 9.19

$$n_A \, dX_A + n_B \, dX_B + \cdots = 0 \tag{9.20}$$

and, dividing throughout by $(n_A + n_B + n_C + \cdots)$:

$$x_A \, dX_A + x_B \, dX_B + \cdots = 0 \tag{9.21}$$

The corresponding equation for relative molar quantities is:

$$x_A \, d\Delta X_A + x_B \, d\Delta X_B + \cdots = 0 \tag{9.22}$$

These are forms of the *Gibbs–Duhem equation*. The significance of Equation 9.21 is that it shows that the partial molar quantity of one component of a solution cannot change independently of the others. In a binary solution, if one partial molar quantity increases the other must decrease.

Integration of Equation 9.21 for a binary solution provides a means for determining the partial molar quantity of one component if the variation with composition of the partial molar quantity of the other component is known. From Equation 9.21, for a binary solution,

$$x_A \, dX_A + x_B \, dX_B = 0$$

Therefore,

$$dX_B = -\frac{x_A}{x_B} dX_A$$

and, integrating

$$X_B\left(\text{at } x_B''\right) - X_B\left(\text{at } x_B'\right) = - \int_{X_A\left(\text{at } x_B'\right)}^{X_A\left(\text{at } x_B''\right)} \frac{x_A}{x_B} \, dX_A$$

A value of X_B must be known, and usually the value is known for the pure substance. Then,

$$X_B\left(\text{at } x_B\right) - X_B^0 = - \int_{X_A\left(x_B=1\right)}^{X_A\left(\text{at } x_B\right)} \frac{x_A}{x_B} \, dX_A = - \int_{X_A\left(x_B=1\right)}^{X_A\left(\text{at } x_B\right)} \frac{x_A}{1-x_A} \frac{dX_A}{dx_A} \, dx_A$$

For partial molar quantities other than entropy and volume, Gibbs–Duhem integrations involving relative partial quantities must be used. The integral may be evaluated graphically or analytically (if analytical dependence of X_A with composition is known).

Graphical determination requires a plot of x_A/x_B *versus* X_A (Figure 9.3). The value of the integral is the area under the curve between the nominated limits. However, the area under the curve may not be bounded well with the given limits since the curve is asymptotic to both curves. To determine the area accurately, either alternative limits need to be used, or ways to resolve the problems of tails

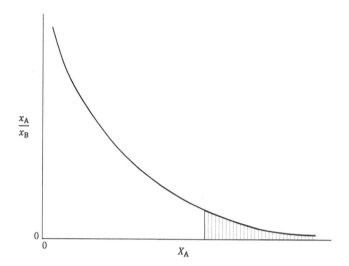

FIGURE 9.3 Graphical integration of the Gibbs–Duhem equation to obtain the value of one partial quantity X_B knowing the value of the other X_A in a binary solution.

to infinity need to be found. The latter aspect is beyond the scope of this book. Analytical determination involves the following steps: Express X_A as a function of one of the composition variables; take the derivative of X_A, and place it into the integral; arrange the function inside the integral so as to leave only one variable, and change, if necessary, the limits of the integral to make it in accord with the derivative of the integral; integrate the function.

9.4   GAS MIXTURES

Although we refer to systems containing several different gases as gas mixtures, gas mixtures are solutions because they are a single phase consisting of several components. Gas mixtures are the simplest solutions to consider because their thermodynamic behaviour can be quantified using the ideal gas law or a modified form for non-ideal gases. We start the discussion of the thermodynamics of solutions with a consideration of the thermodynamics of ideal and non-ideal gas mixtures and then in Section 9.5 extend these ideas to solid and liquid solutions.

9.4.1   IDEAL GAS MIXTURES

Since for 1 mole of ideal gas, $pV = RT$, it follows from Equation 8.12 that

$$\left(\frac{dG}{dp}\right) = \frac{RT}{p} \tag{9.23}$$

Therefore

$$dG = RT \frac{dp}{p} \tag{9.24}$$

which on integrating yields

$$G_2 - G_1 = RT \ln \frac{p_2}{p_1} \tag{9.25}$$

where G_1 and G_2 are Gibbs energy of 1 mole of ideal gas at temperature T and pressures of p_1 and p_2, respectively. $G_2 - G_1$ is the change in Gibbs energy of the gas when its pressure is changed from p_1 to p_2. If the gas is originally in its standard state, Equation 9.25 becomes

$$G - G^0 = RT \ln \frac{p}{p^0} \qquad (9.26)$$

Note the Gibbs energy of an ideal gas is proportional to its pressure p and equal to G^0 when its pressure is p^0 since then $\ln p/p^0 = \ln p^0/p^0 = 0$ (see Figure 9.4).

For a mixture of ideal gases, Equation 9.25 applies to each component. For component B, the change in Gibbs energy of 1 mole of B at temperature T when the partial pressure of the gas is changed from $p_{B,1}$ to $p_{B,2}$ is

$$G_{B,2} - G_{B,1} = RT \ln \frac{p_{B,2}}{p_{B,1}} \qquad (9.27)$$

where $G_{B,1}$ and $G_{B,2}$ are the partial molar Gibbs energies of component B in the gas mixture at temperature T and partial pressures p_1 and p_2, respectively. If the units of $p_{B,1}$ and $p_{B,2}$ are the same the ratio $p_{B,2}/p_{B,1}$ has the same value irrespective of the unit used. If component B is originally in its standard state, then

$$G_B - G_B^0 = RT \ln \frac{p_B}{p_B^0} \qquad (9.28)$$

Since the standard state for ideal gases is 1 bar pressure at temperature T, Equation 9.26 can be written as:

For 1 mol of pure gas: $$\Delta G_m = G - G^0 = RT \ln p \qquad (9.29)$$

where ΔG_m is the *relative molar Gibbs energy* of the gas and p is the dimensionless pressure, numerically equal to the pressure of the gas expressed in bars. Similarly,

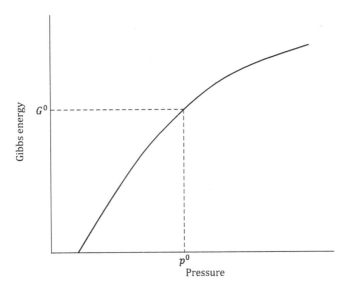

FIGURE 9.4 Variation of the Gibbs energy of an ideal gas with pressure. It is in its standard state when $p = p^0$.

For component B in 1 mol of a gas mixture: $\qquad \Delta G_B = G_B - G_B^0 = RT \ln p_B \qquad$ (9.30)

where ΔG_B is the *relative partial molar Gibbs energy* of component B in the gas mixture and p_B is the dimensionless partial pressure of B (numerically equal to the partial pressure of B expressed in bars). Note that the molar Gibbs energy of a gas G_B is proportional to its partial pressure and temperature. Since values of G_B^0 are known for most substances (Section 8.6), numerical values for G_B at any partial pressure in any ideal gas mixture can be calculated using Equation 9.30.

For historical reasons, the partial molar Gibbs energy of a substance is also known as the *chemical potential** μ_B of the substance, and the two names can be used interchangeably. Thus

$$G_B = \mu_B$$

and

$$G_B^0 = \mu_B^0$$

and, therefore

$$\Delta \mu_B = \mu_B - \mu_B^0 = G_B - G_B^0 = \Delta G_B = RT \ln p_B \qquad (9.31)$$

The term $\Delta \mu_B$ is the *relative chemical potential* of B but in some texts is also referred to as the chemical potential of B. There is, therefore, the possibility of confusion between μ_B and $\mu_B - \mu_B^0$, and the context in which the term chemical potential is used must be understood in order to know which is being referred to.

EXAMPLE 9.2   The relative partial molar Gibbs energy (chemical potential) of a gas

i. Calculate the relative chemical potential of oxygen in the atmosphere at 25°C.

SOLUTION

Oxygen makes up approximately 21% by volume of the atmosphere. Therefore, from Equation 3.7, $p_{O_2} = 0.21 \times 1.0132 = 0.2128$ bar.

$$\Delta \mu_{O_2} = \mu_{O_2} - \mu_{O_2}^0 = G_{O_2} - G_{O_2}^0 = RT \ln p_{O_2} = 8.314 \times 298.15 \times \ln(0.2128) = -3836 \text{ J mol}^{-1}$$

ii. Calculate the chemical potential of oxygen in the atmosphere at 25°C.

SOLUTION

$$\Delta \mu_{O_2} = \mu_{O_2} - \mu_{O_2}^0 = G_{O_2} - G_{O_2}^0$$

According to the data of Barin, $G_{O_2}^0 (298.15 \text{ K}) = -61165$ J. Therefore,

$$\mu_{O_2} = \Delta \mu_{O_2} + \mu_{O_2}^0 = -3836 + (-61165) = -65001 \text{ J mol}^{-1}$$

---

* The term was introduced by J. W. Gibbs in 1876.

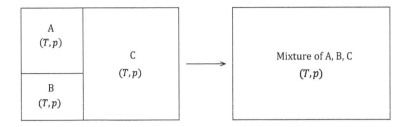

FIGURE 9.5 Illustration of the mixing of gases. The initial state consists of three systems each containing a quantity of one of the components at temperature T and pressure p. The partitions are removed, and the components mix to form a mixture at temperature T and total pressure p, but the partial pressures of the components are each now less than p.

For a gas mixture of components A, B, … of composition x_A, x_B, … the relative integral molar Gibbs energy (or more simply, the *Gibbs energy of mixing*) is given by Equation 9.10 as:

$$\Delta_{mix}G_m = G_m - G_m^0 = x_A \, \Delta G_A + x_B \, \Delta G_B + \cdots \text{J mol}^{-1}$$

Substituting for ΔG_A, ΔG_B, … from Equation 9.27,

$$\Delta_{mix}G_m = x_A RT \ln \frac{p_A}{p} + x_B RT \ln \frac{p_B}{p} + \cdots \text{J mol}^{-1}$$

where p is both the initial pressure of A, B, … and also the pressure of the gas mixture, and p_A, p_A, … are the partial pressures of A, B, … in the gas mixture (see Figure 9.5). Since $(p_A/p) = x_A$, $(p_B/p) = x_B$ … then, if p is set equal to p^0,

$$\Delta_{mix}G_m = G_m - G_M^0 = RT\left(x_A \ln x_A + x_B \ln x_B + \cdots\right) \text{J mol}^{-1} \qquad (9.32)$$

$\Delta_{mix}G$ is the difference between the sum of the Gibbs energies of the individual components at T and p and the sum of their Gibbs energies when present in the mixture. This is the same as Equation 8.30 derived previously using the entropy of mixing equation derived from statistical mechanics principles.

EXAMPLE 9.3   The relative integral molar Gibbs energy of a gas mixture

Calculate the relative integral molar Gibbs energy of air at 25°C.

SOLUTION

Air consists of approximately 21 volume percent O_2 and 79 volume percent N_2. Therefore $x_{O_2} = 0.21$; $x_{N_2} = 0.79$. From Equation 9.32,

$$\Delta_{mix}G_m = 8.314 \times 298.15 \times (0.21 \times \ln 0.21 + 0.79 \times \ln 0.79) = -1274 \text{ J}$$

COMMENT

This is the equivalent of calculating the Gibbs energy of mixing of 0.21 moles oxygen and 0.79 moles of nitrogen each at 1 atm pressure and 25°C to form a mixture at 1 atm pressure and 25°C.

Since, at constant pressure $dG/dT = -S$ (Equation 8.13), then

$$\Delta_{\text{mix}}S_m = -\left(\frac{d\Delta_{\text{mix}}G_m}{dT}\right)_p$$

where $\Delta_{\text{mix}}S_m$ is the *relative integral molar entropy* of the mixture or *entropy of mixing*. Substituting from Equation 9.32

$$\Delta_{\text{mix}}S_m = -\frac{d}{dT}\left[RT(x_A \ln x_A + x_B \ln x_B + \cdots)\right]_p$$

Therefore,

$$\Delta_{\text{mix}}S_m = -R(x_A \ln x_A + x_B \ln x_B + \cdots) \qquad (9.33)$$

Note this has the same form as Equation 7.12. The relative integral molar Gibbs energy, enthalpy and entropy of a mixture are related through the equation

$$\Delta_{\text{mix}}G = \Delta_{\text{mix}}H - T\Delta_{\text{mix}}S$$

It follows from Equations 9.32 and 9.33, therefore, that for an ideal gas mixture the *relative integral molar enthalpy* of the mixture, or *enthalpy of mixing*, is (as expected) zero:

$$\Delta_{\text{mix}}H_m = 0 \qquad (9.34)$$

9.4.2 NON-IDEAL GAS MIXTURES

Equation 9.26 applies to ideal gases, but the same type of equation can be used to express the Gibbs energy of real gases. To do this the actual dimensionless pressure p is replaced by an effective pressure, called the fugacity. The *fugacity* f* of 1 mole of a gas is defined as follows:

$$G - G^0 = RT \ln \frac{f}{p^0} \qquad (9.35)$$

Fugacity has the same units as pressure. To make Equation 9.35 useful we need to relate fugacity to pressure, which can be measured. Thus we write

$$f = \varphi\, p \qquad (9.36)$$

where φ is the *fugacity coefficient*. For example, nitrogen gas (N_2) at 0°C and 100 atm has a fugacity of 97.03 atm, and the fugacity coefficient is $97.03/100 = 0.970$. For an ideal gas, fugacity and pressure are equal so φ is 1. Fugacities are determined experimentally or estimated from models such as van der Waals equation, as discussed below.

Substituting Equation 9.36 into Equation 9.35 and separating the terms,

$$G - G^0 = RT \ln \frac{p}{p^0} + RT \ln \varphi \qquad (9.37)$$

---

* The term fugacity is derived from the Latin for fleetness, which is interpreted as the tendency to flee or escape. The concept of fugacity was introduced by Gilbert N. Lewis in 1901.

Comparing Equation 9.35 with Equation 9.26 (for an ideal gas), it is clear that the term $RT \ln \varphi$ must express the effect of the intermolecular forces between the gas molecules. Since all gases approach ideal behaviour as pressure approaches zero, from Equation 9.36

$$f \to p \text{ as } p \to 0$$

and

$$\mu \to 1 \text{ as } p \to 0$$

The same form of equation as Equation 9.35 applies to components of a gas mixture,

$$G_B - G_B^0 = RT \ln \frac{f_B}{p_B^0} \qquad (9.38)$$

and has the same form as Equation 9.28 for ideal gas mixtures.

As might be expected, the fugacity of a gas is related to its compressibility factor Z. For 1 mole of gas,

$$\left(\frac{dG}{dp} \right)_T = -V_m \qquad (8.12)$$

Integrating from p' to p,

$$G - G' = \int_{p'}^{p} V_m \, dp$$

Also, from Equation 9.35,

$$G - G' = (G - G^0) - (G' - G^0) = RT \ln \frac{f}{p^0} - RT \ln \frac{f'}{p^0} = RT \ln \frac{f}{f'}$$

where f is the fugacity of the gas at pressure p and f' is the fugacity at p'. Therefore,

$$\int_{p'}^{p} V_m \, dp = RT \ln \frac{f}{f'}$$

For an ideal gas, the corresponding equation is:

$$\int_{p'}^{p} V_m^i \, dp = RT \ln \frac{p}{p'}$$

where the superscript i denotes ideal behaviour. Subtracting the two equations and rearranging gives

$$\int_{p'}^{p} \left(V_m - V_m^i \right) dp = RT \left(\ln \frac{f}{f'} - \ln \frac{p}{p'} \right)$$

or

$$\ln \frac{fp'}{pf'} = \frac{1}{RT} \int_{p'}^{p} \left(V_m - V_m^i \right) dp$$

As p' approaches zero the gas begins to behave ideally, and f' approaches p', that is, $p'/f' \to 1$ as $p' \to 0$. If we set this limit for p', then the equation becomes

$$\ln \frac{f}{p} = \frac{1}{RT} \int_0^p \left(V_m - V_m^i \right) dp$$

and, since $\varphi = f/p$,

$$\ln \varphi = \frac{1}{RT} \int_0^P \left(V_m - V_m^i \right) dp$$

For an ideal gas, $V_m^i = RT/p$, and for a real gas, $V_m = RTZ/p$ (Equation 3.8). Therefore,

$$\ln \varphi = \int_0^P \left(\frac{Z-1}{p} \right) dp \tag{9.39}$$

To calculate φ, and through Equation 9.36 the fugacity, values of Z are required from very low pressure up to the pressure of interest. These can be obtained from experimental measurements or from equations for Z derived from an equation of state such as the van der Waals or Virial equation.

9.5   LIQUID AND SOLID SOLUTIONS

The concept of partial molar Gibbs energy of components of gas mixtures can be extended to establish the partial molar Gibbs energy of components in solid or liquid solutions. This is done through linking the partial molar Gibbs energy of the components of the vapour in equilibrium with a solid or liquid solution to a new property of the components of solutions, called activity.

In a closed system, when substances are mixed to form a solution, each component exerts a unique vapour pressure which is less than the vapour pressure of the pure substance at the same temperature. It is an experimental observation that the partial pressures of components of solutions increase as their concentration in a solution increases. This is illustrated for three types of binary solutions in Figure 9.6. Note, in each case the partial pressure of component B increases from a value of zero at $x_B = 0$ to a value equal to its saturated vapour pressure at $x_B = 1$ while the partial pressure of component A decreases from its saturated vapour pressure to zero. The total vapour pressure is the sum of the individual vapour pressures (Equation 3.5).

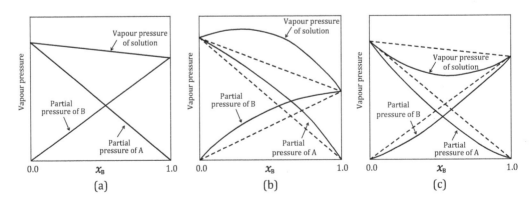

FIGURE 9.6   The variation with concentration of the vapour pressure of components in solution and total vapour pressure for a binary system: (a) an ideal solution, (b) a solution showing positive deviation from ideality, (c) a solution showing negative deviation from ideality.

The solution illustrated in Figure 9.6(a) is an *ideal solution*. In an ideal solution, the partial pressure of each component is directly proportional to its concentration. This is known as *Raoult's law* which can be stated more precisely as follows: The partial pressure of each component of an ideal solution is equal to the product of the vapour pressure of the pure component and its mole fraction in the solution, that is,

$$\text{Raoult's law:} \qquad p_B = x_B \, p_B^* \qquad\qquad (9.40)$$

where p_B^* is the vapour pressure of pure B at the relevant temperature.

Consider a solution composed of the components A and B. For the solution to be ideal the following criteria must be met:

- The atoms or molecules of A and B must be of similar size; and
- The attractive force between A and B molecules (A–B) must be the same as the attractive forces between the A molecules (A–A) and the B molecules (B–B).

Note the difference between ideal solutions and ideal gas mixtures – in the latter there is no interaction between the molecules. No real solutions are truly ideal, although some real solutions approximate ideal behaviour when they contain atoms or molecules that are structurally similar. This includes mixtures of isotopes (for example, H_2O–D_2O), homologous series (for example, methanol–ethanol–propanol) and some molten metallic solutions (for example, Fe–Cr).

The two other cases in Figure 9.6 illustrate situations where the molecules of A and B are still assumed to be of similar size and the vapour behaves as an ideal gas, but the attraction A–B is greater or less than the attraction between A molecules and B molecules. If A–B attraction is weaker than either A–A or B–B the vapour pressure of A and B will be higher than for an ideal solution, and this leads to positive deviation from Raoult's law (Figure 9.6(b)). Conversely, if A–B attraction is stronger than either A–A or B–B the vapour pressure of A and B will be lower than for an ideal solution. This leads to negative deviation from Raoult's law (Figure 9.6(c)).

9.5.1 The concept of activity

Consider a system at temperature T consisting of a solution of A, B, … in equilibrium with its vapour (Figure 9.7 (a)). It will be shown in Chapter 13 that for a system at equilibrium the partial molar Gibbs energy of a component is the same in each phase. Accepting that for the present, then

$$G_A(\text{sln}) = G_A(g); \quad G_B(\text{sln}) = G_B(g)$$

where (sln) indicates a component in a solution. For component B in the gas phase in equilibrium with the solution, from Equation 9.31,

$$G_B(g) = G_B^0(g) + RT \ln p_B$$

and, therefore

$$G_B(\text{sln}) = G_B^0(g) + RT \ln p_B \qquad\qquad (9.41)$$

Now consider another solution of A, B, … in equilibrium with its vapour also at temperature T in which component B is present at a different concentration x_B' (see Figure 9.7 (b)). A similar equation can be written for component B in the gas phase in equilibrium with the solution:

$$G_B'(g) = G_B^0(g) + RT \ln p_B'$$

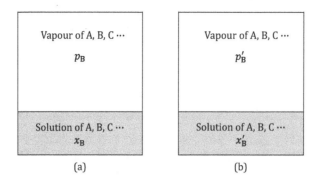

FIGURE 9.7   The concept of activity of components of solutions.

Therefore,

$$G'_B(\text{sln}) = G^0_B(\text{g}) + RT \ln p'_B \qquad (9.42)$$

Subtracting Equation 9.42 from Equation 9.41

$$G_B(\text{sln}) - G'_B(\text{sln}) = \left(G^0_B + RT \ln p_B \right) - \left(G^0_B + RT \ln p'_B \right)$$

and,

$$G_B(\text{sln}) - G'_B(\text{sln}) = RT \ln \frac{p_B}{p'_B} \qquad (9.43)$$

where $G_B(\text{sln}) - G'_B(\text{sln})$ is the Gibbs energy difference between 1 mole of B in the two solutions at constant temperature. Similar equations can be written for the other components of the solution.

If both solutions behave ideally, so Raoult's law applies (Equation 9.40),

$$G_B(\text{sln}) - G'_B(\text{sln}) = RT \ln \frac{x_B}{x'_B} \qquad (9.44)$$

This equation relates the partial molar Gibbs energy of B to a measurable property of the solution (the mole fraction of B) rather than to a property of the vapour phase and hence is more useful. For non-ideal (or real) solutions the form of this equation can be retained by defining the *activity* of component B in such a way that the Gibbs energy difference between 1 mole of B in the two solutions is given by

$$G_B(\text{sln}) - G'_B(\text{sln}) = RT \ln \frac{a_B}{a'_B} \qquad (9.45)$$

Similar equations can be written for the other components of the solution.

Equation 9.45 does not define the actual (or absolute) activity but rather the ratio of the activities of the particular component at two concentrations. In order to assign a value to the activity of components of a solution it is necessary to choose for each component a standard state in which the activity is arbitrarily taken as one. The activity of a component of a solution, therefore, is the ratio of its value in the particular solution to that in the chosen standard state (taken to be one). Since activity is a ratio it is dimensionless. The actual standard state chosen for each component is the most convenient for the purpose. If the solution containing B at a concentration of x'_B is taken as representing the standard state, then Equation 9.45 may be written in the general form

$$G_B(\text{sln}) - G'_B(\text{sln}) = RT \ln a_B \qquad (9.46)$$

It is apparent from the previous discussion that the activity of a component of a solution will vary with its concentration.

The deviation of a component of a solution from ideal behaviour is represented by a quantity called the *activity coefficient* which may be expressed in terms of the various standard states. It is defined by an equation of the general form:

$$\text{Activity of B} = \text{Activity coefficient of B} \times \text{Dimensionless concentration of B} \qquad (9.47)$$

where the terms of the equation depend on the standard state chosen. Thus, the activity of a component can be thought of as its effective (but dimensionless) concentration. The symbol γ is used for activity coefficient. The activity of components of solutions can be determined experimentally by a variety of methods (Section 9.7).

9.5.2 PURE SUBSTANCE STANDARD STATE

The most commonly used standard state is chosen such that the activity coefficient of the relevant component approaches a value of one at high concentrations of that component, that is,

$$\gamma_B \to 1 \text{ as } x_B \to 1$$

Therefore, from Equation 9.47

$$a_B \to x_B \text{ as } x_B \to 1$$

This implies that the standard state of component B is pure B at the relevant temperature in the form (solid or liquid) which is stable at that temperature and 1 bar pressure. Equation 9.46 then becomes:

$$\Delta\mu_B^0 = \mu_B - \mu_B^0 = \Delta G_B^0 = G_B - G_B^0 = RT \ln a_B \qquad (9.48)$$

This is consistent with the standard state introduced in Section 2.5 and used consistently in subsequent chapters:

Standard state for solid or liquid substances, mixtures of substances and for solvents: The standard state of a pure substance or solvent is the state of the substance or solvent at a pressure of 1 bar.

Activities based on this standard state are often referred to informally as *Raoultian activities*. From Equation 9.47,

$$a_B = \gamma_B x_B \qquad (9.49)$$

Values of activity or activity coefficients of components of solutions must be determined by experimental measurement or from mathematical models of solutions. Since values of G_B^0 are known for most common substances (Section 8.6), numerical values for $G_B(\text{sln})$ at any composition can be calculated using Equation 9.48 from known values of activity.

For an ideal solution, and since $x_B' = 1$, it follows from Equation 9.44 that

Ideal solution: $$\Delta G_B^0 = G_B - G_B^0 = \mu_B - \mu_B^0 = RT \ln x_B \qquad (9.50)$$

and, comparing this with Equation 9.46:

Ideal solution: $\quad\quad\quad\quad\quad\quad\quad\quad\quad a_B = x_B$ $\quad\quad\quad\quad\quad\quad\quad\quad\quad\quad$ (9.51)

The variation of activity of the components of a two-component solution using Raoultian standard states for both components is illustrated in Figure 9.8. It is seen that the activity of component B increases from 0 to 1 and that the activity of A decreases from 1 to 0 as the mole fraction of B increases from 0 to 1 (or mole fraction of A decreases from 1 to 0). The explanation for the deviations from ideality due to molecular interactions between the A and B atoms or molecules is the same as discussed previously for the deviation of vapour pressures above the solutions (see Figure 9.6).

Equation 9.43 is applicable to all solutions, ideal or non-ideal, and comparison with Equation 9.45 shows that

$$\frac{p_B}{p_B'} = \frac{a_B}{a_B'}$$

Adopting the pure substance standard state, $p_B' = p_B^0$ and $a_B' = 1$. Therefore,

$$a_B = \frac{p_B}{p_B^0}$$ $\quad\quad\quad\quad\quad\quad\quad\quad\quad\quad$ (9.52)

Equation 9.52 provides a means of experimentally measuring the activity of components of solutions, but in practice it is useful only for components which have a measurable vapour pressure at the relevant temperature. In the above discussion it was assumed that gases behave ideally. While this will be true in many situations, activity should strictly be expressed in terms of fugacity:

$$a_B = \frac{f_B}{f_B^0}$$ $\quad\quad\quad\quad\quad\quad\quad\quad\quad\quad$ (9.53)

9.5.3 The infinitely dilute standard state

Dilute solutions are an important class of solutions in many applications, particularly in aqueous chemistry and in metal refining (since most metals are produced in an impure molten form in which small quantities of undesirable elements are present). By definition, components present in small

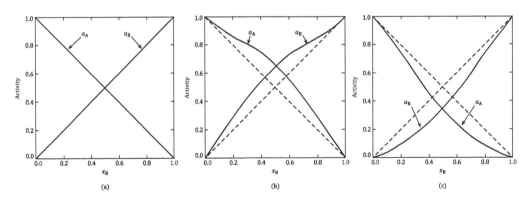

FIGURE 9.8 The variation of activity of components of a binary solution for the case of complete miscibility over the entire composition range: (a) ideal solution behaviour, (b) solution exhibiting positive deviation from Raoult's law, (c) solution exhibiting negative deviation from Raoult's law.

amounts in a solution are called solutes while the dominant component is the solvent. Thus in aqueous chemistry water is the solvent, and in steelmaking iron is the solvent and carbon, silicon, manganese, phosphorus, etc., are solutes. The concentration of solutes in dilute solutions may range from only a few parts per million (or less) up to several mass percent. The chemical behaviour of solutes can be important (for example, if their concentration has to be controlled within specified limits).

It has been found experimentally that on increasing the degree of dilution of a solute its activity coefficient approaches a constant, finite value, that is,

$$\gamma_B \rightarrow \gamma_B^0 \text{ as } x_B \rightarrow 0.$$

γ_B^0 is called the *activity coefficient at infinite dilution*. In practice over some, usually small, composition range, γ_B^0 is constant, and a linear relation applies:

$$a_B = \gamma_B^0 x_B \tag{9.54}$$

This relationship is known as *Henry's law* and is illustrated in Figure 9.9. γ_B^0 is the slope of the Henry's law line and is equal to the intercept on the a_B axis at $x_B = 1$.

Henry's law behaviour for a solute at low dilution is understandable at the microscopic level. In a very dilute solution the environment of component B is constant since B atoms/molecules will be entirely surrounded by A atoms/molecules (see Figure 9.10). As the concentration of B is increased the probability of it having a like neighbour is initially small, and its activity coefficient will remain constant until the B atoms/molecules reach a concentration at which they start to interact with each other.

The most common standard state for activities is the pure substance in its stable form at 1 bar pressure at the temperature of interest. In that case the activity of a component varies from zero when its mole fraction is zero to a value of one when its mole fraction is one. It is convenient to adopt an alternative standard state for solutes in dilute solutions so that the activity of the solute will be equal to one at some point within the composition range of interest. Since Henry's law applies for

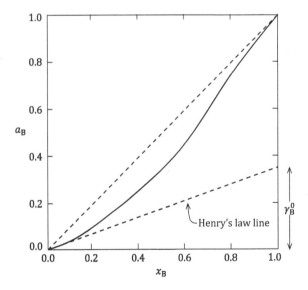

FIGURE 9.9   Schematic representation of Henry's law behaviour of solute B in a binary solution (in this case for negative deviation from ideal behaviour).

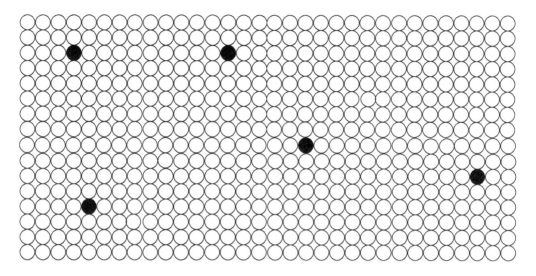

FIGURE 9.10 Illustration of the environment of solute atoms in a very dilute solution. Molecules of the solute will be entirely surrounded by solvent molecules.

all solutes at sufficiently low concentrations, the standard state adopted for solutes in dilute solutions is the *infinitely dilute solution standard state*:

> *Standard state for solutes*   The (hypothetical) state of the solute at a concentration of 1
> *in dilute solutions:*   unit (mole fraction, molality, mass percent or mole percent,
> as appropriate) and 1 bar assuming infinitely dilute solution
> behaviour.

In a solution exhibiting infinitely dilute solution behaviour a solute will have an activity of one at a concentration of 1 unit (Figure 9.11(a)). The word 'hypothetical' is used in this definition because a real solution at a concentration of 1 unit will not necessarily obey Henry's law up to

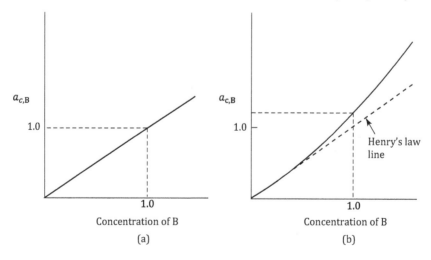

FIGURE 9.11 Variation of the activity of B on the infinitely dilute solution standard state: (a) solution exhibiting infinitely dilute solution behaviour, (b) a dilute real solution.

that concentration (Figure 9.11(b)). This standard state introduces a new activity, informally called *Henrian activity*, which is designated $a_{c,B}$, where the subscript c (for concentration) indicates the infinitely dilute solution standard state. In practice, the c is replaced with x (for mole fraction), m (for molality), % (for mass percent) or %x (for mole percent), as appropriate. When writing a chemical equation, the standard state of a solute in a dilute solution is indicated by placing (m), (%), or (%x), as appropriate, after the symbol or formula of each of the reactants and products that are present in solution. The suffix (aq) is used to indicate the molality standard state for solutes in aqueous solutions.

The Henrian activity coefficient is defined, after the form of Equation 9.47, as:

$$a_{c,B} = \gamma_{c,B}\, c_B \qquad (9.55)$$

where c_B is the dimensionless concentration of B ($c_B = c_B/c_B^0 = c_B/1$). According to Henry's law, and adopting the infinitely dilute standard state, the activity coefficient of a solute approaches a value of one as the concentration of the solute approaches zero, that is,

$$\gamma_{c,B} \to 1 \text{ as } c_B \to 0$$

and, therefore, from Equation 9.47

$$a_{c,B} \to c_B \text{ as } c_B \to 0$$

Activity is defined as a ratio (Equation 9.45),

$$G_{c,B}(\text{sln}) - G'_{c,B}(\text{sln}) = RT \ln \frac{a_{c,B}}{a'_{c,B}} \qquad (9.56)$$

and for the infinitely dilute standard state $a'_{c,B} = 1$. Therefore,

$$\Delta\mu^0_{c,B} = \mu_{c,B} - \mu^0_{c,B} = \Delta G^0_{c,B} = G_{c,B} - G^0_{c,B} = RT \ln a_{c,B} \qquad (9.57)$$

Henrian activity, like Raoultian activity, is a dimensionless quantity.

In the concentration range in which Henry's law is obeyed by component B, called the Henry's law region, $\gamma_{c,B}$ is equal to one, and

Henry's law region: $\qquad\qquad a_{c,B} = c_B \qquad (9.58)$

where c_B is the dimensionless concentration of B. $\gamma_{c,B}$ is less than or greater than one at concentrations outside the Henry's law region. The Henrian activity scale is open-ended, and Henrian activities can have values greater than one. Clearly, the numerical value of the Henrian activity of a component is different from the value of the Raoultian activity at the same concentration, and it is necessary to always distinguish between them when writing equations for chemical reactions.

9.5.4 CONVERSION BETWEEN STANDARD STATES

In terms of Raoultian activities,

$$G_B - G_B^0 = RT \ln a_B \qquad (9.46)$$

The equivalent relation in terms of Henrian activities, and choosing mole fraction as the concentration unit, is:

$$G_B - G_{x,B}^0 = RT \ln a_{x,B} \tag{9.59}$$

Subtracting

$$\Delta G_{x,B}^0 = G_{x,B}^0 - G_B^0 = RT \ln \frac{a_B}{a_{x,B}} = RT \ln \frac{\gamma_B \, x_B}{\gamma_{x,B} \, x_B}$$

This relation applies at all concentrations and hence also for infinite dilution. At infinite dilution, $\gamma_B = \gamma_B^0$ and $\gamma_{x,B} = 1$. Hence,

$$\Delta G_{x,B}^0 = G_{x,B}^0 - G_B^0 = RT \ln \gamma_B^0 \tag{9.60}$$

$\Delta G_{x,B}^0$ is the Gibbs energy change accompanying the change of state from Raoultian activity to Henrian activity for the case where the concentration of solute is expressed in terms of mole fraction. It is the difference in the Gibbs energy of 1 mole of B at infinite dilution and 1 mole of pure B, both at the same temperature and 1 bar pressure. Equation 9.60 enables conversion of the standard state; for the transformation reaction B(s or 1)=B(x), $\Delta_r G^0 = \Delta G_{x,B}^0 = G_{x,B}^0 - G_B^0$.

The conversion to Henrian activity on the molality scale ($c_B = m_B$) is:

$$G_{m,B}^0 - G_B^0 = RT \ln \frac{a_B}{a_{m,B}} = RT \ln \frac{\gamma_B \, x_B}{\gamma_{m,B} \, m_B}$$

At infinite dilution, $\gamma_B = \gamma_B^0$ and $\gamma_{m,B} = 1$; therefore,

$$G_{m,B}^0 - G_B^0 = RT \ln \frac{\gamma_B^0 \, x_B}{m_B} \tag{9.61}$$

Similarly, the conversion of the mass percent scale ($c_B = \%B$) is:

$$G_{\%,B}^0 - G_B^0 = RT \ln \frac{a_B}{a_{\%,B}} = RT \ln \frac{\gamma_B \, x_B}{\gamma_{\%,B} \, \%B}$$

At infinite dilution, $\gamma_{x,B} = \gamma_B^0$ and $\gamma_{\%,B} = 1$; therefore,

$$G_{\%,B}^0 - G_B^0 = RT \ln \frac{a_B}{a_{\%,B}} = RT \ln \frac{\gamma_B^0 \, x_B}{\%B} \tag{9.62}$$

For binary solutions, at low concentrations of B, $x_B/\%B$ is very nearly equal to $M/(100 \times M_B)$, where M is the molar mass of the solvent and M_B is the molar mass of solute B. Hence

$$G_{\%,B}^0 - G_B^0 = RT \ln \frac{\gamma_B^0 \, M}{100 \times M_B} \tag{9.63}$$

It is not uncommon to use mixed standard states. For a dilute solution the activity of the solutes will usually be expressed in Henrian activity while the activity of the solvent will be in Raoultian activity.

9.5.5 THE GIBBS–DUHEM EQUATION

If the activity of one component of a solution has been determined experimentally, the activity of the other component can be calculated using the Gibbs–Duhem equation. Since (Equation 9.22)

$$x_A \, d\Delta G_A^0 + x_B \, d\Delta G_B^0 + \cdots = 0$$

and $\Delta G_A^0 = RT \ln a_A$ and $\Delta G_B^0 = RT \ln a_B$, then

$$x_A RT \, d \ln a_A + x_B RT \, d \ln a_B + \cdots = 0$$

Dividing by RT,

$$x_A d \ln a_A + x_B d \ln a_B + \cdots = 0 \qquad (a)$$

Therefore, for a binary solution

$$d \ln a_B = -\frac{x_A}{x_B} d \ln a_A$$

$x_B = 1 - x_A$, so that $dx_B = -dx_A$, and

$$\frac{x_B}{x_B} dx_B = \frac{x_A}{x_A} dx_A$$

Since $d \ln x = dx / x$, then

$$x_B d \ln x_B = -x_A d \ln x_A \qquad (b)$$

Subtracting Equation (b) from Equation (a) gives

$$x_B d \ln a_B - x_B d \ln x_B + x_A d \ln a_A - x_A d \ln x_A = 0$$

or,

$$x_B \, d \ln \frac{a_B}{x_B} + x_A d \ln \frac{a_A}{x_A} = 0$$

and, since $a = \gamma x$,

$$x_B d \ln \gamma_B + x_A d \ln \gamma_A = 0$$

where γ_A and γ_B are the Raoultian activity coefficients of A and B, respectively. Integrating gives:

$$\ln \gamma_B = -\int_1^{x_B} \frac{x_A}{x_B} d \ln \gamma_A \qquad (9.64)$$

The integral may be evaluated graphically, by plotting x_A/x_B *versus* $\ln \gamma_A$ (Figure 9.12), or analytically. Because γ_A is always finite there is no difficulty in integrating in the region $x_B = 1$, but there is still difficulty when x_B is small because x_A/x_B approaches infinity. Several mathematical procedures have been developed to overcome this problem, but discussion of these is beyond the scope of this book.

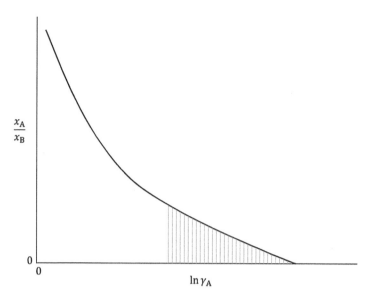

FIGURE 9.12 Graphical integration of the Gibbs–Duhem equation to obtain the activity coefficient of one component of a binary solution knowing the activity coefficient of the other.

9.6 PROPERTIES OF SOLUTIONS

It follows from Equation 8.12 that the relative partial molar volume of component B in a solution is given by

$$\Delta V_B = \left(\frac{d\Delta G_B}{dp} \right)_T$$

and at any temperature, ΔV_B is the slope of the tangent to the ΔG_B v p curve. Since temperature is constant then, from Equation 9.46

$$\Delta V_B = RT \left(\frac{d \ln a_B}{dp} \right)_T \tag{9.65}$$

Similarly, from Equation 8.13

$$\Delta S_B = - \left(\frac{d\Delta G_B}{dT} \right)_p \tag{9.66}$$

Thus at any temperature, ΔS_B is the negative slope of the tangent to the ΔG_B versus T curve at constant pressure. Since $\Delta G_B = \Delta H_B - T\Delta S_B$, at constant pressure

$$\Delta G_B = \Delta H_B + T \left(\frac{d\Delta G_B}{dT} \right)_p$$

Manipulation of this equation yields

$$\Delta H_B = \left(\frac{d\Delta G_B / T}{d(1/T)} \right)_p$$

and, since $\Delta G_B = -RT \ln a_B$,

$$\Delta H_B = \left(\frac{d\Delta G_B / T}{d(1/T)} \right)_p = R \left(\frac{d \ln a_B}{d(1/T)} \right)_p \tag{9.67}$$

Equations 9.65 to 9.67 are needed in the following discussion of ideal and non-ideal solutions.

9.6.1 IDEAL SOLUTIONS

In ideal solutions $a_B = x_B$; therefore Equation 9.65 can be rewritten as

$$\Delta V_B = RT \left(\frac{d \ln x_B}{dp} \right)_T$$

Since mole fraction is independent of pressure the differential in the above equation is equal to zero. Therefore, for an ideal solution:

Relative partial molar volume: $\Delta V_B = 0$ \hspace{2cm} (9.68)

that is, component B has the same volume in the solution as an equal quantity of pure B at the same temperature and pressure. The integral molar volume of an ideal solution therefore is (Equation 9.6):

Integral molar volume: $V_m = x_A V_A + x_B V_B + \cdots$ \hspace{1.5cm} (9.69)

Similarly, Equation 9.67 can be written as

$$\Delta H_B = R \left(\frac{d \ln x_B}{d(1/T)} \right)_p$$

and, again, since mole fraction is independent of temperature the differential is zero. Hence,

Relative partial molar enthalpy: $\Delta H_B = H_B - H_B^0 = 0$ \hspace{1.5cm} (9.70)

For an ideal solution

Relative partial molar Gibbs energy: $\Delta G_B = G_B - G_B^0 = RT \ln x_B$ \hspace{1cm} (9.71)

and, since $\Delta G_B = \Delta H_B - T\Delta S_B$

$$\Delta S_B = -\left(\frac{\Delta G_B - \Delta H_B}{T} \right) = -\left(\frac{RT \ln x_B - 0}{T} \right)$$

therefore,

Relative partial molar entropy: $\Delta S_B = S_B - S_B^0 = -R \ln x_B$ \hspace{1cm} (9.72)

We can now use these partial molar values to calculate the respective integral molar properties of ideal solutions using Equation 9.10:

$$\Delta_{mix}X_m = X_m - X_m^0 = x_A\,\Delta X_A + x_B\,\Delta X_B + \cdots$$

Relative integral molar Gibbs energy (or Gibbs energy of mixing):

$$\Delta_{mix}G_m = G_m - G_m^0 = x_AG_A + x_BG_B + \cdots = x_ART\ln x_A + x_BRT\ln x_B + \cdots$$

(9.73)

$$= RT\left(x_A\ln x_A + x_B\ln x_B + \cdots\right)$$

Relative integral molar entropy (or entropy of mixing):

$$\Delta_{mix}S_m = S_m - S_m^0 = x_AS_A + x_BS_B + \cdots = -x_AR\ln x_A - x_BR\ln x_B + \cdots$$

(9.74)

$$= -(x_AR\ln x_A + x_BR\ln x_B + \cdots)$$

Relative integral molar enthalpy (or enthalpy of mixing):

$$\Delta_{mix}H_m = H_m - H_m^0 = x_AH_A + x_BH_B + \cdots = 0$$

(9.75)

Note Equations 9.74, 9.75 and 9.47 are identical to the corresponding equations for ideal gas mixtures, namely Equations 9.32, 9.33 and 9.34 respectively. This is as expected. In ideal gases there is no interaction between the atoms or molecules comprising the gas. In ideal solutions there is attraction between the atoms or molecules comprising it, but all attractions are of equal strength.

EXAMPLE 9.4   The Gibbs energy of mixing of an ideal solution

One mole of benzene (C_6H_6) is mixed with 2 moles of toluene (C_7H_8) at 25°C. Assuming they form an ideal solution (which in this case is a good assumption) calculate the Gibbs energy change on mixing the two substances.

SOLUTION

$$n = 1+2 = 3; \quad x_{C_6H_6} = 0.333; \quad x_{C_7H_8} = 0.667$$

Substituting into Equation 9.73,

$$\Delta_{mix}G = 3 \times 8.3148 \times 298.25$$

$$\times (0.333 \times 0.333 + 0.667 \times \ln 0.667) = -4732\ \text{J}$$

and

$$\Delta_{mix}G_m = -\frac{4732}{3} = -1577\ \text{J mol}^{-1}$$

9.6.2 NON-IDEAL SOLUTIONS

Relative partial Gibbs energy: $\qquad \Delta G_B = RT \ln a_B \qquad$ (9.46)

and, since $\Delta G_B = \Delta H_B - T\Delta S_B$, then

Relative partial molar entropy: $\quad \Delta S_B = \dfrac{\Delta H_B - \Delta G_B}{T} = \dfrac{\Delta H_B - RT \ln a_B}{T} = \dfrac{\Delta H_B}{T} - R \ln a_B$ (9.76)

Values of a_B and ΔH_B are normally determined experimentally.

The corresponding relative integral molar values are given by Equation 9.10. Therefore,

Relative integral molar Gibbs energy (or Gibbs energy of mixing):

$$\Delta_{mix}G_m = G_m - G_m^0 = x_A G_A + x_B G_B + \cdots = x_A RT \ln a_A + x_B RT \ln a_B + \cdots \qquad (9.77)$$

$$= RT \left(x_A \ln a_A + x_B \ln a_B + \cdots \right)$$

The value of $\Delta_{mix}G_m$ is always negative, indicating forming a solution is a spontaneous process and in binary systems is zero at each extreme (pure A and pure B) as illustrated in Figure 9.13.

Relative integral molar entropy (or entropy of mixing):

$$\Delta_{mix}S_m = S_m - S_m^0 = x_A S_A + x_B S_B + \cdots \qquad (9.78)$$

$$= -x_A \left(\frac{\Delta H_A}{T} - R \ln a_A \right) - x_B \left(\frac{\Delta H_B}{T} - R \ln a_B \right) + \cdots$$

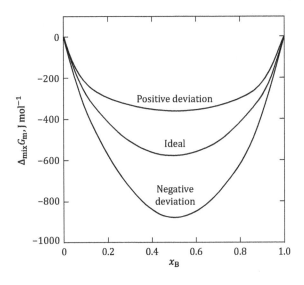

FIGURE 9.13 The Gibbs energy of mixing at 100 K for the solutions with activities varying as shown in Figure 9.8.

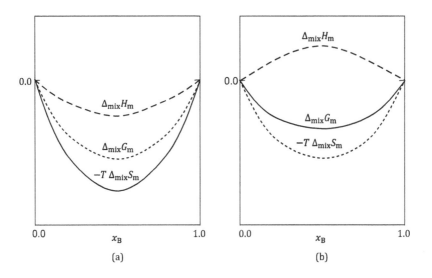

FIGURE 9.14 Typical variation of Gibbs energy, enthalpy and entropy of mixing with composition in a binary system at constant temperature and pressure: (a) $\Delta_{mix}H < 0$, (b) $\Delta_{mix}H > 0$.

The value of $\Delta_{mix}S_m$ is always positive indicating an increase in entropy (randomness) due to forming a solution from the pure components.

Relative integral molar enthalpy (or enthalpy of mixing):

$$\Delta_{mix}H_m = H_m - H_m^0 = x_A H_A + x_B H_B + \cdots \tag{9.79}$$

The relationship between $\Delta_{mix}H_m$, $\Delta_{mix}S_m$ and $\Delta_{mix}G_m$ is illustrated in Figure 9.14.

9.6.3 EXCESS MOLAR QUANTITIES

To express the relative deviation of real solutions from ideal behaviour excess functions which measure the difference between the value of a property in the real solution and the value in an ideal solution of the same composition have been found to be useful. Thus, the *excess relative partial molar quantity* for component B of a solution, ΔX_B^X, is defined as:

$$\Delta X_B^X = \Delta X_B - \Delta X_B^{id} \tag{9.80}$$

and the *excess integral molar quantity* for a solution, $\Delta_{mix}X_m^X$, is defined as:

$$\Delta_{mix}X_m^X = \Delta_{mix}X_m - \Delta_{mix}X_m^{id} \tag{9.81}$$

Therefore,

Excess relative partial molar Gibbs energy:

$$\Delta G_B^X = \Delta G_B - \Delta G_B^{id} = RT \ln a_B - RT \ln x_B = RT \ln \frac{a_B}{x_B} = RT \ln \gamma_B \tag{9.82}$$

Excess integral molar Gibbs energy (or excess Gibbs energy of mixing):

$$\Delta_{mix}G_m^X = \Delta_{mix}G_m - \Delta_{mix}G_m^{id} = RT\left(x_A \ln \gamma_A + x_B \ln \gamma_B + \cdots\right) \tag{9.83}$$

Excess integral molar enthalpy (or excess enthalpy of mixing):

$$\Delta_{mix}H_m^X = \Delta_{mix}H_m - \Delta_{mix}H_m^{id} = \Delta_{mix}H_m - 0 = \Delta_{mix}H_m \tag{9.84}$$

Excess integral molar entropy (or excess entropy of mixing):

$$\Delta_{mix}S_m^X = \frac{\Delta_{mix}H_m^X - \Delta_{mix}G_m^X}{T} \tag{9.85}$$

9.7 EXPERIMENTAL MEASUREMENT OF ACTIVITIES

- *Measurement of the vapour pressure* of a component in equilibrium with the phase containing the component of concern. The activity is then calculated using Equation 9.52. This is feasible if the component of interest has a sufficiently high vapour pressure to be measured experimentally.
- *Measurement of heterogeneous equilibria.* This method involves establishing equilibrium between the phase containing the component of interest and another phase in which the activity of the component has previously been determined. There are two methods. In the first the component becomes distributed between the phases without reacting with either (non-reactive systems). In the second, the component is present in different forms in each phase, and their activities are related through a chemical reaction (reactive systems). This is discussed in Chapters 13 and 14.
- *Measurement of the electrical potential* of electrochemical cells for reactions which can be performed galvanically. The Nernst equation (Equation16.9) is then used to calculate the unknown activity.

9.8 SOURCES OF ACTIVITY DATA

A great deal of experimental work was done throughout the 20th century on measuring the activities of components of solutions of theoretical and practical interest, and a great amount of data has been accumulated. The work has largely been performed by experimental chemists, metallurgists and geochemists. There are some compilations of data, but these are not comprehensive, and reference frequently has to be made to publications of original research or critical reviews. Other than for relatively simple binary and ternary systems it is now more common to use mathematical models to estimate activities of components, particularly in complex systems. These models range from simplistic correlation models to structurally based models making use of atomic-level understanding of the nature of the particular solvents and solutes. Experimental work is still performed, mainly to obtain parameters required for the models and to validate models for complex systems at selected compositions. Discussion of solution models, other than the ideal solution model, is beyond the scope of this book.

PROBLEMS

9.1   Calculate the vapour pressure of water above a solution prepared by dissolving 40 g glycerol ($C_3H_8O_3$) in 150 g of water at 65°C assuming the solution is ideal. The vapour pressure of pure water at 65°C is 25.0 kPa.

9.2   Calculate the mass of ethylene glycol ($C_2H_6O_2$) that must be added to a kilogram of ethanol (C_2H_5OH) to reduce its vapour pressure by 1.3 kPa at 35°C. The vapour pressure of pure ethanol at 35°C is 13.33 kPa. Assume glycol and ethanol form an ideal solution.

9.3   Calculate the vapour pressure of solder consisting of 60 mass% Sn and 40 mass% Pb at 700°C given

$$\log p_{Sn}(\text{atm}) = 5.262 - \frac{15332}{T}$$

$$\log p_{Pb}(\text{atm}) = 4.911 - \frac{9701}{T}$$

Assume Sn and Pb form an ideal solution. At what temperature will the solder boil?

9.4   The volume changes when molten zinc and tin are mixed to form a molten alloy at 420°C are listed in the table below. Plot these values, and estimate the partial molar volume of mixing of tin in an alloy of composition 0.30 mole fraction of zinc.

x_{Zn}	0.10	0.20	0.30	0.40	0.50	0.60	0.70	0.80	0.90
ΔV (mL mol^{-1})	0.0539	0.0964	0.1274	0.1542	0.1763	0.1888	0.1779	0.1441	0.0890

9.5   Calculate values of the relative partial molar Gibbs energies at 25°C of components A and B in an ideal binary solution of A and B at mole fractions of B of 0.1, 0.2, 0.4, 0.6, 0.8 and 0.9. Also calculate the relative integral molar Gibbs energy of the solution at the same concentrations. Plot the values on a graph. Draw a tangent to the integral curve at $x_B = 0.4$, and confirm the values previously calculated for ΔG_A^0 and ΔG_B^0 using the method of intercepts. This can be done either analytically or numerically. Calculate the corresponding entropy values (ΔS_A^0, ΔS_B^0 and $\Delta_{mix}S_m$), and plot on a graph. Using the calculated values confirm that ΔH_A^0, ΔH_B^0 and $\Delta_{mix}H_m$ are zero at all concentrations for an ideal solution.

9.6   The following vapour pressure values of zinc have been determined experimentally for Cu–Zn alloys at 1333 K.

x_{Zn}	1.000	0.800	0.600	0.450	0.300	0.200	0.150	0.100	0.050
p_{Zn} (bar)	4.053	3.200	2.186	1.293	0.608	0.240	0.120	0.060	0.0294

i.   Calculate the activity of zinc at each composition, and plot the values as a function of composition. Also, show the ideal solution line. Given that zinc melts at 419.5°C, what is the standard state for the zinc activity values?

ii.   Estimate the approximate composition ranges over which zinc obeys Henry's law and Raoult's law.

iii. Calculate values of $\ln \gamma_{Zn}$, and plot them as a function of x_{Zn}/x_{Cu}. By graphical or numerical integration using the Gibbs–Duhem equation, estimate the activity coefficient of Cu over the composition range. Plot the values on the same graph as in part i.

iv.   Using the activity data, calculate the Gibbs energy change when 1 mole of Zn is dissolved in a very large quantity of Cu–Zn alloy at 1333 K in which $x_{Zn} = 0.3$.

v.   Calculate the integral molar Gibbs energy of mixing of Cu and Zn at the compositions in the table, and present them graphically. Show, either graphically or numerically, how the relative partial molar Gibbs energy of Cu at $x_{Zn}=0.3$ can be determined. The value should agree with that calculated in part iv.

vi. Calculate (a) the excess partial molar Gibbs energy of Zn; (b) the excess partial molar Gibbs energy of Cu; (c) the integral excess Gibbs energy of the solution. Present the results on one graph.

9.7 The activity coefficient of Zn in Al–Zn alloys has been found to be given by the relation $T \ln \gamma_{Zn} = -880(1 - x_{Zn})^2$. Calculate the activity coefficient of Al at 477 K for $x_{Zn}=0.4$ by solving analytically the Gibbs–Duhem equation. What is the activity of Zn?

9.8 The activities of components A and B in a solution at 80°C determined experimentally are as follows:

x_B	0.000	0.062	0.133	0.276	0.453	0.616	0.828	0.935	0.969	1.000
a_B	0.000	0.216	0.403	0.631	0.770	0.835	0.908	0.960	0.970	1.000
a_A	1.000	0.963	0.896	0.800	0.720	0.656	0.524	0.318	0.180	0.000

i.   Plot these values graphically, and also show the lines for ideal behaviour for each component.

ii. Calculate the value of ΔG_A^0 and ΔG_B^0 at each composition, and plot them graphically.

iii. Calculate the value of γ_A and γ_B at each composition, and plot them graphically.

iv. Calculate the value of $\Delta_{mix}G^0$ at each composition, and plot them graphically.

v.   Repeat the above for the excess functions ΔG_A^{ex}, ΔG_B^{ex} and $\Delta_{mix}G_m^{ex}$.

9.9 ZnS and FeS form a solid solution over the entire composition range. Values of the activity of ZnS and FeS in (Zn,Fe)S solid solution at 850°C are:*

x_{ZnS}	0.1	0.2	0.3	0.4	0.5	0.6	0.7	0.8	0.9
a_{ZnS}	0.165	0.290	0.400	0.501	0.597	0.685	0.767	0.844	0.920
a_{FeS}	0.912	0.828	0.744	0.660	0.570	0.480	0.387	0.290	0.169

i.   Calculate the activity coefficients of ZnS and FeS at the compositions listed, and show graphically how the values vary with composition.

ii. Calculate the excess Gibbs energy of mixing and the molar Gibbs energy of mixing, and show the results graphically.

9.10 As originally formulated in 1803, Henry's law states that the solubility of a gas in a liquid is directly proportional to the partial pressure of the gas above the liquid. Under what conditions will this be true?

9.11 The solubility of argon in a molten mixture of Li, Na and K fluorides at 800°C has been determined as follows:

Solubility (mol mL^{-1})	6.84×10^{-8}	5.16×10^{-8}	3.40×10^{-8}
Pressure (atm)	2.03	1.51	1.00

Does argon obey Henry's law in this system?

---

* Fleet, M. E. 1975. *American Mineralogist*. 60, 466–470.

9.12 The activity coefficient of silver at infinite dilution in liquid lead at 1000°C is 2.3 relative to
      pure liquid silver as standard state. Calculate the Gibbs energy change from the Raoultian
      to the Henrian standard state (mass% scale).

9.13 At 1873 K the activity coefficient of silicon in Fe–Si solution relative to pure liquid silicon
      is 0.0014 at $x_{Si} = 0.01$. Calculate the Henrian activity and activity coefficient of silicon on
      the mass% scale at this concentration. Data: Si(l) = Si(%); $\Delta G^0 = -119250 - 24.3\,T$ J.

10 Reactive systems – single reactions

SCOPE

This chapter examines the equilibrium of single reactions occurring under non-standard state conditions.

LEARNING OBJECTIVES

1. *Understand the criteria for equilibrium of reactions involving reactants and products in solutions and gas mixtures.*
2. *Understand the concept of equilibrium constant of a reaction.*
3. *Be able to use the van't Hoff Isotherm equation to predict the spontaneity of a reaction and the equilibrium position of a reaction.*
4. *Be able to predict qualitatively and quantitatively the effect of changes in temperature, pressure and composition on the equilibrium position of a reaction.*

10.1 INTRODUCTION

We saw in Chapter 8 that when solid or liquid reactants and products are in their standard state (pure solid or liquid substances and gases at 1 bar pressure) a reaction is thermodynamically feasible if the value of the standard Gibbs energy of the reaction is negative. In that case a reaction will proceed until one of the reactants is fully consumed. If the value of the standard Gibbs energy is positive, the reverse reaction is thermodynamically feasible and can proceed until one of the reactants is fully consumed.

In this chapter we are concerned with reactions in which one or more of the reactants or products is present in a solution or gas mixture,* that is, at activities less than one. When solutions are present, reactions between components can never go entirely to completion; that is, the reactants can never be completely consumed. As they react their concentrations become smaller, their activities decrease, and they become increasingly less available for reaction until an equilibrium between the reactants and products is reached. The concepts of Gibbs energy change for reactions and activities of components of solutions are brought together in the van't Hoff Isotherm, the essential equation for predicting the feasibility and extent of a reaction occurring in solutions and/or gas phase. It is this that we explore in this chapter.

Reactant and product species may be present entirely in one phase (homogeneous reactions) or in two or more phases (heterogeneous reactions), for example, a gas phase and a liquid or solid phase, or two liquid or two solid phases. Homogeneous reactions occur throughout the phase whereas heterogeneous reactions occur only at the interfaces between phases since reactants and products need to be in contact for reaction to occur. The reactants diffuse from the bulk phases to the interface where reaction occurs, and the products diffuse away from the interface into the bulk phases.

---

* A gas mixture, being a single phase of several components, is a solution.

10.2   THE FEASIBILITY OF CHEMICAL REACTIONS

The chemical reaction

$$k\,A + l\,B = m\,C + n\,D \tag{4.31}$$

can be considered to be a system consisting of k moles of A, l moles of B, m moles of C and n moles of D. The total Gibbs energy of the system at any time during the reaction is the sum of the Gibbs energies of A, B, C and D at that time:

$$G = m\,G(\mathrm{C}) + n\,G(\mathrm{D}) + k\,G(\mathrm{A}) + l\,G(\mathrm{B}) \tag{10.1}$$

For reactions occurring between species in solutions (solid, liquid or gaseous), Equation 10.1 is written as

$$G = m\,G_{\mathrm{C}} + n\,G_{\mathrm{D}} + k\,G_{\mathrm{A}} + l\,G_{\mathrm{B}}$$

where G_{C}, G_{D}, G_{A} and G_{B} are the partial molar Gibbs energies of species A, B, C and D. If a reaction is spontaneous then the total Gibbs energy of the system will decrease as the reaction occurs. For the case where species A, B, C and D are pure substances, the reaction will continue, and G will continue to decrease, until one of the reactants is fully consumed. In the case where A, B, C and D are in solution the reaction will continue until a minimum value is reached. This is illustrated in Figure 10.1 which shows the variation in G with the relative extent of reaction.* The composition

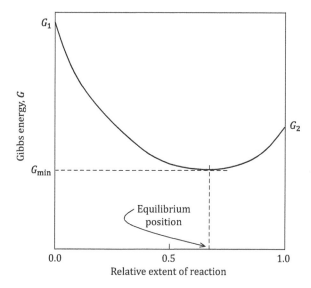

FIGURE 10.1   The change in G for a closed system as a chemical reaction proceeds to equilibrium at constant temperature and pressure. The horizontal axis represents the relative extent of the reaction. Points G_1 and G_2 represent the total Gibbs energy of the reactants and products, respectively.

---

* The *extent of reaction* is an extensive quantity describing the progress of a chemical reaction. It is defined as $\xi = (n_{\mathrm{B}} - n_{\mathrm{B,0}})/\nu_{\mathrm{B}}$ where B is a reactant, ν_{B} its stoichiometric coefficient and $n_{\mathrm{B,0}}$ and n_{B} the moles of species B at the beginning of the reaction and when the extent of reaction ξ is reached, respectively. The sign of ν_{B} is positive for the products and negative for the reactants. The *relative extent of reaction* is the ratio of the extent of reaction at any stage to the extent of reaction assuming all reactant B has been consumed.

of the system at this minimum is the equilibrium composition. Calculation of the integral Gibbs energy of a system as a function of composition is a particularly useful way of determining the equilibrium composition of complex systems such as when several parallel reactions occur. This is examined further in Chapter 15. For the present, however, consideration is restricted to systems in which we are interested in single reactions only.

For single reactions, it is more convenient to deal with changes in Gibbs energy rather than actual values of total Gibbs energy so the basis for establishing the criteria for the equilibrium of reactions occurring in solutions or gas mixtures is the Gibbs energy change of the relevant reaction. For Reaction 4.31,

$$\Delta G = m\,G_C + n\,G_D - k\,G_A - l\,G_B$$

Substituting for the partial molar Gibbs energies of A, B, C and D from Equation 9.48 gives

$$\Delta G = m\left(G_C^0 + RT\ln a_C\right) + n\left(G_D^0 + RT\ln a_D\right)$$

$$- k\left(G_A^0 + RT\ln a_A\right) - l\left(G_B^0 + RT\ln a_B\right)$$

$$= \left(mG_C^0 + nG_D^0 - kG_A^0 - lG_B^0\right) + RT\left(m\ln a_C + n\ln a_D - k\ln a_A - l\ln a_B\right)$$

and, therefore,

$$\Delta G = \Delta_r G^0 + RT\ln\frac{a_C^m \times a_D^n}{a_A^k \times a_B^l} = \Delta_r G^0 + RT\ln Q \tag{10.2}$$

where Q is the *reaction quotient*. Equation 10.2 is the *van't Hoff Isotherm*, named after J. H. van't Hoff[*] who first derived it.

If any of the species of the reaction are gases, then the expression for partial molar Gibbs energy for gases (Equation 9.29 for a pure component or Equation 9.30 for a component in a gas mixture) is substituted in the derivation, that is, $RT\ln p$ or $RT\ln p_B$, where p is the dimensionless pressure and p_B is the dimensionless partial pressure of B (numerically equal to the pressure or partial pressure of B expressed in bars).

The values of a_A and a_B are the activities of reactants A and B at any particular time during the reaction, and the corresponding values of a_C and a_D are the activities of the products C and D at that time. These values change as the reaction proceeds, changing the value of Q. If the reactants and products are not initially at equilibrium, the reaction proceeds in the direction that decreases the Gibbs energy difference (ΔG) between the reactants and products as illustrated in Figure 10.2. If $\Delta G < 0$, the reaction will proceed spontaneously to the right, and if $\Delta G > 0$, the reaction will proceed spontaneously to the left. This difference decreases until finally it becomes zero and the reaction reaches equilibrium. At this point, the activities must be such that the $RT\ln Q$ term cancels the ΔG^0 term (since $\Delta G = 0$ at equilibrium). The system can no longer decrease

---

[*] Jacobus Henricus van't Hoff (1852–1911) was a professor at the University of Amsterdam. In 1884, he published the book *Études de Dynamique Chimique* (*Studies in Chemical Dynamics*) in which he used the principles of thermodynamics to explain mathematically the rates of chemical reactions based on changes in the concentration of reactants with time. He showed how the previously independently developed concepts of dynamic equilibrium (that chemical equilibrium results when the rates of forward and reverse reactions are equal), the law of mass action (that the concentration of substances affects the rate of reaction) and the equilibrium constant (the ratio of the concentrations of starting materials to products at equilibrium) together provide a coherent explanation of the nature of chemical reactions. He was the first winner of the Nobel Prize for Chemistry (in 1901).

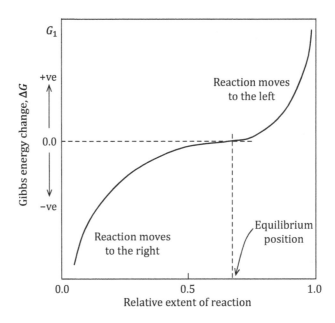

FIGURE 10.2 The change in ΔG for a closed system as a chemical reaction proceeds to equilibrium at constant temperature and pressure. The horizontal axis represents the relative extent of the reaction.

its Gibbs energy by transforming reactants into products, and G is at its minimum value. When $\Delta G = 0$, therefore,

$$\Delta_r G^0 = -RT \ln \frac{\left(a_C^*\right)^m \times \left(a_D^*\right)^n}{\left(a_A^*\right)^k \times \left(a_B^*\right)^l} \qquad (10.3)$$

where a_A^*, a_B^*, a_C^* and a_D^* are the activities of A, B, C and D at equilibrium of the reaction.

10.3   THE EQUILIBRIUM CONSTANT

We now introduce a new term, the *equilibrium constant K*, which is defined as follows:

For solid and liquid reactants and products: $K = \dfrac{a_C^m \times a_D^n}{a_A^k \times a_B^l}$ (10.4)

For gaseous reactants and products: $K = \dfrac{p_C^m \times p_D^n}{p_A^k \times p_B^l}$ (10.5)

If the gases are non-ideal, then fugacities would be used in Equation 10.5.

Since K is a ratio of dimensionless parameters, it is itself dimensionless – it has no unit. Equilibrium values of activities and partial pressures are implied in Equations 10.4 and 10.5 because the equations apply only at equilibrium. For convenience, in future the superscripted asterisks (*) on the activities in the equilibrium constant expression are omitted.

It follows from Equation 10.3 that

$$\Delta_r G^0 = -RT \ln K \qquad (10.6)$$

Substituting back into Equation 10.2, the van't Hoff Isotherm can be rewritten as:

$$\Delta G = -RT \ln K + RT \ln Q \tag{10.7}$$

Clearly the equilibrium constant for a reaction has a unique value at any particular temperature; that is, K is a function of temperature only.

Since

$$\ln K = -\frac{\Delta_r G^0}{RT}$$

a reaction with a large negative value of $\Delta_r G^0$ will have a large equilibrium constant. Then, according to Equation 10.4

$$a_C^m \times a_D^n \gg a_A^k \times a_B^l$$

The reaction will be strongly displaced to the right, and at equilibrium the concentration of the reactants will be small and that of the products large; that is, most of the reactants will be converted into products. Conversely, a reaction with a large positive value of $\Delta_r G^0$ will have a small equilibrium constant and

$$a_C^m \times a_D^n \ll a_A^k \times a_B^l$$

This reaction will be strongly displaced to the left, and at equilibrium the concentration of the reactants will be high and that of the products small; that is, most of the reactants will remain unconverted into products. A reaction with $\Delta_r G^0$ around zero will have an equilibrium constant of around one. In this case,

$$a_C^m \times a_D^n \approx a_A^k \times a_B^l$$

and at equilibrium the concentrations of reactant and products will all be significant.

Expressing the Gibbs–Helmholtz equations (Equations 8.27 and 8.28) in terms of ΔG^0 and ΔH^0,

$$\frac{d(\Delta G^0 / T)}{dT} = \frac{-\Delta H^0}{T^2} \tag{10.8}$$

$$\frac{d(\Delta G^0 / T)}{d(1/T)} = \Delta H^0 \tag{10.9}$$

Substituting for ΔG^0 from Equation 10.6 gives

$$\frac{d \ln K}{dT} = \frac{\Delta_r H^0}{RT^2} \tag{10.10}$$

$$\frac{d \ln K}{d(1/T)} = -\frac{\Delta_r H^0}{R} \tag{10.11}$$

Equations 10.10 and 10.11 are forms of the *van't Hoff Isochore* (isochore means constant pressure). These equations show how the equilibrium constant of a reaction varies with temperature. If a reaction is endothermic, $\Delta_r H^0$ is positive, and it follows that the equilibrium constant of the reaction increases with temperature. Similarly, if a reaction is exothermic, $\Delta_r H^0$ is negative, and the equilibrium constant of the reaction decreases with temperature.

EXAMPLE 10.1   Application of the van't Hoff Isotherm

A piece of silver is placed in an electric oven and heated in air. Will the silver oxidise?

SOLUTION
The relevant data are:

T, K	298.15	300	400	500
$\Delta_f G^0 (Ag_2O), kJ\,mol^{-1}$	−11,184	−11,060	−4432	2065

The reaction of interest is the formation of Ag_2O from its constituent elements:

$$2\,Ag(s) + 0.5\,O_2(g) = Ag_2O(s)$$

The silver and its oxide are pure; therefore their activities are one. The partial pressure of oxygen in air is 0.21 atm $= 1.01325 \times 0.21 = 0.213$ bar. Therefore (and remembering to convert kJ mol^{-1} to J mol^{-1}),

$$\Delta G = \Delta_f G^0 + RT \ln \frac{a_{Ag_2O}}{a_{Ag} \times p_{O_2}^{0.5}} = \Delta_f G^0 + 8.314 \times T \ln \frac{1}{p_{O_2}^{0.5}}$$

and using the values in the table above:

T, K	298.15	300	400	500
ΔG, J mol^{-1}	−9266	−9130	−1859	+5281

The graphical relation is shown in Figure 10.3 from which it is seen that ΔG changes from being negative to positive at a temperature between 400 and 450 K. The relationship is linear, and

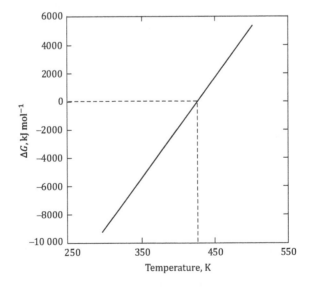

FIGURE 10.3   The variation of ΔG with temperature of the reaction $2\,Ag(s) + 0.5\,O_2(g) = Ag_2O(s)$ when silver is heated in air.

when an equation is fitted to the points and solved for $\Delta G = 0$ the temperature is 426 K (153°C). Therefore, at temperatures below 153°C silver will oxidise in air (and be completely converted to oxide), but at temperatures above 153°C it will not. Conversely, if oxidised silver is placed in an oven at temperatures above 153°C the oxide will decompose back to metallic silver and oxygen.

COMMENT

Interestingly, silver is not stable in air at normal ambient temperatures and should oxidise. That silver jewellery and other items do not oxidise in air is due to the kinetic barrier at room temperature. However, silver does tarnish readily at room temperature by reaction with sulfur compounds in the air to form Ag_2S.

EXAMPLE 10.2   Calculation of the equilibrium constant of a reaction

Calculate the equilibrium constant for the reaction

$$C(s) + CO_2(g) = 2\,CO(g)$$

at temperatures between 300 and 1300 K.

SOLUTION

The relevant data are:

	300	500	700	900	1100	1300
$\Delta_f G^0$ (CO), kJ mol^{-1}	−137.345	−155.43	−173.534	−191.431	−209.056	−226.482
$\Delta_f G^0$ (CO$_2$), kJ mol^{-1}	−394.369	−394.912	−395.369	−395.716	−395.918	−396.079

First calculate $\Delta_r G^0$ using the $\Delta_f G^0$ values for CO and CO_2: $\Delta_r G^0 = 2 \times \Delta_f(CO) - \Delta_f(CO_2)$, then calculate K from Equation 10.6:

$$K = \exp\left(\frac{-\Delta_r G^0}{RT}\right)$$

T, K	300	500	700	900	1100	1300
$\Delta_r G^0$, kJ mol^{-1}	119.679	84.052	48.301	12.854	−22.194	−56.885
K	1.450×10^{-21}	1.655×10^{-9}	2.487×10^{-4}	1.794×10^{-1}	1.132×10^{1}	1.931×10^{7}

COMMENT

Carbon forms two oxides, CO and CO_2, both of which are gases. When carbon is reacted with oxygen both oxides can form (since their Gibbs energy of formation is negative), and, because they are gases, they form a mixture in which each exerts a partial pressure. Since K is very small at low temperatures, the equilibrium of the reaction is displaced towards the left (CO_2 is the dominant oxide). At high temperatures, K is large, and the reaction is displaced to the right (CO is the dominant oxide). The equilibrium between CO and CO_2, namely

$$CO_2(g) + C(s) = 2\ CO(g)$$

is called the *Boudouard reaction*,* and it is an important reaction in many industrial situations including smelting of iron ore, smelting of aluminium and producing graphite flakes, filamentous graphite, lamellar graphite and carbon nanotubes.

EXAMPLE 10.3   Find the equilibrium state of a reaction

i. Calculate the composition of a $CO - CO_2$ gas mixture in equilibrium with carbon over the temperature range 300 to 1300 K.

SOLUTION

The equilibrium is expressed by the Boudouard reaction:

$$CO_2(g) + C(s) = 2\ CO(g)$$

$$K = \frac{p_{CO}^2}{a_C p_{CO_2}}$$

If the total pressure of the gas mixture is p, then $p_{CO} + p_{CO_2} = p$ and

$$K = \frac{p_{CO}^2}{a_C(p - p_{CO})}$$

Rearranging

$$p_{CO}^2 + K\ a_C p_{CO} - K\ a_C p = 0$$

Using the K values from Example 10.2 and setting $a_C = 1$ (since the gases are in equilibrium with solid carbon), the equilibrium partial pressures of CO can be calculated by solving the equation analytically or by using the Goal Seek function in Excel™. The corresponding partial pressure of CO_2 is $p_{CO_2} = p - p_{CO}$. The respective volume fractions are directly proportional to the partial pressures (Equation 3.7). The results are shown in Figure 10.4 for three values of total pressure. At any pressure, the equilibrium composition of the gas mixtures lies on the appropriate line at the relevant temperature.

COMMENT

The approach used here is not a general method to find the equilibrium composition of a system and will work only in specific cases. A general approach which applies for all reactions is described in Section 10.6.1.

ii. What will happen at a particular temperature and pressure if the composition of a gas mixture lies below the curve?

The composition represented by point A in Figure 10.4 is below the curve for 1 bar pressure at 900°C. The gas mixture has a lower concentration of CO than the equilibrium composition. Therefore, at constant temperature, CO_2 in the mixture will react with carbon, and the composition will move vertically upwards until it lies on the curve.

---

* Named after Octave Leopold Boudouard (1872–1923), a professor at the Conservatoire National des Arts et Métiers in Paris who worked in areas of applied chemistry. He discovered the Boudouard reaction in 1905.

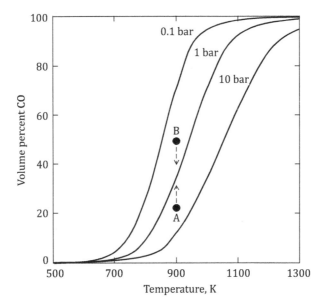

FIGURE 10.4   The composition of a CO–CO$_2$ gas mixture in equilibrium with graphite.

iii. What will happen at a particular temperature and pressure if the composition of a gas mixture lies above the curve?

The composition represented by point B in Figure 10.4 is above the curve for 1 bar pressure at 900°C.
The gas mixture has a higher concentration of CO than the equilibrium composition. Therefore, at
constant temperature, CO in the mixture will dissociate into CO$_2$ and carbon by means of the
reverse Boudouard reaction. Carbon will precipitate out of the gas mixture, and the gas composition
will move vertically downwards until it lies on the curve.

COMMENT

The reverse Boudouard reaction is very slow and usually only occurs at a significant rate when a
catalyst is present. The type of catalyst, the temperature and pressure all influence the morphol-
ogy of the carbon product.

EXAMPLE 10.4   Find the conditions for a reaction to occur

At what temperature will limestone begin to decompose if heated in air?

SOLUTION

The decomposition of limestone occurs by means of the reaction:

$$CaCO_3(s) = CaO(s) + CO_2(g)$$

for which

$$K = \frac{a_{CaO}\, p_{CO_2}}{a_{CaCO_3}} = p_{CO_2}, \quad \text{for pure } CaCO_3 \text{ and } CaO$$

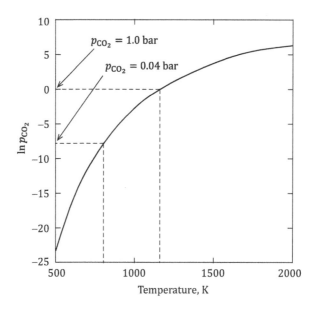

FIGURE 10.5 The pressure of CO_2 in equilibrium with solid $CaCO_3$ and CaO as a function of temperature.

For a particular temperature (value of K), p_{CO_2} is the partial pressure of CO_2 in equilibrium with a mixture $CaCO_3$ and CaO. If the ambient partial pressure of CO_2 is less than p_{CO_2}, $CaCO_3$ will decompose; if the ambient pressure of CO_2 is greater than p_{CO_2}, CaO will be unstable relative to $CaCO_3$, and the reverse reaction will occur.

The concentration of CO_2 in the atmosphere is about 0.04% (by volume); therefore, the partial pressure of CO_2 in the atmosphere is 0.0004 atm (Equation 3.4). We need to find the temperature at which the partial pressure of CO_2 generated by the decomposition reaction equals 0.0004 atm. To do this we need to know the variation of K with temperature. This was calculated in Example 8.8.

The results are shown graphically in Figure 10.5 from which, assuming for simplicity that 1 bar ≈ 1 atm, it is evident that limestone will start to decompose in air when the temperature reaches approximately 803 K (530°C). We can refer to this as the onset of decomposition. Above this temperature CO_2 formed by the reaction can diffuse away from the limestone particles into the atmosphere.

Also marked on the figure is the temperature at which $p_{CO_2} = 1$ atm. At this point, 1160 K (887°C), the pressure of the CO_2 generated by the reaction is equal to the external pressure, the CO_2 can now expand against the atmosphere, and rapid decomposition will commence. This temperature corresponds to the boiling point of liquids at atmospheric pressure.

10.4 CHOICE OF STANDARD STATE

Because of the different forms of activity – Raoultian, Henrian (molality), Henrian (mass percent) – care is needed when performing calculations that the appropriate value of $\Delta_r G^0$ is used. The activity values used in calculations involving the Gibbs energy change, and the equilibrium constant of the reaction, *must* be referred to the standard states implied by the value of $\Delta_r G^0$ used. For example, for the reaction between carbon and FeO to form Fe and CO at 1350°C, the standard states would normally be taken as the stable form of the species at 1350°C. If the stable form is a gas then the pressure of the gas in its standard state is 1 bar. The standard states would, therefore, be:

C	pure, solid carbon at 1350°C
Fe	pure, solid iron at 1350°C
FeO	pure, molten FeO at 1350°C
CO	CO gas at 1 bar and 1350°C

All this is implied simply by writing the reaction as:

$$C(s) + FeO(l) = Fe(s) + CO(g); \quad \Delta_r G^0 (1623.15)$$

for which the equilibrium constant expression is:

$$K = \frac{a_{Fe}\ p_{CO}}{a_C\ a_{FeO}}$$

The standard Gibbs energy change of the reaction is then calculated from the standard Gibbs energies, or standard Gibbs energies of formation, of the individual components at 1350°C:

$$\Delta G^0 = G^0_{Fe(s)} + G^0_{CO(g)} - G^0_{C(s)} - G^0_{FeO(l)}$$

or

$$\Delta G^0 = \Delta G^0_{Fe(s)} + \Delta G^0_{CO(g)} - \Delta G^0_{C(s)} - \Delta G^0_{FeO(l)} = \Delta G^0_{CO(g)} - \Delta G^0_{FeO(l)}$$

The form of a component (solid, liquid or gas) selected as the standard state does not have to correspond to the actual form of the component in the reaction system. For example, if the above reaction was carried out at 1350°C it is likely that the iron would actually be present as liquid even though pure iron melts at 1536°C because carbon is soluble in iron and lowers its melting point to below 1350°C. The activity of iron, however, would usually still be expressed relative to pure solid iron at 1350°C since that is the stable form of iron at that temperature.

Furthermore, the same chemical reaction can sometimes be expressed with different stoichiometric numbers. For example, the reactions

$$0.5\,N_2(g) + 1.5\,H_2(g) = NH_3(g) \tag{1}$$

$$N_2(g) + 3\,H_2(g) = 2\,NH_3(g) \tag{2}$$

are equivalent reactions, but their $\Delta_r G^0$ values will be different, as will the expressions for their equilibrium constants:

$$K_1 = \frac{p_{NH_3}}{p_{N_2}^{0.5}\ p_{H_2}^{1.5}}$$

$$K_2 = \frac{p_{NH_3}^2}{p_{N_2}\ p_{H_2}^3}$$

The same principles apply to the use of Henrian and mixed Raoultian and Henrian standard states. For example, for the reaction

$$CO_2(g) + H_2O(l) = H_2CO_3(aq)$$

Raoultian activities are implied for CO_2 and H_2O and Henrian activities (molality) for H_2CO_3, and this indicates the appropriate values of G required to calculate $\Delta_r G^0$, and hence K.

SUMMARY

Values and expressions for $\Delta_r G^0$ and K should always be associated with the relevant chemical reaction, and the chemical reaction should be written to show the standard state for each component.

EXAMPLE 10.5   The solubility of CO_2 in water

What mass of carbon dioxide is dissolved in 1 L of soda water if the water was bottled at a pressure of 2.7 bar at 10°C? Assume CO_2 in water obeys Henry's law. The solubility of CO_2 in water at 10°C and 1 atmosphere pressure is 2.5 g kg⁻¹.

SOLUTION

CO_2 dissolves in water predominantly in the molecular form, and the equilibrium between CO_2 gas and CO_2 dissolved in water is:

$$CO_2(g) = CO_2(aq)$$

Here, we have adopted Henrian activity for CO_2 in water and Raoultian activity for gaseous CO_2. Since CO_2 obeys Henry's law in solution in water,

$$K = \frac{a_{\%,CO_2}}{p_{CO_2}} = \frac{\%CO_2}{p_{CO_2}}$$

2.5 g kg⁻¹ ≡ 0.025 mass percent; 1 atm = 1.01325 bar. Therefore,

$$K = \frac{0.25}{1.01325} = 0.2467$$

At 2.7 bar,

$$\%CO_2 = K \times p_{CO_2} = 0.2467 \times 2.7 = 0.67$$

Therefore, the equilibrium concentration of CO_2 in water at 10°C and 2.7 bar is 0.67 mass percent or 0.67 g per hundred grams. The mass of 1 L of water is 1000 g; therefore, a litre of the bottled soda water contains 6.7 g CO_2.

10.5   THE EFFECT OF TEMPERATURE, PRESSURE AND CONCENTRATION ON EQUILIBRIUM

The well-known *Le Chatelier's* Principle* (1884) states that:

> If a reactive system in equilibrium is subjected to a change which alters the equilibrium (such as addition or removal of heat, increase or decrease in pressure, addition of more of a reactant or removal of

---

* Henry Louis Le Châtelier (1850–1936) was a French chemist who taught at the École des Mines in Paris, the Collège de France and later at the Sorbonne. He is best known for his research on chemical equilibrium though he also did much applied research, particularly in the field of metallurgy.

a product) the direction of the reaction which takes place is such as to tend to reduce the effect of the change.

This a useful qualitative rule for predicting the effect of a change in conditions on the equilibrium of a reaction. Its theoretical basis is now explained.

10.5.1 THE EFFECT OF TEMPERATURE

For endothermic reactions, K increases with temperature. From Equation 10.3 it is apparent that as K increases, the term $(a_C)^m \times (a_D)^n$ increases relative to the term $(a_A)^k \times (a_B)^l$, and the equilibrium position of a reaction is displaced further to the right. For exothermic reactions, K decreases with temperature, and the equilibrium position is displaced further to the left.

10.5.2 THE EFFECT OF PRESSURE

Consider the reaction

$$k\,A(g) + l\,B(s) = m\,C(g) + n\,D(s)$$

for the case where there is a gaseous reactant and a gaseous product. Then

$$K = \frac{p_C^m \times a_D^n}{p_A^k \times a_B^l}$$

If we assume B and D are pure, their activities are one, and

$$K = \frac{p_C^m}{p_A^k}$$

applying Dalton's law of partial pressures (Equation 3.6),

$$K = \frac{\left(p\,x_C\right)^m}{\left(p\,x_A\right)^k} = \frac{x_C^m}{x_A^k} \times p^{m-k} \tag{10.12}$$

Since p^{m-k} is a constant and K is a function of temperature only (Equation 10.6), it follows that at constant temperature the equilibrium position of a reaction is independent of the pressure of the system if $m = k$ (since then $m - k$ is zero) but depends on the total pressure if $m \neq k$. The value of x_C^m / x_A^k must increase with pressure if $m < k$ and decrease with pressure if $m > k$.

10.5.3 THE EFFECT OF COMPOSITION

Consider the reaction

$$k\,A + l\,B = m\,C + n\,D$$

for which

$$K = \frac{a_C^m \times a_D^n}{a_A^k \times a_B^l}$$

Let us assume that the activities of B and D are held constant; for example, they may be solids or immiscible liquids. Then

$$K = c \times \frac{a_C^m}{a_A^k} \tag{10.13}$$

where c is a constant. Since K is constant at a given temperature, if the activity of reactant A is decreased, for example by diluting it, then the equilibrium of the reaction will be displaced to the left, and less of reactant A will be consumed (and less of product C will be formed). Conversely, if the activity of product C is decreased the equilibrium of the reaction will be displaced to the right, and more of reactant A will be consumed (and more of product C formed).

EXAMPLE 10.6   The effect of temperature, pressure and composition on the equilibrium of a reaction

Ammonia gas is made industrially from H_2 and N_2 using the Haber process. The key reaction is:

$$N_2(g) + 3 H_2(g) = 2 NH_3(g); \Delta_r H^0 \approx -92 \text{ kJ}$$

High temperature (typically 400–500°C) and pressure (typically 150–250 bar) are required to obtain a high degree of conversion to ammonia, and the reaction is accelerated by means of a catalyst.

$$K = \frac{\left(p_{NH_3}\right)^2}{p_{N_2}\left(p_{H_2}\right)^3} = \frac{\left(x_{NH_3}\right)^2 p^2}{x_{N_2} p \left(x_{H_2}\right)^3 p^3} = \frac{\left(x_{NH_3}\right)^2}{x_{N_2}\left(x_{H_2}\right)^3 p^2}$$

where p is the total pressure of the system. Rearranging gives

$$\left(x_{NH_3}\right)^2 = K \, p^2 \times x_{N_2}\left(x_{H_2}\right)^3$$

This relation shows how the concentration of ammonia in a mixture of H_2, N_2 and NH_3 in equilibrium varies with the pressure of the system, the temperature (since K is a function of T) and concentration of N_2 and H_2. As pressure increases, the proportion of NH_3 must increase relative to H_2 and N_2 in order to maintain the equality. Since the reaction is exothermic, K will decrease as the temperature increases, and the proportion of NH_3 will increase relative to H_2 and N_2. If additional H_2 or N_2 is added to the system to increase its concentration, the proportion of NH_3 will again increase relative to H_2 and N_2.

The effects of temperature, pressure and composition illustrated in Example 10.6 can be quantified by taking into account the stoichiometric relations among the reactants and products. This is discussed for single reaction systems in Section 10.6.1. Methods for doing this for more complex systems, including those in which multiple reactions can occur, are discussed in Chapter 15.

10.6   THE EQUILIBRIUM COMPOSITION OF A SYSTEM

10.6.1   SINGLE REACTIONS

Calculation of the equilibrium composition for a single reaction system is relatively easy and was done for the Boudouard reaction in Example 8.8.8ii. In this section, a general method of calculation is described which can be applied to any reaction in which the activity of reactants and products can be related to concentration, for example, partial pressure for gases or mole fraction for solids and liquids.

Suppose the reaction of interest is:

$$k \, A + l \, B = m \, C + n \, D$$

TABLE 10.1

Stoichiometric relations for the reaction $k\,A + l\,B = m\,C + n\,D$

	A	B	C	D	Total
Initial	n_A	n_B	0	0	$n_A + n_B$ mole
Change	$-\alpha$	$\dfrac{-\alpha l}{k}$	$\dfrac{\alpha m}{k}$	$\dfrac{\alpha n}{k}$	2α mole
Equilibrium	$n_A - \alpha$	$n_B - \dfrac{\alpha l}{k}$	$\dfrac{\alpha m}{k}$	$\dfrac{\alpha n}{k}$	Variable, depending on number of phases

The stoichiometry of the reaction tells us that α moles of A will react with $\alpha(l/k)$ moles of B to form $\alpha(m/k)$ moles of C and $\alpha(n/k)$ moles of D. If we start with n_A, moles of A, n_B, moles of B, 0 moles of C and 0 moles of D, then we can set up Table 10.1.

The Equilibrium line is the sum of the Initial and Change lines and gives the moles of each species at equilibrium. These can be related to the equilibrium state of the system through the equilibrium constant expression:

$$K = \frac{a_C^m\, a_D^n}{a_A^k\, a_B^l}$$

If the species is a gas, the activity term can be replaced by the partial pressure, $p \times x_B$ where p is the total pressure of the system; if the species is in solution the activity can be replaced by $\gamma \times x_B$. In both cases, we need the mole fractions of the species at equilibrium.

Single phase reactions

If all the species are gases, there is one phase, and the total number of moles at equilibrium is:

$$\left(n_A - \alpha\right) + \left(n_B - \frac{\alpha l}{k}\right) + \frac{\alpha m}{k} + \frac{\alpha n}{k} = n_A + n_B - \alpha - \frac{\alpha}{k}\left(m+n\right)$$

The mole fraction of each species can now be calculated, and if the total pressure of the system is p, the partial pressures can be calculated and summarised in the form of an Initial–Change–Equilibrium (or ICE) table, as shown in Table 10.2.

TABLE 10.2

The Initial–Change–Equilibrium table for the reaction $k\,A + l\,B = m\,C + n\,D$ if all species are gases

	A	B	C	D	Total
Initial	n_A	n_B	0	0	$n_A + n_B$ mole
Change	$-\alpha$	$\dfrac{-\alpha l}{k}$	$\dfrac{\alpha m}{k}$	$\dfrac{\alpha n}{k}$	2α mole
Equilibrium	$n_A - \alpha$	$n_B - \dfrac{\alpha l}{k}$	$\dfrac{\alpha m}{k}$	$\dfrac{\alpha n}{k}$	$n = n_A + n_B - \alpha - \dfrac{\alpha}{k}\left(m+n\right)$
x_i	$\dfrac{n_A - \alpha}{n}$	$\dfrac{n_B - (\alpha l/k)}{n}$	$\dfrac{(\alpha m/k)}{n}$	$\dfrac{(\alpha n/k)}{n}$	1.0
p_i	$\dfrac{(n_A - \alpha)p}{n}$	$\dfrac{(n_B - (\alpha l/k))p}{n}$	$\dfrac{(\alpha m/k)p}{n}$	$\dfrac{(\alpha n/k)p}{n}$	p bar

The partial pressures are substituted into the equilibrium constant expression:

$$K = \frac{p_C^m \, p_D^n}{p_A^k \, p_B^l}$$

Then if n_A, n_B and p are known, values of α can be calculated at any desired temperature by solving the resulting equation. The values of α are then substituted into the expressions for mole fraction and partial pressure to obtain the equilibrium composition of the system.

If the reaction occurs within one solid or liquid phase the same approach applies only the activity terms in the equilibrium constant expression are replaced by $\gamma \times x$ terms rather than partial pressures. If ideal solution behaviour applies or is assumed, $\gamma = 1$.

Heterogeneous reactions

The calculation procedure is slightly different when the reaction involves more than one phase, for example, gas–solid, gas–liquid, liquid–solid, solid–solid, etc. In this case, the mole fractions of the species are expressed in terms of the relevant phase rather than of the system as a whole. For the reaction

$$k\,A + l\,B = m\,C + n\,D$$

if species A and C are present in one phase (Phase I), and species B and D in another phase (Phase II),

$$n(\text{I}) = (n_A - \alpha) + \frac{\alpha\,m}{k} = n_A - \alpha\left(1 - \frac{m}{k}\right)$$

$$n(\text{II}) = n_B - \frac{\alpha\,l}{k} + \frac{\alpha\,n}{k} = n_B - \frac{\alpha}{k}(l + n)$$

The Initial–Change–Equilibrium table then takes the form shown in Table 10.3.

TABLE 10.3

The Initial–Change–Equilibrium table for the reaction $k\,A + l\,B = m\,C + n\,D$ where A and C are in one phase and B and D are in a second phase

	A(I)	B(II)	C(I)	D(II)	Total
Initial	n_A	n_B	0	0	$n_A + n_B$ mole
Change	$-\alpha$	$\dfrac{-\alpha\,l}{k}$	$\dfrac{\alpha\,m}{k}$	$\dfrac{\alpha\,n}{k}$	2α mole
Equilibrium	$n_A - \alpha$	$n_B - \dfrac{\alpha\,l}{k}$	$\dfrac{\alpha\,m}{k}$	$\dfrac{\alpha\,n}{k}$	$n(\text{I}) = n_A - \alpha\left(1 - \dfrac{m}{k}\right)$ $n(\text{II}) = n_B - \dfrac{\alpha}{k}(l + n)$
x_i	$\dfrac{n_A - \alpha}{n(\text{I})}$	$\dfrac{n_B - (\alpha\,l/k)}{n(\text{II})}$	$\dfrac{(\alpha\,m/k)}{n(\text{I})}$	$\dfrac{(\alpha\,n/k)}{n(\text{II})}$	1.0 (Phase I); 1.0 (Phase II)
p_i (if i is a gas)	$\dfrac{(n_A - \alpha)p}{n(\text{I})}$	$\dfrac{(n_B - (\alpha\,l/k))p}{n(\text{II})}$	$\dfrac{(\alpha\,m/k)p}{n(\text{I})}$	$\dfrac{(\alpha\,n/k)p}{n(\text{II})}$	p bar

---

EXAMPLE 10.7   Calculate the equilibrium of a single-reaction system

Calculate the equilibrium composition of the C–O system at 900 K assuming the only reaction is:

$$C(s) + CO_2(g) = 2\ CO(g)$$

and the system is saturated with carbon (solid carbon is in equilibrium with CO and CO_2).

SOLUTION

Suppose the initial state of the system was 1 mol of CO_2 (any amount could be chosen). When equilibrium is established, part of the CO_2 will have dissociated into C and CO_2 to satisfy the stoichiometry of the reaction and the equilibrium condition:

$$K = \frac{p_{CO}^2}{a_C\ p_{CO_2}}$$

Let α be the amount of CO_2 that dissociates. Then α moles of C and 2α moles of CO will be formed, and $1 - \alpha$ moles of CO_2 will remain. This is summarised in the Initial and Change rows of the ICE table:

	CO_2	CO	C	Total
Initial	1	0	0	1 mole
Change	$-\alpha$	2α	α	2α mole
Equilibrium	$1-\alpha$	2α	α	$n_C = \alpha$
				$n_{gas} = 1+\alpha$
x_B (gas phase)	$\dfrac{1-\alpha}{1+\alpha}$	$\dfrac{2\alpha}{1+\alpha}$		1
p_B	$\dfrac{(1-\alpha)p}{1+\alpha}$	$\dfrac{2\alpha\ p}{1+\alpha}$		p

The Equilibrium row is the sum of the first two rows and is the composition of the system at equilibrium. The amount of gas phase formed is $1 - \alpha + 2\alpha = 1 + \alpha$ moles. The fourth row is the composition of the gas phase. Carbon forms a separate phase, and its vapour pressure is negligible. The fifth row is the partial pressures of CO and CO_2.

Substituting the partial pressures into the equilibrium constant expression, and setting $a_C = 1$,

$$K = \frac{p_{CO}^2}{p_{CO_2}} = \frac{4\alpha^2 p^2}{(1+\alpha)^2} \times \frac{(1+\alpha)}{(1-\alpha)p} = \frac{4\alpha^2 p}{(1-\alpha^2)}$$

Rearranging,

$$4\alpha^2 p + K\alpha^2 = K$$

$$\alpha = \left(\frac{K}{4p+K}\right)^{1/2}$$

The equation can be solved at any pressure p and temperature.

At 900 K, $K = 0.17945$ and at 1 bar pressure, $\alpha = 0.207$. Therefore, the equilibrium composition of the system is:

$$n_{CO_2} = 1 - \alpha = 1 - 0.207 = 0.793 \text{ moles}$$

$$n_{CO} = 2\alpha = 2 \times 0.207 = 0.414 \text{ moles}$$

$$n_C = \alpha = 0.207 \text{ moles}$$

The composition of the gas phase is:

$$x_{CO_2} = \frac{(1-\alpha)}{1+\alpha} = 0.343 \quad \text{and} \quad x_{CO_2} = \frac{2\alpha}{1+\alpha} = 0.657$$

or 34.3 vol% CO_2 and 65.7 vol% CO, which is as calculated in Example 8.8(ii)

10.6.2 MULTIPLE REACTIONS WITHIN A SYSTEM

In many systems of practical interest, more than one reaction can occur between the reactants. Sometimes it is possible to ignore all except the main reaction if the amounts of products formed by the other reactions are small; in other situations small amounts of products may be important, or there may be several reactions all of which occur to a significant extent. Calculation of the equilibrium composition of a system which cannot be defined by a single reaction is mathematically difficult and will be discussed further in Chapter 15.

PROBLEMS

10.1 Write the equilibrium constant expression for the following equations, and calculate its value from the given values of $\Delta_r G^0$:

$$2\,SO_3(g) = 2\,SO_2(g) + O_2(g); \quad \Delta_r G^0(573) = 89.9 \text{ kJ}$$

$$Zn(s) + H_2SO_4(aq) = ZnSO_4(aq) + H_2(g); \quad \Delta_r G^0(298) = -160.3 \text{ kJ}$$

$$SiO_2 + 3\,C = SiC + 2\,CO(g); \quad \Delta_r G^0(1873) = -25.9 \text{ kJ}$$

10.2 Using Le Chatelier's Principle, predict the effect of the following changes on the equilibrium of the reaction.
  i.   An increase in the amount of one of the reactants
  ii.   Removal of a gaseous product from the system (if applicable)
  iii. Removal of a solid product from the system (if applicable)
  iv.   An increase in temperature of the system
  v.   A decrease in pressure of the system
  Reactions:

$$2\,SO_3(g) = 2\,SO_2(g) + O_2(g); \quad \Delta_r H^0(573) = 198 \text{ kJ}$$

$$H_2(g) + 0.5\,O_2(g) = H_2O(g); \quad \Delta_r H^0(573) = -489 \text{ kJ}$$

$$Zn(s) + H_2SO_4(aq) = ZnSO_4(aq) + H_2(g); \quad \Delta_r H^0(298) = -138 \text{ kJ}$$

$$CaCO_3 = CaO + CO_2; \quad \Delta_r H^0(1073) = 168 \text{ kJ}$$

$$SiO_2 + 3\,C = SiC + 2\,CO(g); \quad \Delta_r H^0(1873) = 588 \text{ kJ}$$

$$Si + C = SiC; \quad \Delta_r H^0(1273) = -72 \text{ kJ}$$

10.3 The solubility of O_2 and N_2 in water at 1 atm pressure and 0°C is, respectively, 0.029 g kg^{-1} and 0.069 g kg^{-1}. Calculate the concentration of O_2 and N_2 in ice water when it is in equilibrium with air at 1 atm. Oxygen and nitrogen dissolve in water in their molecular form and obey Henry's law.

10.4 The solubility of oxygen in molten silver at 1075°C has been measured as follows:

Pressure of oxygen, kPa	17.065	65.061	101.325	160.386
Dissolved oxygen, mass%	0.116	0.224	0.277	0.364

i. Determine whether oxygen dissolves in silver in the atomic or molecular form.

ii. What mass percent oxygen does silver absorb from the air if it is melted in an open crucible at 1075°C?

10.5 It is required to produce argon gas containing less than 10^{-8} volume percent oxygen. This can be achieved by passing a stream of argon over heated copper turnings in a tube furnace. Assuming equilibrium is achieved what is the maximum temperature of the copper? Cu_2O is the stable oxide of copper at high temperatures.

10.6 In a laboratory experiment tin oxide is mixed with powdered graphite (C), placed in a crucible and heated slowly to 1000 K in a muffle furnace.

i. At what temperature will the reaction $SnO_2(s) + C(s) = Sn(l) + CO_2(g)$ begin to occur spontaneously?

ii. If the atmosphere in the furnace is air, is there any advantage in adding a salt mixture cover over the SnO_2/C mixture? Assume the salt mixture melts to form a seal over the top of the mixture.

10.7 When copper sulfate ($CuSO_4.5H_2O$) is heated it releases the water of crystallisation to form anhydrous copper sulfate: $CuSO_4.5H_2O = CuSO_4 + 5H_2O$. Over the range 350–550 K, the Gibbs energy of reaction as a function of temperature is given by the approximate relation: $\Delta_r G^0 = 110.51 - 0.243\,T$ kJ.

i. What is the average enthalpy and entropy of reaction over the temperature range? Comment on the significance of these values.

ii. At what temperature will $CuSO_4.5H_2O$ begin rapid decomposition when heated in an open container?

iii. If the air contained 0.25 volume percent moisture, at what temperature would $CuSO_4.5H_2O$ begin to slowly decompose? $CuSO_4.5H_2O$ is actually stable at ambient temperatures. How can this be explained in the light of the answer just obtained?

10.8 In an experiment in a sealed container it was found that the equilibrium pressure above a sample of ammonium chloride was 726 kPa at 425°C and 1172 kPa at 450°C. Estimate the equilibrium constant of the reaction, the standard Gibbs energy of reaction, the standard enthalpy of reaction and the standard entropy of reaction, all at 425°C. Assume the vapour behaves ideally. Ammonium chloride dissociates according to the reaction: $NH_4Cl(s) = NH_3(g) + HCl(g)$.

10.9 The vapour pressure of zinc above molten zinc is given by the relation:

$$\ln p_{Zn}(\text{atm}) = -\frac{15250}{T} - 1.255 \ln T + 21.79$$

i. Calculate the vapour pressure of molten zinc at 50 K intervals between 700 and 1300 K, and plot them on a graph as a function of temperature.

ii.   Estimate the normal boiling point of zinc by interpolation, and mark the point on the graph.

iii. Calculate the enthalpy of volatilisation of zinc near the boiling point of zinc.

10.10 Magnesium metal can be made from magnesium oxide using silicon as the reductant: $2\,MgO + Si = 2\,Mg(g) + SiO_2$.

i.   Calculate the temperature required for the reaction to occur at atmospheric pressure.

ii.   If the reaction is to be performed at 1500 K, to what maximum pressure must the system be reduced for the reaction to occur?

iii. If CaO is added to the system, Ca_2SiO_4 rather than SiO_2 is formed: $2\,MgO + Si + 2\,CaO = 2\,Mg(g) + 2\,CaO.SiO_2$. Calculate the pressure required at 1500 K.

iv.   In practice, silicon in the form of ferrosilicon (an alloy of iron and silicon) containing about 80 mass% Si is used as the reductant. Assuming the activity coefficient of Si in Fe at this concentration is 0.9, calculate the pressure required.

10.11 It is proposed to remove zinc contamination from a batch of recycled lead by passing chlorine gas through the molten lead at 700 K to form $ZnCl_2$. Some $PbCl_2$ will also be produced and will form a melt consisting of $ZnCl_2$ and $PbCl_2$. If at the end of the process the mole fraction of $ZnCl_2$ in the melt is 0.90 to what concentration (mass%) can the Zn be lowered? The activity coefficient of Zn in Pb at low concentrations is about 30. Assume the $ZnCl_2 - PbCl_2$ melt behaves ideally.

10.12 A mixture of 1 mole each of iron sulfide and zinc metal powders is sealed in an inert container and heated in a furnace.

i.   Assuming no solid solutions form, will the reaction $FeS(s) + Zn(g) = ZnS(s) + Fe(s)$ occur at 1300°C and, if so, to what extent?

ii.   In fact ZnS and FeS form a solid solution. Assuming the solution is ideal, calculate its equilibrium composition at 1300°C. Assume there is excess zinc in the mixture.

a.   Data: At 1300°C, $G^0(Fe) = -87.025\ kJ$, $G^0(Zn, g) = -150.743\ kJ$, $G^0(FeS) = -305.625\ kJ$, $G^0(ZnS) = -362.897\ kJ$. The vapour pressure of zinc is 18.16 bar at 1300°C. Zinc is a vapour at 1300°C.

10.13 Trichlorosilane (HCl_3Si), a colourless and volatile liquid, is the principal precursor for the production of ultrapure silicon for the semi-conductor and photovoltaic industries. Trichlorosilane is produced by reacting impure, powdered metallurgical grade silicon with hydrogen chloride at around 300°C: $Si(s) + 3\,HCl(g) = HCl_3Si(g) + H_2(g)$. To produce silicon for photovoltaic cells, purified trichlorosilane can be decomposed by heating it (in the presence of silicon as nuclei) at around 1150°C: $2\,HCl_3Si(g) = Si(s) + 2\,HCl(g) + SiCl_4(g)$

i.   Using data from the NIST–JANAF database explain the thermodynamic basis for the temperatures for these reactions.

ii.   If trichlorosilane is heated in the presence of silicon at 1400 K, what would be the equilibrium amount of silicon deposited, and what would be the composition of the gas? Assume the total pressure is 1 bar. What is the percent recovery of silicon from the trichlorosilane?

In the following two problems, the objective is to use knowledge gained in this chapter to make general observations about the nature of the reactions. A detailed thermodynamic analysis of the proposed processes is possible, but it requires the use of Gibbs energy minimisation to investigate all the possible species that may form and the conditions that best optimise the process. Gibbs energy minimisation is introduced in Chapter 15, but to tackle these problems Gibbs energy minimisation software is needed.

10.14 The combustion of methane (natural gas) in air to generate heat occurs by the following reaction: $CH_4(g) + 2\,O_2(g) = CO_2(g) + 2\,H_2O(g)$. The product gas is diluted by nitrogen, and this would make recovery of the CO_2, for sequestration, expensive. This problem could be overcome by carrying out the reaction in two stages using an intermediate as the carrier of oxygen between the stages. The intermediate would be regenerated and recycled. Such a concept is called chemical looping. For the above reaction, iron (or another metal) can be used as the carrier. In step 1, metallic iron is oxidised in air to form magnetite: $Fe(s) + 2\ O_2(g) = Fe_3O_4(s)$. The magnetite is then transferred to a second reactor, step 2, where it reacts with methane $Fe_3O_4(s) + CH_4(g) = 3\ Fe(s) + CO_2(g) + 2\ H_2O(g)$. In this reaction a mixture of CO_2 and H_2O is produced. H_2O is easily separated from CO_2 by condensation leaving a pure stream of CO_2. Assume the reactions are carried out at 1 bar pressure.

    i.   What would be the composition (volume percent) of the combustion gas if methane was burned in air? Assume the stoichiometric amount of oxygen is used.

    ii.   What would be a suitable temperature for step 1? Explain.

    iii. What would be a suitable temperature for step 2? Explain.

    iv. Using the selected temperatures what is the approximate heat produced by the two-step process? Assume all heat produced in step 1 can be transferred to step 2. Compare this to the heat produced by burning methane in air.

    v.   The assumption that all heat produced in step 1 can be transferred to step 2 is unrealistic. Discuss this in terms of the limitations imposed by the second law.

10.15   Investigate whether it is thermodynamically feasible to produce a stream of pure hydrogen from methane by a looping process, again using iron as the intermediate. The potential reactions are:

$$Fe_3O_4(s) + CH_4(g) = 3\ Fe(s) + CO_2(g) + 2\ H_2O(g)$$

$$3\ Fe(s) + 4\ H_2O(g) = Fe_3O_4(s) + 4\ H_2(g)$$

    i.   Write the net equation. Is the overall process thermodynamically feasible?

    ii.   What would be a suitable temperature for step 1? Explain.

    iii. What would be a suitable temperature for step 2? Explain.

    iv. Is the process a net producer or consumer of thermal energy?

11 Gibbs energy applications to metal production

SCOPE

This chapter examines one application of Gibbs energy, namely the theory of the extraction of metals from naturally occurring oxide minerals.

LEARNING OBJECTIVES

1. *Understand the concept of stability of a metal oxide and the criteria for its reduction to metal.*
2. *Be able to calculate the thermodynamic conditions required to reduce a metallic oxide to metal using carbon, CO, H_2 or a more reactive metal as the reductant.*
3. *Be able to determine whether reduction of an oxide by carbon under specified conditions will produce the metal or metal carbide.*
4. *Understand the principles of fire-refining of impure metals.*

11.1 INTRODUCTION

As an example of the practical application of thermodynamics to reaction equilibria, this chapter examines its application to the production and refining of metals using high-temperature processing. This is but one area of application, and it was chosen because of the importance of metals to modern society and because the application of thermodynamic principles has made major impacts on the development and improvement of metal production processes.

Most metals occur naturally as oxides or sulfides (Table 11.1), and these compounds must be reduced to produce metals; that is, the cationic form of the metal in the mineral compound must be reduced to the atomic form, for example

$$Fe^{2+} - 2\,e^- = Fe$$

or

$$FeO = Fe + 0.5\,O_2$$

There are many ways to produce metals through the application of aqueous chemistry, electrochemistry and high-temperature processing. The route selected for a particular metal depends upon, among other factors, the stability and nature of the host mineral, the thermodynamically feasible reactions to produce the metal and the cost of processing. Thus, steel, copper, lead, magnesium, titanium and tin are produced predominately by high-temperature processing, aluminium by a combination of aqueous extraction and high-temperature electrochemistry and zinc by a combination of high-temperature processing, aqueous extraction and aqueous electrochemistry.

After an ore has been mined, if the concentration of the desired mineral is small the ore is processed using physical operations to liberate (by crushing and grinding), then separate (by froth flotation or gravity separation, for example) the unwanted waste minerals (gangue) to produce a *concentrate* with a high concentration of the wanted mineral. This concentrate is then processed chemically to produce metal of the required purity.

TABLE 11.1

Some common ore minerals from which metals are extracted

Metal	Oxide Minerals	Sulfide Minerals
Aluminium	Bauxite, $AlO(OH) + Al(OH)_3$	
	Alunite, $K_2O.3Al_2O_3.4SO_4.6H_2O$	
Chromium	Chromite, $(Fe,Mg)(Cr,Al)_2O_4$	
Copper	Cuprite, Cu_2O	Chalcopyrite, $CuFeS_2$
	Azurite, $2CuCO_3.Cu(OH)_2$	Chalcocite, Cu_2S
Iron	Hematite, Fe_2O_3	
	Magnetite, Fe_3O_4	
Magnesium	Dolomite, $(Ca,Mg)CO_3$	
Nickel	Garnierite, $(Mg,Ni)_6Si_4O_{10}(OH)_8$	Pentlandite, $(Fe,Ni)S$
Lead	Cerussite, $PbCO_3$	Galena, PbS
Silicon	Quartz, SiO_2	
Tin	Cassiterite, SnO_2	
Titanium	Rutile, TiO_2	
	Ilmenite, $FeTiO_3$	
Zinc		Sphalerite, ZnS

11.2   STABILITY OF OXIDES

The stability of a metal oxide relative to a mixture of the metal and oxygen is given by the magnitude of the Gibbs energy of formation for reactions of the type

$$2/x \, M + O_2 = 2/x \, MO_x \tag{11.1}$$

where M represents a metal, MO_x its oxide and x the stoichiometric amount of oxygen in the oxide (which is most commonly 1 but can be 0.5, 1.5, 2 and 2.5). From Equation 10.3

$$\Delta_f G^0 = -RT \ln K = -RT \ln \left(\frac{a_{MO_x}^{2/x}}{a_M^{2/x} \, p_{O_2}} \right) = RT \ln p_{O_2} - RT \ln \left(\frac{a_{MO_x}^{2/x}}{a_M^{2/x}} \right) \tag{11.2}$$

For the special case of pure metal and oxide,

$$\Delta_f G^0 = RT \ln p_{O_2} \tag{11.3}$$

The variation of $\Delta_f G^0$ with temperature for some important metal/metal oxide systems is shown in Figure 11.1. Also included are lines for the formation of gaseous CO and CO_2 from carbon and oxygen. Many of the relations are nearly linear, and, since $\Delta_f G^0 = \Delta_f H^0 - T\Delta_f S^0$, the slope of the line is the average value of $\Delta_f S^0$, and the intercept at 0 K is the average value of $\Delta_f H^0$ of the respective reactions.

The slopes of metal oxide lines in Figure 11.1 are similar; that is, the entropies of formation of metal oxides are negative and generally of comparable magnitude. The decrease in entropy (negative value of $\Delta_f S^0$) results mainly from the decrease in disorder due to the consumption of 1 mol of gaseous oxygen in forming the oxide from the metal. When metals and oxides melt there is a slight increase in their entropy. Therefore, when a metal or oxide melts the slope of the lines changes slightly. When a metal boils and turns to vapour its entropy increases greatly, and the line shows a correspondingly larger change in slope (for example, the lines for Zn/ZnO and Mg/MgO).

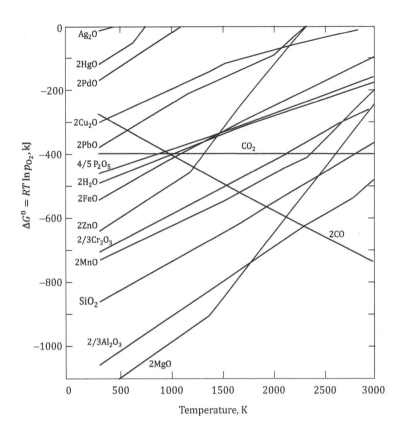

FIGURE 11.1 The variation with temperature of $\Delta_f G^0$ for reactions of the type $2 / x \, M + O_2 = 2 / x \, MO_x$. Note, $\Delta_f G^0$ is expressed on the basis of per mole of oxygen rather than per mole of metal oxide.

The lines for formation of CO and CO_2 have quite different slopes from the others. The reaction

$$C(s) + O_2(g) = CO_2(g) \tag{11.4}$$

has equal volumes of gas on each side of the equation, and, since the entropy of solid carbon is small compared with that of gases, there is little change in entropy on forming CO_2 from carbon and oxygen, and the line is nearly horizontal. For the reaction

$$2 \, C(s) + O_2(g) = 2 \, CO(g) \tag{11.5}$$

there is an increase in volume of gases. Therefore, entropy increases when CO is produced from carbon and oxygen, and the line for the reaction has an opposite slope to that of the lines for metal oxides. That the lines for CO and CO_2 intersect the lines for the metal oxides has important consequences for the extraction of metals from oxides.

The position of the lines in Figure 11.1 indicates the relative stability of oxides. The lower the line on the diagram the more negative is the standard Gibbs energy of formation of the oxide and the more stable is the oxide or, equivalently, the more reactive is the metal. Magnesium, for example, is a highly reactive metal, and magnesium oxide is a highly stable oxide. The order of stability of oxides can be demonstrated by estimating the *decomposition temperature* (or *dissociation temperature)* of oxides at 1 bar pressure of oxygen by reading the temperature at which the standard Gibbs energy of formation is equal to zero. At its decomposition temperature, an oxide will decompose spontaneously into the metal and oxygen. The approximate temperatures of dissociation of the relatively unstable oxides Ag_2O and HgO, for example, are 205 and 585°C, respectively, but most metallurgically important oxides must be heated to in excess of 2000°C before they will decompose. Clearly, this is not a practical way of producing metals from their oxides.

If the reactants and products are not in their standard states but in solution* in which their activities are less than unity (Equation 11.2),

$$\Delta_r G^0 = RT \ln p_{O_2} - RT \ln \left(\frac{a_{MO_x}^{2/x}}{a_M^{2/x}} \right)$$

$\Delta_r G^0$ varies approximately linearly with temperature (Equation 8.35),

$$\Delta_r G^0 = p + tT$$

where p and t are constants. Combining these equations,

$$p + tT = RT \ln p_{O_2} - RT \ln \left(\frac{a_{MO_x}^{2/x}}{a_M^{2/x}} \right)$$

and, therefore,

$$RT \ln p_{O_2} = p + \left[t + R \ln \left(\frac{a_{MO_x}^{2/x}}{a_M^{2/x}} \right) \right] T \qquad (11.6)$$

This is also a linear relation, with slope equal to $\left[t + R \ln \left(\frac{a_{MO_x}^{2/x}}{a_M^{2/x}} \right) \right]$. The effect of changing the activity of M or MO_x is to rotate the line of $RT \ln p_{O_2}$ *versus* T about the point $(0, p)$. If the activity of MO_x is less than one the slope decreases and the line is rotated clockwise. Similarly, a reduction in the activity of M results in an anti-clockwise rotation of the line. These rotations can change the apparent order of stability of oxides and metals. This has important implications as discussed later in the chapter.

11.3   REDUCTION REACTIONS

For an oxide to be reduced to metal, ΔG of Reaction 11.1 must be positive, since the reaction is required to move to the left. In terms of the van 't Hoff Isotherm (Equation 10.7), for the case of pure metal and pure oxide

$$\Delta G = -RT \ln \frac{1}{p_{O_2}^*} + RT \ln \frac{1}{p_{O_2}} = RT \ln p_{O_2}^* - RT \ln p_{O_2} > 0 \qquad (11.7)$$

where $p_{O_2}^*$ is the pressure of oxygen at which M and MO_x are at equilibrium (that is, they co-exist) and p_{O_2} is the actual pressure of oxygen in the system. Therefore, for reduction to take place the following condition must be met:

$$RT \ln p_{O_2} < RT \ln p_{O_2}^*$$

or, at constant temperature, $p_{O_2} < p_{O_2}^*$. In other words, for reduction to occur the oxygen potential or (oxygen partial pressure) of the surroundings of the oxide must be less than the oxygen potential due to the equilibrium of Reaction 11.1.

---

* In high-temperature processes (smelting processes), the solutions are usually molten metallic solutions (alloys), molten oxide solutions (slags), molten sufhide solutions (mattes) or solid-state solutions (Section 9.2).

The reduction of an oxide could in principle be achieved either physically or chemically. The pressure of the system could be lowered and/or the temperature raised until the oxygen potential of the oxide exceeds that of the surroundings. This is not a practical way except perhaps for a few relatively unstable oxides (such as Ag_2O and HgO) since the temperatures are too high or the pressures too low to be attained in large-scale processes. A better method is to react the oxide chemically with a substance which has a greater affinity for oxygen than the metal and which, therefore, maintains the oxygen potential at a lower value than that created by the M/MO_x equilibrium. Substances with a greater affinity for oxygen than the metal are called *reducing agents*. For large-scale reduction of oxides, relatively cheap and plentiful reducing agents are required, the major ones being carbon (usually in the form of coal or coke), carbon monoxide, hydrogen and reactive metals.

11.3.1 REDUCTION USING CARBON

Since the slopes of the lines for C/CO and C/CO_2 in Figure 11.1 are different from the metal/metal oxide lines there is a temperature for each metal oxide above which the lines for C/CO and C/CO_2 lie below the line for the metal oxide. At temperatures equal to or greater than the temperature at which they intersect a metal/metal oxide line the metal oxide can be reduced by carbon. This class of reactions is called *carbothermic reduction*. The situation is made more complicated by the fact that carbon forms two oxides,

$$2\,C(s) + O_2(g) = 2\,CO(g); \quad \Delta_f G^0 = -228\,780 - 171.6\,T \text{ Joules} \tag{11.8}$$

$$C(s) + O_2(g) = CO_2(g); \quad \Delta_f G^0 = -395\,350 - 0.54\,T \text{ Joules} \tag{11.9}$$

and when metal oxides are reduced by carbon the oxidation product is a mixture of CO and CO_2:

$$MO + C = M + CO \tag{11.10}$$

$$2\,MO + C = 2\,M + CO_2 \tag{11.11}$$

Here, for simplicity we've assumed the metals are divalent ($x = 2$).

Equations 11.8 and 11.9 can be combined to eliminate oxygen as follows:

$$C(s) + CO_2(g) = 2\,CO(g); \quad \Delta_r G^0 = 166570 - 170.6\,T \text{ Joules} \tag{11.12}$$

This is the *Boudouard reaction* encountered previously in Example 10.3. Changing the pressure by an order of magnitude has a considerable effect on the equilibrium of the Boudouard reaction, but changing the temperature has a much greater effect (the relation is shown graphically in Figure 10.4). At 1 atm pressure, the gas is effectively pure CO_2 at temperatures less than about 400°C while it is effectively pure CO at temperatures greater than about 1000°C.

The minimum temperature for the reduction of an oxide by carbon is given at the point of intersection of the carbon and metal lines in Figure 11.1. The method is approximate, since each carbon line assumes only one oxide of carbon is formed, and it can be used only in the limiting cases, at temperatures up to about 400°C and greater than about 1000°C. Approximate values for some common oxides are 75 and 315°C for Cu_2O and PbO, respectively (in the low-temperature region) and 1225, 1415, 1615 and 2000°C for Cr_2O_3, MnO, SiO_2 and Al_2O_3, respectively (in the high-temperature region).

The enthalpy of formation of CO is -114.4 kJ mol^{-1} O_2 while that of most oxides is between -300 and -1200 kJ. The latter is apparent from Figure 11.1 from the intercepts of the lines at $T = 0$ K. Furthermore, the enthalpy of formation of oxides is more negative the more stable is the oxide. Therefore, the carbothermic reduction of most oxides is endothermic and is more endothermic the more stable the oxide. Also, the temperature required for the reduction of oxides by carbon increases downwards in Figure 11.1; that is, the more stable the oxide the higher is the minimum temperature required for its reduction. The situation arises for the reduction of an oxide by carbon, therefore, that the more stable the oxide the more heat is required and at a higher temperature than for a less stable oxide. This has an important implication for the way heat is supplied for carbothermic reduction of the more stable oxides. As discussed in Section 6.5 electrical heating becomes more attractive the higher the temperature of a process. In practice, when oxides of the more reactive metals (typically those below iron in Figure 11.1) are reduced carbothermically the process is performed in electrically heated reactors rather than in furnaces using a fossil fuel for heating.

Many elements form stable carbides at high temperatures (Figure 11.2), and when oxides of these are reduced with carbon the reaction

$$y\,MO_x + y(x+y)\,C = y\,MC_y + xy\,CO \qquad (11.13)$$

can occur under appropriate conditions, and the carbide will form in preference to the metal. The relative stability of a metal and its carbide under the conditions for reducing an oxide by carbon is determined by the thermodynamics of the reaction:

$$y\,MO_x + x\,MC_y = (x+y)\,M + xy\,CO \qquad (11.14)$$

If the Gibbs energy of the reaction is negative the carbide is unstable relative to the oxide, and metal is the reduction product. If the Gibbs energy is positive the metal carbide will form in preference to the metal.

The variation of $\Delta_r G^0$ for Equation 11.14 for some common metals is shown in Figure 11.3. $\Delta_r G^0$ decreases for all metals as the temperature increases and attains a value of zero at temperatures attainable in practice (say a maximum of 1850°C) for Co, Fe, W, Mo, Cr, Mn and Si (in order of increasing temperature). These, therefore, can be produced as metals at a sufficiently high temperature by reduction of their oxides with carbon. However, if they are molten at the temperature of production they will contain dissolved carbon up to the level of saturation at that temperature, and for some of these metals the solubility of carbon is considerable. Of the more commercially important metals, V, Al, Ca, Ti and Zr can be produced only as carbides at realistically attainable temperatures when their oxides are reduced by carbon.

If the activity of M or partial pressure of CO is lowered the lines in Figure 11.3 are rotated clockwise and the reaction to form metal becomes thermodynamically more favoured. If the activity of MO_x and/or MC_y is lowered the lines are rotated anti-clockwise and the reaction is less favoured. Carbide-forming metals, therefore, can be produced as metals if they are formed at a sufficiently low activity, for example, by dissolving the metal as it forms in another metal to form a dilute alloy. Aluminium, for example, has been produced this way in the laboratory at practical temperatures by dissolving it as it forms in another metal. A separate process would then be required to separate the aluminium from the other metal.

11.3.2   REDUCTION WITH CARBON MONOXIDE AND HYDROGEN

Low oxygen potentials can be obtained by means of the reaction

$$2\,CO(g) + O_2(g) = 2\,CO_2(g); \quad \Delta_r G^0 = -561\,920 + 170.54\,T \text{ Joules} \qquad (11.15)$$

which is a combination of Reactions 11.8 and 11.9. Now,

$$K = \frac{p_{CO_2}^2}{p_{CO}^2\,p_{O_2}} \qquad (11.16)$$

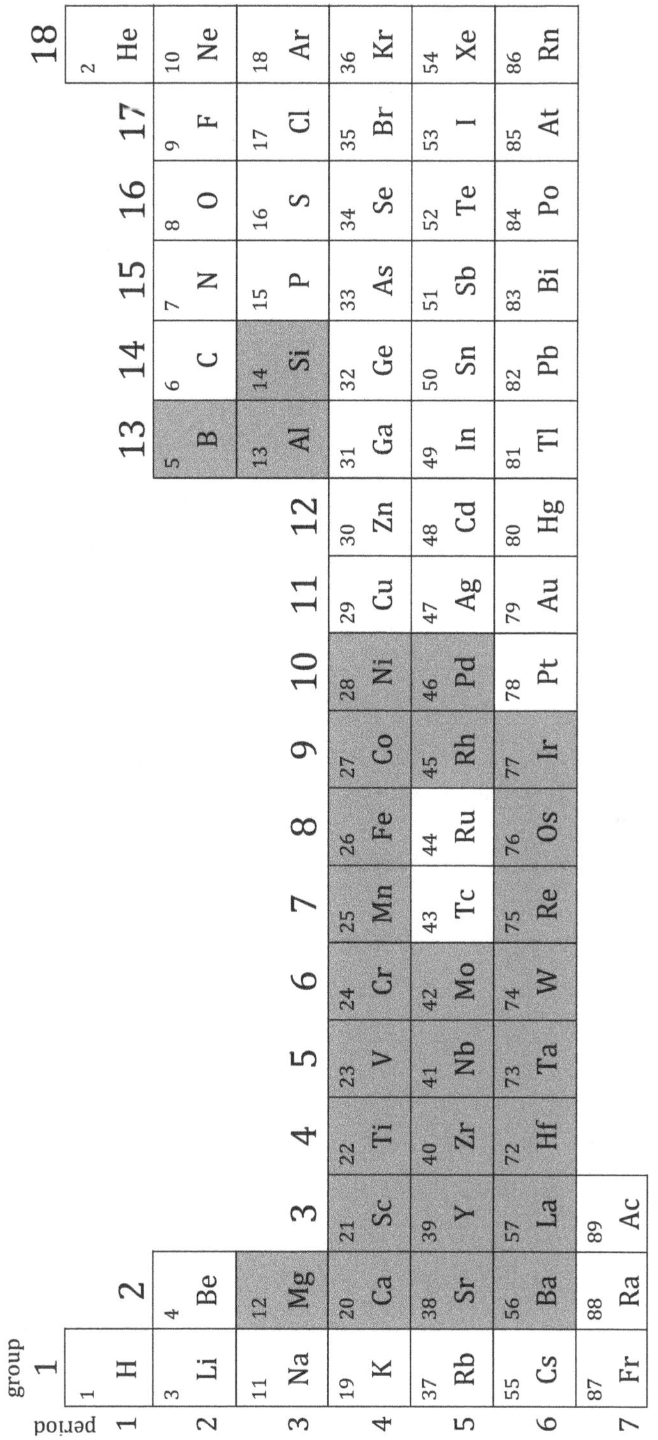

FIGURE 11.2 Elements with an affinity for carbon at high temperatures (shaded); the element either forms a solid carbide or a melt having a significant carbon solubility.

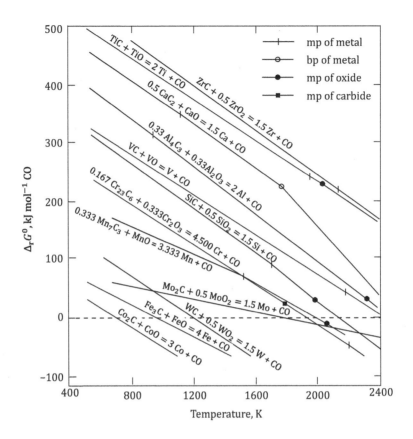

FIGURE 11.3 The variation with temperature of the standard Gibbs energy of the reaction $y\,MO_x + x\,MC_y = (x+y)\,M + xy\,CO$ for some common metals.

and taking logarithms, rearranging then multiplying through by RT gives

$$RT \ln p_{O_2} = 2\,RT \ln \frac{p_{CO_2}}{p_{CO}} - RT \ln K$$

$$= 2\,RT \ln \frac{p_{CO_2}}{p_{CO}} + \Delta G^0$$

$$= 2\,RT \ln \frac{p_{CO_2}}{p_{CO}} - 561\,920 + 70.54\,T$$

Therefore,

$$RT \ln p_{O_2} = \left(2\,R \ln \frac{p_{CO_2}}{p_{CO}} + 170.54 \right) T - 561\,920 \qquad (11.17)$$

The oxygen potential and, therefore, the partial pressure of oxygen, due to Reaction 11.15 is a function of the temperature and the ratio of the partial pressures of CO_2 and CO. The relation is shown graphically in Figure 11.4 in the form of lines of constant oxygen partial pressure. Also shown is the curve for the Boudouard reaction at 1 bar. Gas compositions within the area under the Boudouard

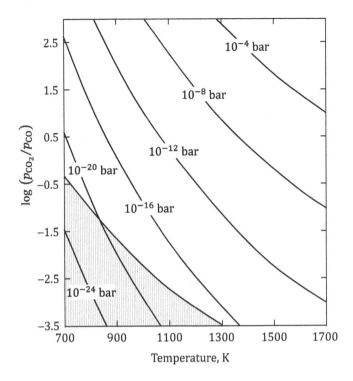

FIGURE 11.4 The variation of the partial pressure of oxygen with temperature and ratio of the partial pressures of CO and CO_2 calculated using Equation 11.17. The equilibrium of the Boudouard reaction at 1 bar is shown by the dashed line. Gas compositions below the Boudouard line are unstable and will precipitate carbon until the composition lies on the line.

curve are thermodynamically unstable, and a mixture in this region will react to precipitate carbon (the 'sooting' reaction) until its composition moves to the Boudouard curve. Under most conditions the reaction is sluggish, and it is actually possible for unstable gas mixtures to persist for some time, and, at room temperature, thermodynamically unstable mixtures of CO and CO_2 may be stored for considerable periods without reaction occurring.

In an exactly similar manner low oxygen potentials can be obtained by means of the reaction

$$2\,H_2(g) + O_2(g) = 2\,H_2O(g); \quad \Delta_r G^0 = -495\,000 + 111.8\,T \text{ Joules} \qquad (11.18)$$

Now,

$$K = \frac{p_{H_2O}^2}{p_{H_2}^2\,p_{O_2}} \qquad (11.19)$$

and proceeding as before,

$$RT \ln p_{O_2} = \left(2\,R \ln \frac{p_{H_2O}}{p_{H_2O}} + 111.8\right)T - 495\,000 \qquad (11.20)$$

The partial pressure of oxygen is again independent of the total pressure and varies only with temperature and the ratio of H_2 to H_2O. The relation is shown in Figure 11.5 in a similar form to that for CO and CO_2.

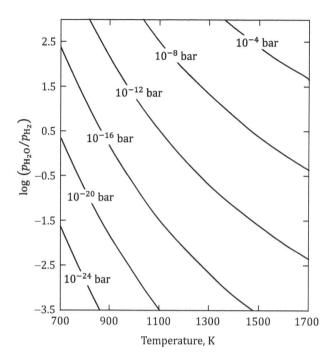

FIGURE 11.5 The variation of the partial pressure of oxygen with temperature and ratio of the partial pressures of H_2 and H_2O calculated using Equation 11.20.

EXAMPLE 11.1   Reduction of nickel oxide by gaseous CO and H_2

i. Calculate the maximum concentration of CO_2 allowable in gaseous CO so that NiO will be reduced to metal at 1200°C.

SOLUTION

The reduction of NiO by CO is given by the reaction:

$$NiO(s) + CO(g) = Ni(s) + CO_2(g)$$

which is a combination of the two reactions:

$$NiO(s) = Ni(s) + 0.5\,O_2(g); \ \Delta_f G^0 = 235\,600 - 86.10\,T \text{ Joules}$$
$$CO(g) + 0.5\,O_2(g) = CO_2(g); \ \Delta_r G^0 = -280\,960 + 85.26\,T \text{ Joules}$$
$$\overline{NiO(s) + CO(g) = Ni(s) + CO_2(g); \ \Delta_r G^0 = -453\,600 - 0.84\,T \text{ Joules}}$$

Taking the activities of NiO and Ni as unity, then

$$K = \frac{p_{CO_2}}{p_{CO}}$$

Therefore,

$$\Delta_r G^0 = -453\,600 - 0.84\,T = -RT \ln \frac{p_{CO_2}}{p_{CO}}$$

and, at 1473.15 K

$$\frac{p_{CO_2}}{p_{CO}} = 44.9$$

The significance of this value is that if the ratio of the partial pressure of CO_2 to the partial pressure of CO is less than 44.9, NiO will be reduced to nickel metal at 1200°C and that at ratios greater than 44.9 metallic nickel will oxidise to NiO. Since $p_B \propto x_B$ (Equation 3.6) and $x_B = \phi_B$ (Equation 3.4),

$$\frac{\phi_{CO_2}}{\phi_{CO}} = 44.9$$

But $\phi_{CO_2} + \phi_{CO} = 1$ since p_{O_2} is extremely small; therefore solving the equations simultaneously

$$\phi_{CO} = \frac{1}{(44.9+1)} = 0.0218$$

Thus the equilibrium volumetric composition of the gas is 2.18 vol% CO and 97.82 vol% CO_2.

ii. Calculate the maximum concentration of H_2O allowable in gaseous hydrogen so that NiO will be reduced to metal at 1200°C.

The reduction of NiO by CO is given by the reaction:

$$NiO(s) + H_2(g) = Ni(s) + H_2O(g)$$

which is a combination of the two reactions:

$NiO(s) = Ni(s) + 0.5\,O_2(g);$ $\qquad \Delta_f G^0 = 235\,600 - 86.10\,T$ Joules

$H_2(g) + 0.5\,O_2(g) = H_2O(g);$ $\qquad \Delta_f G^0(J) = -247\,500 + 55.90\,T$ Joules

$NiO(s) + H_2(g) = Ni(s) + H_2O(g);$ $\quad \Delta_r G^0(J) = -11\,900 - 30.20\,T$ Joules

Therefore,

$$\Delta_r G^0 = -11\,900 - 30.20\,T = -RT \ln \frac{p_{H_2O}}{p_{H_2}}$$

At 1473.15 K, $\dfrac{p_{H_2O}}{p_{H_2}} = 99.89$. Proceeding as before, the equilibrium composition of the gas is 99.01 vol% H_2O and 0.99 vol% H_2.

COMMENTS

In the above examples, if pure CO or H_2 were used to reduce NiO, then the respective equilibrium CO and H_2O concentrations represent the unusable amounts of CO and H_2. These would be lost unless the equilibrium gases were collected, purified to remove the CO_2 and H_2O, then reused.

If equilibrium values of p_{CO_2}/p_{CO} and p_{H_2O}/p_{H_2} are calculated over a range of temperatures then the relations shown in Figure 11.6 are obtained. At high temperatures, CO and H_2 are slightly less efficient at reducing NiO than at lower temperatures because the proportion of unusable CO or H_2 in the gas mixture is higher. To take an extreme case, at 1400°C the equilibrium value of p_{CO_2}/p_{CO} is 29 and the volume of CO in the gas mixture is 3.33 vol%. At low temperatures, H_2 is a more efficient reducing agent than CO, but at temperatures above about 860°C, CO becomes more efficient.

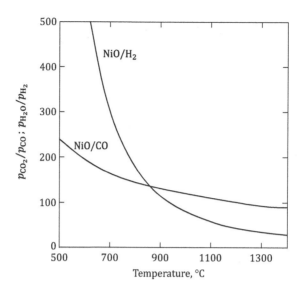

FIGURE 11.6 The variation of p_{CO_2}/p_{CO} and p_{H_2O}/p_{H_2} with temperature for NiO/CO and NiO/H$_2$ equilibrium, respectively.

11.3.3 Reduction using another metal

Metal M$_1$ will reduce the oxide of another metal, M$_2$, which lies above it in Figure 11.1 since $\Delta_r G^0$ of the reaction

$$M_1 + M_2O = M_1O + M_2 \tag{11.21}$$

will in that case be negative. This type of reaction is called *metallothermic reduction*. For example, the line for Si/SiO$_2$ lies below that of Cr/Cr$_2$O$_3$ at all temperatures, and the reaction

$$Si + 2/3\ Cr_2O_3 = SiO_2 + 4/3\ Cr$$

is thermodynamically favourable at all temperatures though, in reality, for kinetic reasons the reaction will proceed only when a certain minimum temperature is exceeded. The standard Gibbs energy of the reaction can be obtained by combining the standard Gibbs energies for the reactions for formation of Cr$_2$O$_3$ and SiO$_2$:

2/3 Cr$_2$O$_3$(s) = 4/3 Cr(s) + O$_2$(g);	$\Delta_f G^0 = 746\ 840 - 173.2\ T$ J	(11.22)
Si(s) + O$_2$(g) = SiO$_2$(s);	$\Delta_f G^0 = -902\ 070 + 173.6\ T$ J	(11.23)
Si(s) + 2/3 Cr$_2$O$_3$(s) = SiO$_2$(s) + 4/3 Cr(s);	$\Delta_r G^0 = -155\ 230 + 0.4\ T$ J	(11.24)

Clearly, $\Delta_r G^0$ is negative at all temperatures of practical interest. The average enthalpy of reaction is $-155\ 230$ J mol^{-1} of silicon (since $\Delta_r G^0 = \Delta_r H^0 - T\Delta_r S^0$), and, therefore, the reaction is exothermic. Metallothermic reactions are always exothermic, and many reactions of practical interest are self-sustaining and require no additional heat other than that to raise the reactants to the ignition temperature. Metallothermic reduction is an expensive process as it consumes large quantities of reactive metal, and metallothermic reduction can be used economically to produce only those metals that are considerably more valuable than the reactive metal used for the reduction.

11.4   OXIDATION REACTIONS

Oxidation is the reverse of reduction, for example,

$$Fe - 2\,e^- = Fe^{2+}$$

or

$$Fe + 0.5\,O_2 = FeO$$

From the point of view of the high temperature extraction of metals, the importance of oxidation is that it provides a method for removing dissolved impurity elements from metals. This is done at high temperatures (at which the metal is molten) and is called *fire-refining*. The most common oxidising agent used is oxygen, but other reactants are also used, such as chlorine. Oxygen may be in a variety of forms: Air, gaseous oxygen, CO/CO_2 or H_2/H_2O mixtures, the oxide of a less reactive metal or some other compound. For oxidation of an element to occur, ΔG of Reaction 11.1 must be negative; that is, the oxygen potential of the surroundings must exceed the equilibrium oxygen potential for the metal/metal oxide.

In fire-refining, impurity elements are converted to oxides which separate from the metal as solids or liquids (and dissolve in the slag phase) or as gas. Any impurity element in molten iron which lies below iron in Figure 11.1 can be removed to very low concentrations because, being more reactive, they react in preference to the iron with oxygen. In steelmaking, elements such as carbon, silicon and manganese can be oxidised from molten iron for this reason. When the line for an impurity element lies close to or even a little above the line for the impure metal it is still possible to oxidise the impurity selectively from the host metal by adjusting the activities of the components of the reaction.

The selective oxidation of phosphorus from molten iron, an important reaction in steelmaking, demonstrates the effect of changing the activities. It is apparent from Figure 11.1 that the selective oxidation of phosphorus is not possible when the metal and metal oxide are in their standard states because the line for phosphorus lies above that of iron at the temperatures at which the reaction would be performed (~1600°C). In steelmaking, the phosphorus is actually present at a low concentration in the iron, and the formation of P_2O_5 from a 1 mass% solution of phosphorus in iron is represented by line (a) in Figure 11.7. In practice, a slag is formed in which the P_2O_5 and FeO can dissolve and have an activity of less than one. In typical steelmaking slags the activity of P_2O_5 is of the order 10^{-20}. When this value is substituted in Equation 11.6 it has the effect of rotating line (a) to position (b) so that it now lies well below line (x) for pure Fe/pure FeO. This means that the phosphorus can now be oxidised selectively from the iron. A complication is that FeO also dissolves in the slag. However, the activity coefficient of FeO in the slag is much greater than that for P_2O_5, and the activity of FeO is typically around 0.2. Substitution of this value in Equation 11.6 causes line (x) to rotate clockwise slightly to position (y) but without changing the overall picture. The effect of oxidising the phosphorus and lowering its concentration in iron below 1 mass% causes line (b) to rotate anti-clockwise to positions (c) and (d) at 0.1 and 0.01 mass%P, respectively. Total elimination of phosphorus from the iron is not possible since a concentration is reached at which the line crosses that for FeO formation at the oxidation temperature and from then on the iron is oxidised preferentially.

This example illustrates the important point that an impurity element can be eliminated down to a certain concentration but can never be entirely eliminated. This is examined further in Section 14.4.

If it was attempted to refine an impure form of aluminium by oxidation reactions, elements below the line for aluminium could be removed by oxidation. However, most elements lie above aluminium (see Figure 11.1), which is a reactive metal, and oxidation is not a practical method for refining in this case. One way around this problem for reactive metals is to first produce a pure oxide, or

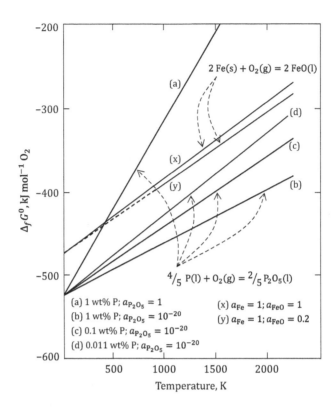

FIGURE 11.7 The oxidation of phosphorus from iron.

some other compound of the metal, then reduce it, for example by electrowinning (Section 17.1.1) to produce a metal sufficiently pure that it needs little or no further refining.

11.5 METAL PRODUCTION STRATEGY

We can now draw the above theory together to give a thermodynamic basis for metal production. For this purpose, metals can be divided into three classes: The reactive metals, the noble (inert) metals and those in between (see Figure 11.8). The compounds of the reactive metals are the most stable and the most difficult to reduce. Usually, the production of a metal includes refining steps, in addition to reduction, to remove unwanted impurities in gangue minerals or in the crystal structure of the ore mineral. In broad terms, either the concentrate or ore is treated chemically to produce a pure compound which is then reduced to a pure metal, or the concentrate is treated chemically to produce an impure metal which, if necessary, is then refined to the required purity. The two approaches are illustrated in Figure 11.9. The refine–reduce sequence is most commonly used for the production of highly reactive metals, and the reduce–refine sequence is mainly used for less reactive metals. The reason for this is that an impure form of a reactive metal is difficult to refine using relatively simple oxidation reactions since the host metal, being more reactive than most of the impurity elements, oxidises preferentially. Hence, it is usually preferable to remove most impurities before the metal is reduced.

Noble metals (Au, Ag, Pt, etc), which usually occur in nature in the elemental form, are extracted using physical separation techniques (such as gravity concentration) or, if they occur with metals such as copper or lead, extracted as dilute solutes in the copper or lead when they are extracted and then separated during the refining step. Gold can also be extracted directly using a sodium cyanide solution to form a soluble gold cyanide complex.

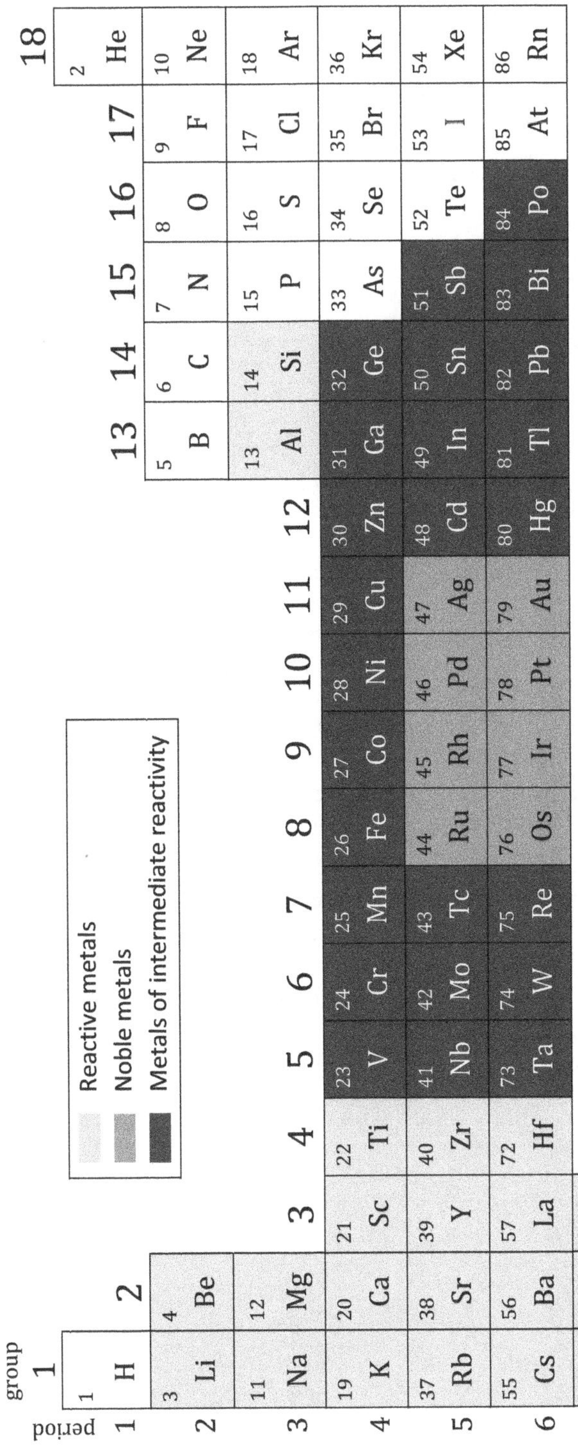

FIGURE 11.8 Periodic table divided into thermodynamically reactive metals, noble metals and metals of intermediate reactivity.

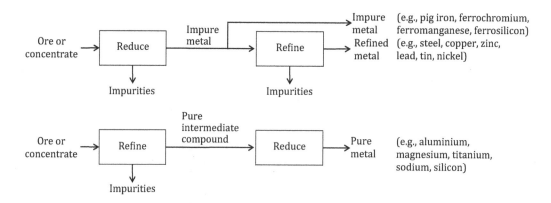

FIGURE 11.9 The reduce–refine and refine–reduce sequences for the production of metals.

PROBLEMS

11.1 Calculate the temperature at which Ag_2O will begin to decompose to metallic silver and oxygen when it is heated in air. At what temperature will decomposition become rapid?

$$2\, Ag(s) + 0.5\, O_2(g) = Ag_2O(s);\ \ \Delta G^0 = -30\, 500 + 66.1\, T \text{ Joules}$$

11.2 Calculate CO/CO_2 and H_2/H_2O equilibria for the reduction of SiO_2 to silicon metal at 2000 K. Why are CO and H_2 not suitable reductants for silica?

11.3 Calculate the maximum allowable concentration of H_2O vapour in a stream of hydrogen used to reduce tungsten oxide (WO_3) to metal at 900 K. Is hydrogen a suitable reductant for WO_3?

11.4 Determine under what condition Cr_2O_3 can be reduced to metal at 1300 K without forming $Cr_{23}C_6$. At 1300 K, $\Delta_f G^0\left(Cr_2O_3\right) = -801.4$ J; $\Delta_f G^0\left(Cr_{23}C_6\right) = -432.9$ J; $\Delta_f G^0(CO) = -226.5$ J.

11.5 Explain the trend illustrated in Figure 11.6 that H_2 is a more efficient reductant for oxides than CO at low temperatures but CO is more efficient at higher temperatures.

11.6 Alumina (Al_2O_3) is reduced to metal with carbon in the presence of molten tin at 2000 K and total pressure of 0.1 bar. What is the equilibrium activity of Al in the tin assuming no other reactions occur? Given that the activity coefficient of Al in tin at 2000 K is about 1.2, what is the concentration of the Al? By examining the Al–Sn phase diagram suggest how the Al and Sn could be separated. Could pure Al be produced?

12 Electrolyte solutions

SCOPE

This chapter examines the thermodynamics of solutions containing ions and shows how the concepts of enthalpy, entropy, Gibbs energy and activity are applied to ionic species in solutions.

LEARNING OBJECTIVES

1. *Understand the nature of electrolytes and electrolyte solutions, in particular aqueous solutions.*
2. *Understand the methodology by which values can be measured and assigned to $\Delta_f H^0$, $\Delta_f S^0$ and $\Delta_f G^0$ for ions in aqueous solutions.*
3. *Understand the concept of the activity of ions and of mean ionic activity.*
4. *Be able to estimate the activity of ions and electrolyte solutes in dilute solutions for fully dissociated and partially dissociated electrolytes.*

12.1 INTRODUCTION

Chapter 9 examined the thermodynamics of solutions of neutral species. In this chapter we consider the thermodynamics of solutions containing charged species. The thermodynamic principles developed in Chapter 9 (chemical potential, activity, etc.) apply to these solutions, but their application is made complicated by the fact that the charged species interact strongly with each other. The charged species we are referring to are atoms or molecules that, through loss or gain of electrons, have a net electrical charge. These are called *ions*. Positively charged ions, with fewer electrons than protons, are called *cations*, while negatively charged ions, with more electrons than protons, are called *anions*. Compounds which dissociate into ions when they are dissolved in solutions, heated above a certain temperature or when they melt are called *electrolytes**. The process by which a compound dissociates, partially or completely, into ions is called *ionisation*. The most familiar electrolyte solutions are solutions of certain compounds in water, and aqueous solutions are the focus of this chapter. However, the general principles presented are applicable to all electrolytes – solid electrolytes, ionic melts (molten salts and slags) and low-temperature ionic liquids (organic compounds).

Ions in solutions or melts do not behave as independent entities because of the forces between them due to electrostatic interactions. As a result, electrolyte solutions deviate from Henrian behaviour (Section 9.5.3) at much lower concentrations than do solutions of neutral solutes. They only begin to behave in a Henrian manner when the solution is sufficiently dilute that individual ions are far enough apart as to not interact with each other. As a result of the many complex interactions, no satisfactory theory of aqueous electrolyte solutions has been developed which enables accurate prediction of the thermodynamic properties of the solutions over wide ranges of concentration, and approximations have to be made.

12.2 AQUEOUS SOLUTIONS

Water dissolves many salts, hydrophilic organic molecules, such as sugars and simple alcohols, and most acids. The two hydrogen atoms in the water molecule are bonded to the oxygen atom at

---

* A solution containing ions is itself sometimes referred to as an electrolyte. What is being referred to is usually clear from the context of the discussion. In this chapter, the term electrolyte refers to compounds.

an angle of 104.45°, and water thus behaves as a polar compound – each molecule is an electric dipole. This results in relatively weak bonding between the molecules (hydrogen bonds) in which the hydrogen atoms function as a bridge between adjacent oxygen atoms. When certain types of solid substances are added to water, the hydrogen bonds are disrupted at the solid–liquid interface thereby releasing water dipoles. These are attracted to the points of charge on the solid and, if these attractions are sufficiently large to overcome the inter–ionic attractions in the solid, the substance dissolves to form ions in solution. These ions are then surrounded by water molecules which form a relatively ordered structure close to the ion, called the inner co-ordination sphere. This structure progressively becomes less ordered towards the bulk of the solvent. The resulting solution remains electrically neutral as the sum of charges on cations is balanced by those on anions.

Strong electrolytes dissociate completely into ions, while *weak electrolytes* dissociate only partially. The principal solute species in a solution of a strong electrolyte are ions, while the principal species in a solution of a weak electrolyte is the un-dissociated compound. All ionic compounds are strong electrolytes. Even nearly insoluble ionic compounds (such as $AgCl$, $PbSO_4$, $CaCO_3$) are strong electrolytes because the small amounts that do dissolve in water do so principally as ions; that is, there is virtually no undissociated form of the compound in solution. Molecular compounds may be non-electrolytes (e.g., sucrose and ethanol), weak electrolytes or strong electrolytes. Strong acids and strong bases are strong electrolytes. Weak electrolytes include weak acids and weak bases. Some examples of aqueous electrolytes are listed in Table 12.1.

Water, itself, is a very weak electrolyte. The ionisation (actually self-ionisation) of water is a reaction in pure water, or in an aqueous solution, in which water molecules deprotonate (lose the nucleus of one of their hydrogen atoms) to become hydroxide ions, OH^-. The hydrogen nuclei then protonates another water molecule to form a hydronium ion, H_3O^+. By convention, the hydronium ion is abbreviated as H^+ with the understanding that both the solvated H^+ and OH^- ions are complexes, stabilised by surrounding water molecules. Thus, for the self–ionisation of water we write the reaction as:

$$H_2O = H^+ + OH^-$$

Being a very weak electrolyte, the equilibrium concentrations of H^+ and OH^- are very low relative to water molecules.

For thermodynamic purposes the concentration of electrolyte solutes and ions in dilute aqueous solutions is expressed in terms of molality (Equation 2.6) rather than mole fraction, and the activities of aqueous species are expressed relative to the infinitely dilute standard state (Section 9.5.3). In aqueous systems these conventions are indicated by adding (aq) after the symbol or formula for the ion, for example, $CuSO_4(aq)$, $H^+(aq)$.

Consider the electrolyte $A_{\nu_+}B_{\nu_-}$ where A is a cation and B is an anion and ν_+ and ν_- are the number of cations and anions per formula unit of the electrolyte. When added to water the electrolyte will dissolve and dissociate according to the equilibrium:

$$A_{\nu_+}B_{\nu_-}(s) = \nu_+ A^{z+}(aq) + \nu_- B^{z-}(aq) \tag{12.1}$$

TABLE 12.1

Some examples of common strong and weak electrolytes in aqueous solutions

Strong electrolytes	Strong acids	HCl, HBr, HI, HNO_3, H_2SO_4
	Strong bases	NaOH, KOH, LiOH, $Ba(OH)_2$, and $Ca(OH)_2$
	Salts	NaCl, KBr, $MgCl_2$
Weak electrolytes	Weak acids	HF, CH_3COOH (acetic acid), H_2CO_3 (carbonic acid), H_3PO_4 (phosphoric acid)
	Weak bases	Water, NH_3 (ammonia), C_5H_5N (pyridine)

where z^+ and z^- are the charge of the cation and anion, respectively and (aq) indicates the infinitely dilute standard state. For example, $CaCl_2$ will dissociate according to the equilibrium:

$$CaCl_2(s) = Ca^{2+}(aq) + 2\,Cl^-(aq)$$

The concentrations of ions in solution are related through the stoichiometry of the dissociation reaction. In the above example, the concentrations of Ca^{2+} and Cl^- ions in water are related by the equation

$$m_{Cl^-} = 2m_{Ca^{2+}}$$

and the concentration of H^+ and OH^- ions by the equation

$$m_{H^+} = m_{OH^-}$$

through the reaction

$$H_2O = H^+ + OH^-$$

Assuming complete dissociation of $CaCl_2$, if the molality of the solution is 0.1 mol kg^{-1} then $m_{CaCl_2} = m_{Ca^{2+}} = 0.5\,m_{Cl^-} = 0.1$. In general,

Strong electrolytes: $\qquad\qquad\qquad m_i = \nu_i\,m_e$ $\qquad\qquad\qquad$ (12.2)

where m_i is the molality of ion i in the solution, ν_i is the number of ions per formula unit of the electrolyte, and m_e is the molality of the electrolyte solution. Equation 12.2 does not apply for partially dissociated electrolytes. This case will be considered later.

Solutions are electrically neutral, so for solutions containing ions the number of positive charges equals the number of negative charges. *Charge balance equations* express this electrical neutrality by equating the molar concentrations of the positive and negative charges:

$$\sum_i m_i z_i = 0 \qquad\qquad\qquad (12.3)$$

where m_i is the molality of ion i and z_i is the charge on ion i. For example, a solution of $CaCl_2$ in water contains Ca^{2+}, Cl^-, H^+ and OH^- ions. The charge balance equation, therefore, is:

$$2m_{Ca^{2+}} + m_{H^+} - m_{Cl^-} - m_{OH^-} = 0$$

12.3   ENTHALPY, GIBBS ENERGY AND ENTROPY OF IONS IN AQUEOUS SOLUTIONS

The standard enthalpy, entropy and Gibbs energy changes of a reaction involving ions in solution can be calculated using values of the standard enthalpies, entropies and Gibbs energies of formation in the same way as for reactions involving compounds (as in respectively, Examples 4.4, 7.3 and 8.4). In this case, however, the values of $\Delta_f H^0$, $\Delta_f S^0$ and $\Delta_f G^0$ refer to the formation of ions from elements, in their relevant reference state, for example, for silver ions,

$$Ag(s) = Ag^+(aq) + e^-$$

$\Delta_f H^0$, $\Delta_f S^0$ and $\Delta_f G^0$ are related in the conventional way through the relation $\Delta_f G^0 = \Delta_f H^0 - T\Delta_f S^0$.

It is not possible to form solutions of single ions since cations and anions are always associated; for example, a solution of hydrochloric acid contains both H^+ and Cl^- ions:

$$HCl(g) = H^+(aq) + Cl^-(aq)$$

Thus, measurement of enthalpy of formation of, for example, Cl^- ions in isolation,

$$0.5\ Cl_2(g) - e^- = Cl^-(aq)$$

is not possible. This problem is overcome by adopting the convention that in aqueous solutions the hydrogen ion at unit activity has zero enthalpy of formation:

Convention: $\Delta_f H^0(H^+, aq) = 0$, at all temperatures

Then, for the reaction

$$0.5\ H_2(g) + 0.5\ Cl_2(g) = H^+(aq) + Cl^-(aq)$$

$$\Delta_r H^0 = \Delta_f H^0(H^+, aq) + \Delta_f H^0(Cl^-, aq) - 0.5\Delta_f H^0(H_2, g) - 0.5\Delta_f H^0(Cl_2, g)$$

Enthalpy of reaction is a measurable quantity (for example by calorimetry; Section 4.10) and at 25°C, $\Delta_r H^0 = -167.16$ kJ mol^{-1}. $\Delta_f H^0(H_2, g)$ and $\Delta_f H^0(Cl_2, g)$, both being the enthalpy of formation of elements, are equal to zero at all temperatures. By convention $\Delta_f H^0(H^+, aq) = 0$; therefore $\Delta_f H^0(Cl^-, aq) = -167.16$ kJ mol^{-1}.

The same approach is used to define the standard Gibbs energy of formation of ions by adopting the convention:

Convention: $\Delta_f G^0(H^+, aq, a = 1) = 0$, at all temperatures

Thus, for the reaction

$$0.5\ H_2(g) + 0.5\ Cl_2(g) = H^+(aq) + Cl^-(aq)$$

$$\Delta_r G^0 = \Delta_f G^0(H^+, aq) + \Delta_f G^0(Cl^-, aq) - 0.5\Delta_f G^0(H_2, g) - 0.5\Delta_f G^0(Cl_2, g)$$

Gibbs energy of reaction is a measurable quantity (Section 8.10), and at 25°C, $\Delta_r G^0 = -131.23$ kJ mol^{-1}. $\Delta_f G^0(H_2, g)$ and $\Delta_f G^0(Cl_2, g)$, both being the Gibbs energy of formation of elements, are equal to zero at all temperatures. By convention $\Delta_f G^0(H^+, aq) = 0$; therefore $\Delta_f G^0(Cl^-, aq) = -131.23$ kJ mol^{-1}.

Since $\Delta_f G^0 = \Delta_f H^0 - T\Delta_f S^0$, and $\Delta_f H^0(H^+, aq)$ and $\Delta_f G^0(H^+, aq)$ are both zero at all temperatures, it follows that:

$$\Delta_f S^0(H^+, aq, a = 1) = 0$$, at all temperatures

In the above example, the entropy of formation of Cl^- ions at 25°C is:

$$\Delta_f S^0 = \frac{\Delta_f H^0 - \Delta_f G^0}{T} = \frac{-167.16 - (-131.23)}{298.15} \times 1000 = -120.41\ J\ mol^{-1}$$

By yet another convention, the entropy of hydrogen ions at unit activity in water is taken as zero:[*]

Convention: $S^0(H^+, aq, a = 1) = 0$, at all temperatures

This enables determination of the entropies of ions relative to the hydrogen ion:

$$0.5\, H_2(g) + 0.5\, Cl_2(g) = H^+(aq) + Cl^-(aq)$$

$$\Delta_r S^0 = S^0(H^+, aq) + S^0(Cl^-, aq) - 0.5\, S^0(H_2, g) - 0.5\, S^0(Cl_2, g)$$

At 25°C, $\Delta_r S^0 = -120.41\,\text{J mol}^{-1}$, $S^0(H_2, g) = 130.68\,\text{J mol}^{-1}$, and $S^0(Cl_2, g) = 223.12\,\text{J mol}^{-1}$. By convention $S^0(H^+, aq) = 0\,\text{J mol}^{-1}$; therefore $S^0(Cl^-, aq) = -56.49\,\text{J mol}^{-1}$. A positive value of entropy means an ion has a higher entropy than hydrogen ions in water, and a negative value of entropy means an ion has a lower entropy than hydrogen ions in water.

These values can now be used to determine formation functions for other ions. To illustrate how this is done, consider the reaction

$$CaCl_2(s) = Ca^{2+}(aq) + 2Cl^-(aq)$$

for which $\Delta_r H^0$ has been determined to be $-81.35\,\text{J mol}^{-1}$. Now

$$\Delta_r H^0 = \Delta_f H^0(Ca^{2+}, aq) + 2\Delta_f H^0(Cl^-, aq) - \Delta_f H^0(CaCl_2, s)$$

Using the known (tabulated) value of $\Delta_f H^0(CaCl_2, s) = -795.80\,\text{kJ mol}^{-1}$ and the value of $\Delta_f H^0(Cl^-, aq)$ obtained above,

$$\Delta_f H^0(Ca^{2+}, aq) = -81.35 - 2 \times (-167.16) + (-795.80) = -542.83\,\text{kJ mol}^{-1}$$

Values for $\Delta_f G^0(Ca^{2+}, aq)$ and $S^0(Ca^{2+}, aq)$ can be determined in similar fashion. The procedure can be extended to include other ions.

By convention the heat capacity of hydrogen ions at unit activity is taken to be zero:

Convention: $C_p(H^+, aq, a = 1) = 0$, at all temperatures

Determination of values of $\Delta_f H^0$, $\Delta_f G^0$ and $\Delta_f S^0$ at higher temperatures is done in the same manner as for pure substances from a knowledge of ΔC_p for the reaction (calculated from the C_p values of the reactants and products) and the value of $\Delta_f H^0$, $\Delta_f G^0$ or $\Delta_f S^0$ at 298.15 K:

$$\Delta_f H^0(T) = \Delta_f H^0(298.18) + \int_{298}^{T} \Delta C_p\, dT$$

$$\Delta_f S^0(T) = \Delta_f S^0(298.18) \int_{298}^{T} \frac{\Delta C_p}{T}\, dT$$

$$\Delta_f G^0(T) = \Delta_f H^0(T) - T\Delta_f S^0(T)$$

---

[*] The absolute (or third law) entropy of hydrogen ions in water has been estimated by structural modelling to be about -21 kJ mol^{-1}. The negative value means that H$^+$ ions increase ordering within the solvent.

12.3.1 Sources of thermodynamic data for aqueous solutions

Published values of $\Delta_f H^0$, $\Delta_f G^0$ and S^0 at 25°C and of C_p for many ionic species are available in tabular format in many compilations. A readily accessible and comprehensive source is:

> Wagman, D. D. et al. 1982. The NBS tables of chemical thermodynamic properties. *Journal of Physical and Chemical Reference Data*, 11, Supplement 2.

An open-access copy (in PDF format) can be downloaded from the NIST website: https://srd.nist.gov/JPCRD/jpcrdS2Vol11.pdf.

The following standard states are used in the NBS tables:

- For undissociated solutes (non-electrolytes) and ions in aqueous solution the standard state is the (hypothetical) state of the solute at a molality of 1 mol kg⁻¹ of water and pressure of 1 bar and exhibiting infinitely dilute solution behaviour. This state is indicated by 'ao' in the State column of the table.
- For electrolytes in solution the standard state is the (hypothetical) state of the solute at a mean ionic molality* of 1 mol kg⁻¹ of water and pressure of 1 bar and exhibiting infinitely dilute solution behaviour. The electrolyte is assumed to be fully dissociated at infinite dilution. This state is indicated by 'ai' in the State column of the table. This has the following implication:

A thermodynamic property of a fully dissociated electrolyte in solution is the sum of the corresponding property of the ions of which the electrolyte is constituted.

For example, for Na_2SO_4,

$$Na_2SO_4(ai) = 2\,Na^+(ao) + SO_4^{2-}(ao)$$

$$\Delta G^0(Na_2SO_4, ai) = 2 \times \Delta G^0(Na^+, ao) + \Delta G^0(SO_4^{2-}, ao)$$

$$= 2 \times -261.905 - 744.53 = 1268.34\ kJ$$

For weak electrolytes, values of thermodynamic properties are frequently quoted for both the fully dissociated and un-dissociated states. For the former, the notation ai is used; for the latter, ao is used.

Another comprehensive open-access source of data for ions in aqueous solutions is the US Geological Survey (USGS) compilation:

> Robie, R. A. and B. S. Hemingway. 1995. Thermodynamic properties of minerals and related substances at 298.15 K and 1 Bar (10^5 pascals) pressure and at higher temperatures. *U. S. Geological Survey Bulletin*, no. 2131. Washington: U.S. Geological Survey.

A scanned copy in PDF format can be downloaded from the USGS website: https://pubs.er.usgs.gov/publication/b2131

---

* Mean ionic molality is defined subsequently (Equation 12.11).

EXAMPLE 12.1 Calculate the Gibbs Energy of Reaction and Equilibrium Constant for an Aqueous Reaction Using Tabulated Data

i. Determine the Gibbs energy change and equilibrium constant at 25°C for the reaction:

$$CuSO_4(aq) = Cu^{2+}(aq) + SO_4^{2-}(aq)$$

SOLUTION

From the NBS tables, $\Delta_f G^0(CuSO_4, ao) = -692.18$ kJ mol^{-1}; $\Delta_f G^0(Cu^{2+}, ao) = 65.49$ kJ mol^{-1}; $\Delta_f G^0(SO_4^{2-}, ao) = -744.53$ kJ mol^{-1}

$$\Delta_r G^0 = \Delta_f G^0(Cu^{2+}) + \Delta_f G^0(SO_4^{2-}) - \Delta_f G^0(CuSO_4)$$

$$= 65.49 + (-744.53) - (-692.18) = 13.14 \text{ kJ mol}^{-1}$$

$$\Delta G^0 = -RT \ln K$$

$$K = \text{Exp}\left(\frac{-\Delta G^0}{RT}\right) = \text{Exp}\left(\frac{-13140}{8.314 \times 298.15}\right) = 4.987 \times 10^{-3}$$

Alternatively, assuming $CuSO_4$ is fully dissociated,

$$\Delta_f G^0(CuSO_4, ai) = \Delta_f G^0(Cu^{2+}) + \Delta_f G^0(SO_4^{2-}) = -779.04 \text{ kJ mol}^{-1}$$

and

$$\Delta_r G^0 = \Delta_f G^0(Cu^{2+}) + \Delta_f G^0(SO_4^{2-}) - \Delta_f G^0(CuSO_4)$$

$$= 65.49 + (-744.53) - (-779.04) = 0$$

ii. Determine the Gibbs energy change at 25°C for the reaction: $CuSO_4(s) = Cu^{2+}(aq) + SO_4^{2-}(aq)$

SOLUTION

From NBS tables, $\Delta_f G^0(CuSO_4, s) = -661.8$ kJ mol^{-1}; $\Delta_f G^0(Cu^{2+}, ao) = 65.49$ kJ mol^{-1}; $\Delta_f G^0(SO_4^{2-}, ao) = -744.53$ kJ mol^{-1}

$$\Delta_r G^0 = \Delta_f G^0(Cu^{2+}) + \Delta_f G^0(SO_4^{2-}) - \Delta_f G^0(CuSO_4) = 65.49 + (-744.53) - (-661.8)$$

$$= -17.24 \text{ kJ mol}^{-1}$$

COMMENT

For the reaction $CuSO_4(s) = CuSO_4(aq)$,

$$\Delta_r G^0 = \Delta_f G^0(CuSO_4, aq) - \Delta_f G^0(CuSO_4, s) = -692.18 - (-661.8) = -30.38 \text{ kJ mol}^{-1}$$

This is the difference in Gibbs energy of 1 mole of solid $CuSO_4$ and 1 mole of $CuSO_4$ at infinite dilution in water.

12.4 ACTIVITIES IN ELECTROLYTE SOLUTIONS

The thermodynamic properties of electrolyte solutions differ in significant ways from the properties of solutions of nonelectrolytes. To illustrate, the variation of the partial pressure of gaseous HCl in equilibrium with aqueous HCl as a function of concentration (see Figure 12.1) shows that the slope of the curve at infinite dilution is zero. If HCl is dissolved in water in the molecular form,

$$HCl(g) = HCl(aq)$$

$$K = \frac{a_{HCl}}{p_{HCl}}$$

At infinite dilution, $\gamma_{HCl} \rightarrow 1$ and $a_{HCl} \rightarrow m_{HCl}$. Therefore, $p_{HCl} \rightarrow m_{HCl} / K$ and the relation between p_{HCl} and m_{HCl} would be linear with slope equal to $1/K$. In fact, HCl is a strong electrolyte and dissociates in aqueous solutions; this dissociation is the reason for the non-Henrian behaviour of aqueous HCl. We will return to this example in Section 12.4.1.

In considering electrolyte solutes, we can consider the solute itself or the individual charged ions that result from dissociation. The same equations for chemical potential, activity coefficient and activity apply to these different species, but only the activity coefficient and activity of the solute as a whole can be determined experimentally.

The equilibrium constant expression for the dissociation reaction

$$A_{\nu_+}B_{\nu_-}(s \text{ or } aq) = \nu_+ A^{z+}(aq) + \nu_- B^{z-}(aq)$$

is:

$$K = \frac{a_{A^{z+}}^{\nu_+} \times a_{B^{z-}}^{\nu_-}}{a_{A_{\nu_+}B_{\nu_-}}} \tag{12.4}$$

K will be large for strong electrolytes and very small for weak electrolytes so the equilibrium will be displaced strongly to the right in the case of strong electrolytes and strongly to the left in the case of

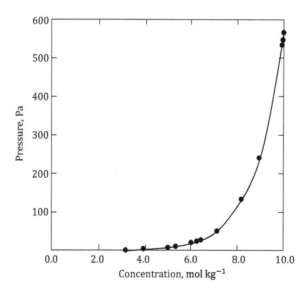

FIGURE 12.1 Variation of the equilibrium vapour pressure of HCl with concentration. Source of data: Bates, S. J. and H. D. Kirschman. 1919. *J. Am. Chem. Soc.*, 41, 1991–2001.

weak electrolytes. In the case of strong electrolytes, the assumption usually made is that the solute compound is completely dissociated.

Since $a_B = \gamma_B\, m_B$ (Equation 9.47),

$$K = \frac{\gamma_{A^{z+}}^{v_+} \times \gamma_{B^{z-}}^{v_-}}{\gamma_{A_{v_+}B_{v_-}}} \times \frac{m_{A^{z+}}^{v_+} \times m_{B^{z-}}^{v_-}}{m_{A_{v_+}B_{v_-}}}. \tag{12.5}$$

By convention, the following standard states are most commonly adopted for aqueous solutions:

- For ions, the standard state is the ion at a molality of 1 mol kg^{-1} of solvent and pressure of 1 bar and exhibiting infinitely dilute solution behaviour.*
- For non-electrolyte solutes, the standard state is the solute at a molality of 1 mol kg^{-1} of solvent and pressure of 1 bar and exhibiting infinitely dilute solution behaviour.
- For electrolyte solutes, the standard state is the solute at a mean ionic molality[†] of 1 mol kg^{-1} of water and pressure of 1 bar and exhibiting infinitely dilute solution behaviour. Alternatively, the pure solid may be taken as the standard state.
- For the solvent (water), the standard state is pure water at a pressure of 1 bar.

Equations 9.47 and 12.5 can be rewritten using simpler notation as follows:

$$a_+ = \gamma_+\, m_+ \text{ and } a_- = \gamma_-\, m_- \tag{12.6}$$

$$K = \frac{\gamma_+^{v_+}\, \gamma_-^{v_-}}{\gamma_e} \times \frac{m_+^{v_+}\, m_-^{v_-}}{m_e} \tag{12.7}$$

where + represents the cation, − represents the anion and e represents the electrolyte. It is important to remember that the molalities in Equations 12.5 to 12.7 are dimensionless since they are actually ratios of the molality in the solution and the molality of the standard state solution, which is one (Equation 9.56). Hence the equilibrium constant K is dimensionless as expected.

While the above equations are thermodynamically valid, as noted earlier the activity coefficients of individual ions in solutions cannot be determined. It is, however, possible to estimate values of the activity coefficient of ions from models of aqueous solutions (this is discussed in Section 12.5) or to make the simplifying assumption that the activity coefficients are unity.

12.4.1   THE UNIT ACTIVITY COEFFICIENT APPROXIMATION

If a solution is infinitely dilute the activity coefficient of an ion will be one and the activity of the ion will be numerically equal to its molal concentration. Alternatively, in some solutions, the value of the fraction $\gamma_+^{v_+}\, \gamma_-^{v_-} / \gamma_e$ may be one. In both cases, then

$$K_m = \frac{m_+^{v_+}\, m_-^{v_-}}{m_e}. \tag{12.8}$$

where K_m is the equilibrium constant based on dimensionless molality. In most solutions these conditions do not apply, but in dilute solutions they may be approximated, and this approximation is frequently made use of in thermodynamic calculations. In general K_m is not equal to K, the true thermodynamic equilibrium constant.

---

* For ease of reading equations, the convention adopted in Section 9.5.3 for writing activity coefficients for solutes in solution, namely $\gamma_{m,B}$, has not been followed in this chapter.
† Mean ionic molality is defined subsequently (Equation 12.11).

Returning now to Figure 12.1, we can explain the variation of p_{HCl} with m_{HCl} observed experimentally in terms of the dissociation reaction

$$HCl(g) = H^+(aq) + Cl^-(aq)$$

$$K = \frac{a_{H^+}\, a_{Cl^-}}{p_{HCl}}$$

At low concentration we can assume $\gamma_{H^+} = \gamma_{Cl^-} = 1$ and, since $m_{H^+} = m_{Cl^-} = m_{HCl}$,

$$K = \frac{m_{HCl}^2}{p_{HCl}}$$

Therefore

$$p_{HCl} = \frac{1}{K} \times m_{HCl}^2$$

Thus, there is a quadratic relation between p_{HCl} and m_{HCl}. The slope of the curve is:

$$\frac{dp_{HCl}}{dm_{HCl}} = \frac{2}{K} \times m_{HCl}$$

which at infinite dilution is equal to zero. The shape of the curve in Figure 12.1 and the slope at infinite dilution, therefore, are explained by the dissociation of HCl into H^+ and Cl^- ions in solution.

Use of the unit activity coefficient approximation can be further illustrated using the concept of solubility product. If a solution is saturated with an electrolyte solute, the activity of the solute will be one.* For example, for the solution of lead chloride in water,

$$PbCl_2(s) = Pb^{2+}(aq) + 2\, Cl^{2-}(aq)$$

$$K = \frac{a_{Pb^{2+}}\, a_{Cl^-}^2}{a_{PbCl_2}}$$

If the solution is saturated, $a_{PbCl_2} = 1$ and if we assume $\gamma_{Pb^{2+}} = \gamma_{Cl^-} = 1$,

$$K_{sol} = m_{Pb^{2+}}\, m_{Cl^-}^2$$

where K_{sol} is the *solubility product* of lead chloride. If lead chloride completely dissociates, $m_{Pb^{2+}} = m_{PbCl_2}$ and $m_{Cl^-} = 2\, m_{PbCl_2}$ so if the concentration of $PbCl_2$ in the solution at saturation is measured experimentally, the value of K_{sol} can be calculated:

$$K_{sol} = m_{Pb^{2+}}\, m_{Cl^-}^2 = m_{PbCl_2} \times \left(2\, m_{PbCl_2}\right)^2 = 4\, m_{PbCl_2}^3$$

At 25°C the measured solubility of $PbCl_2$ is 1.62×10^{-2} mol kg^{-1}, so $K_{sol} = 1.70 \times 10^{-5}$. The value of the thermodynamic equilibrium constant K at 25°C is 1.72×10^{-5} (calculated from the Gibbs energy change of the reaction) so the assumption of Henrian behaviour at these concentration of ions is justified.

---

* In this case the standard state for the solute has been chosen to be the pure solid at 1 bar pressure.

Suppose the solution into which lead chloride was added contained 0.10 mol kg of dissolved NaCl. The NaCl will be present as Na^+ and Cl^- ions, and $m_{Na^+} = m_{Cl^-} = m_{NaCl} = 0.10$ moles kg^{-1}. If α is the solubility (mol kg^{-1} water) of $PbCl_2$ (the amount that dissolves at saturation of the solution), $m_{Pb^{2+}} = \alpha$ and $m_{Cl^-} = 2\alpha + 0.10$ then

$$K_{sol} = m_{Pb^{2+}}\, m^2_{Cl^-} = x(2\alpha + 0.10)^2 = 1.70 \times 10^{-5}$$

Since we know $\alpha \ll 0.10$, we can assume $2\alpha + 0.1 = 0.1$. Therefore

$$\alpha \times 0.1^2 = 1.70 \times 10^{-5} \text{ moles kg}^{-1}$$

and $\alpha = 1.7 \times 10^{-3}$ mol kg^{-1} which is an order of magnitude lower solubility than for lead chloride in pure water. This example illustrates a general effect: The solubility of any salt is less in a solution containing a common ion than in water alone. This is called the *common ion effect*.

12.4.2 MEAN IONIC ACTIVITY

Since it is not possible to separate the $\gamma_+^{\nu_+}\, \gamma_-^{\nu_-}$ term in Equation 12.7 into individual activity coefficients for the cations and anions, the problem is overcome by assigning the non-ideality embodied in the activity coefficients proportionally to both kinds of ions. Because the activities of the cations and anions cannot be measured independently, their geometric mean, weighted according to the number of cations and anions, is taken as the *mean ionic activity coefficient* of the electrolyte,

$$\gamma_\pm = \left(\gamma_+^{\nu_+}\, \gamma_-^{\nu_-}\right)^{1/\nu} \tag{12.9}$$

where

$$\nu = \nu_+ + \nu_- \tag{12.10}$$

In the same manner, the *mean ionic molality* is defined as

$$m_\pm = \left(m_+^{\nu_+}\, m_-^{\nu_-}\right)^{1/\nu} \tag{12.11}$$

and, since $a_B = \gamma_B\, m_B$ (Equation 9.47), the *mean ionic activity* is:

$$a_\pm = \gamma_\pm\, m_\pm \tag{12.12}$$

For a completely dissociated electrolyte $m_+ = \nu_+\, m_e$ and $m_- = \nu_-\, m_e$, where m_e is the molal concentration of the electrolyte. Therefore, substituting into Equation 12.11:

Strong electrolytes: $\qquad m_\pm = (m_e)\left(\nu_+^{\nu_+}\, \nu_-^{\nu_-}\right)^{1/\nu} \tag{12.13}$

In cases where ionisation is not quite complete it is the convention to follow the same procedure, namely to put $m_+ = \nu_+\, m_e$ and $m_- = \nu_-\, m_e$. The expression for ionic activity in Equation 12.12 can still be used, but the value of $\gamma_\pm$ in this case no longer has the significance of being the geometric mean of the activity coefficients of the fully dissociated ions. $\gamma_\pm$ is then referred to as the *stoichiometric activity coefficient* of the electrolyte.

The significance of $\gamma_\pm$ and $a_\pm$ is that they can be measured experimentally, whereas individual ion activities and activity coefficients cannot. Furthermore, they provide a means to calculate the activity of electrolytes in solution as shown in Section 12.4.3.

12.4.3 ACTIVITY OF THE ELECTROLYTE

The relation between the electrolyte and its ions in solution is:

$$A_{\nu_+}B_{\nu_-}(aq) = \nu_+ A^{z+}(aq) + \nu_- B^{z-}(aq)$$

At equilibrium, $\Delta G = 0$. Therefore,

$$\nu_+ G_{A^{z+}}(aq) + \nu_- G_{B^{z-}}(aq) - G_{A_{\nu_+}B_{\nu_-}}(aq) = 0$$

Using the simplified nomenclature and rearranging,

$$G_e(aq) = \nu_+ G_+(aq) + \nu_- G_-(aq) \tag{12.14}$$

Note, Equation 12.14* is consistent with the earlier statement that a thermodynamic property of a fully dissociated electrolyte in solution is the sum of the corresponding property of the ions of which the electrolyte is constituted. For the individual ions, we can write (Equation 9.57)

$$G_+(aq) = G_+^0(aq) + RT \ln a_+ \quad \text{and} \quad G_-(aq) = G_-^0(aq) + RT \ln a_- \tag{12.15}$$

The Gibbs energy of the neutral electrolyte is their sum:

$$G_e(aq) = \nu_+ G_+^0(aq) + \nu_+ RT \ln a_+ + \nu_- G_-^0(aq) + \nu_- RT \ln a_-$$

$$= \nu_+ G_+^0(aq) + \nu_- G_-^0(aq) + RT \ln\left(\gamma_+^{\nu_+}\, \gamma_-^{\nu_-} \times m_+^{\nu_+}\, m_-^{\nu_-}\right)$$

$$= G_e^0(aq) + RT \ln\left(\gamma_+^{\nu_+}\, \gamma_-^{\nu_-} \times m_+^{\nu_+}\, m_+^{\nu_-}\right)$$

$$= G_e^0(aq) + RT \ln\left(\gamma_+^{\nu_+}\, m_+^{\nu_+} \times \gamma_-^{\nu_-}\, m_+^{\nu_-}\right) \tag{12.16}$$

$$= G_e^0(aq) + RT \ln\left(a_+^{\nu_+} \times a_-^{\nu_-}\right) \tag{12.17}$$

Since the partial molar Gibbs energy (or chemical potential) of the electrolyte (Equation 9.57) is

$$G_e(aq) = G_e^0(aq) + RT \ln a_e(aq) \tag{12.18}$$

then,

$$a_e = a_+^{\nu_+} \times a_-^{\nu_-}$$

Furthermore, since (Equation 12.16)

---

* Using the terminology of the NBS tables, Equation 12.14 would be written as: $G_e(ai) = \nu_+ G_+(ao) + \nu_- G_-(ao)$.

$$G_e(\text{aq}) = G_e^0(\text{aq}) + RT \ln\left(\gamma_+^{\nu_+} m_+^{\nu_+} \times \gamma_-^{\nu_-} m_+^{\nu_-}\right)$$

therefore

$$a_e = \gamma_+^{\nu_+} m_+^{\nu_+} \times \gamma_-^{\nu_-} m_-^{\nu_-} = \gamma_+^{\nu_+} \gamma_-^{\nu_-} m_+^{\nu_+} m_-^{\nu_-} = \left(\gamma_\pm m_\pm\right)^\nu = a_\pm^\nu \tag{12.19}$$

Thus, in summary the activity of an electrolyte in solution is given by:

$$a_e = \left(\gamma_\pm m_\pm\right)^\nu = a_\pm^\nu = a_+^{\nu_+} a_-^{\nu_-} \tag{12.20}$$

where the standard state of the electrolyte is the solute at a mean ionic molality of 1 mol kg^{-1} of water and pressure of 1 bar and exhibiting infinitely dilute solution behaviour.

For a particular solution, the value of $m_\pm$ is known from a knowledge of the molality of the solution (Equation 12.11). Thus $\gamma_\pm$ may be calculated using Equation 12.18 if $G_e(\text{aq}) - G_e^0$, or the activity of e, can be determined experimentally. If the electrolyte is sufficiently volatile, its activity in a solution can be evaluated from partial pressure measurements of an equilibrated gas phase. If the solvent is volatile the activity of the electrolyte solute can be calculated from the measured activity of the solvent using the Gibbs–Duhem equation. For systems for which a suitable galvanic cell reaction can be found $\gamma_\pm$ can be calculated from the measured cell potential (Chapter 16). A more general method making use of osmotic coefficients is described in more advanced texts.*

The variation of $\gamma_\pm$ with concentration for some common electrolytes, measured experimentally, is shown in Figure 12.2. In general, due to the changing nature of the interactions between the ions,

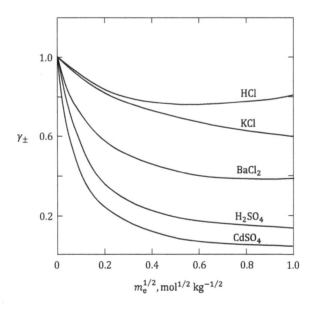

FIGURE 12.2 Variation of mean activity coefficient of some electrolytes in water with the square root of their molality. Source of data: Latimer, W. M. 1952. *The Oxidation States of the Elements and their Potentials in Aqueous Solutions.* New York: Prentice–Hall, Inc.

* For example, Denbigh, K. 1981. *The Principles of Chemical Equilibrium.* Cambridge: Cambridge University Press.

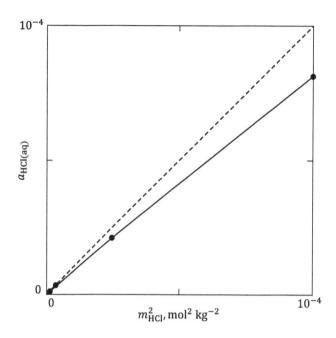

FIGURE 12.3 The activity of HCl as a function of molality squared showing the approach to linearity at low dilution. Source of data: Latimer, W. M. 1952. *The Oxidation States of the Elements and their Potentials in Aqueous Solutions.* New York: Prentice–Hall, Inc.

$\gamma_{\pm}$ decreases initially from a value of one with increasing concentration, passes through a minimum then increases again. When experimental data are not available, values of $\gamma_{\pm}$ may be estimated using the Debye–Hückel law (discussed in Section 12.5).

Returning to the example of HCl in solution, $v=2$, $m_{\pm} = m_{HCl}$ (Equation 12.11) and therefore $a_{HCl} = \left(\gamma_{\pm} m_{\pm} \right)^2$ (Equation 12.20). Thus, at low dilution the activity of HCl in solution should be equal to the square of the HCl molality (since $\gamma_{\pm} \rightarrow 1$). This behaviour is consistent with data for dilute HCl solutions, as shown in Figure 12.3, and also consistent with Figure 12.1. In contrast, the activity of non-electrolyte solutes is equal to the molality of the solute at low dilution (Henry's law, Equation 9.58).

12.4.4 MULTIPLE ELECTROLYTES IN SOLUTION

So far we have considered solutions of a single electrolyte, but often more than one electrolyte will be present. The presence of other electrolytes affects the activity of the electrolyte of interest. For example in a solution of HCl, when electrolytes such as NaCl and H_2SO_4 are also present, the total Cl^- and H^+ concentration is given by the sum of the concentration of ions contributed by the various electrolytes. The relation (Equation 12.19)

$$a_e = \gamma_+^{v_+} \, m_+^{v_+} \times \gamma_-^{v_-} \, m_-^{v_-} = \gamma_{\pm} \, m_+^{v_+} m_+^{v_-}$$

still applies, but the addition of NaCl or H_2SO_4 increases the concentration of Cl^- or H^+ ions and hence increases the activity of the HCl. Similarly, the activity of NaCl in solution will increase if HCl or another sodium salt is added. Under some conditions the increase in activity may be so large that the solution becomes saturated with respect to NaCl. To generalise, *the presence of a common ion raises the activity of the solute containing the common ion.*

EXAMPLE 12.2   Calculation of activities in ionic solutions
using mean ionic activity coefficient

i. Calculate the mean ionic activity and the activity of calcium chloride ($CaCl_2$) in a 0.02 mol kg^{-1} solution at 25°C. Data: $\gamma_{\pm}(CaCl_2) = 0.66$ at $m_{CaCl_2} = 0.02$ mol kg^{-1}.

SOLUTION

$\nu_+ = 1, \nu_- = 2; \nu = 1 + 2 = 3$
$CaCl_2$ is a strong electrolyte; hence

$$m_+ = m_{CaCl_2} = 0.02 \text{ moles kg}^{-1}; m_- = 2\,m_{CaCl_2} = 0.04 \text{ mol kg}^{-1}$$

From Equation 12.11, $m_{\pm} = (0.02^1 \times 0.04^2)^{1/3} = 0.03175$ mol kg^{-1}
(Alternatively, from Equation 12.13, $m_{\pm} = m_{CaCl_2} \times (1^1 \times 2^2)^{1/3} = 0.02 \times 4^{1/3} = 0.03175$ mol kg^{-1})
From Equation 12.12, $a_{\pm}(CaCl_2) = 0.66 \times 0.03175 = 0.021$
From Equation 12.20, $a_{CaCl_2\,(aq)} = (0.66 \times 0.03175)^3 = 9.20 \times 10^{-6}$

ii. Calculate the mean ionic activity and the activity of calcium chloride in a 0.02 mol kg^{-1} solution at 25°C if the solution also contains 0.01 mol kg^{-1} of NaCl.

SOLUTION

As before, for $CaCl_2$, $\nu_+ = 1, \nu_- = 2; \nu = 1 + 2 = 3$
Additionally, since NaCl is also a strong electrolyte, $m_+ = m_{NaCl} = 0.01; m_- = m_{NaCl} = 0.01$
The concentration of Ca^{2+} remains the same as in the previous example, but the concentration of Cl^- is now $0.04 + 0.01 = 0.05$ mole kg^{-1}.
From Equation 12.11, $m_{\pm} = \left(m_+^{\nu_+} m_-^{\nu_-}\right)^{1/\nu} = \left(0.02^1 \times 0.05^2\right)^{1/3} = 0.03685$ mole kg^{-1}
From Equation 12.12, $a_{\pm}(CaCl_2) = \gamma_{\pm} m_{\pm} = 0.66 \times 0.03685 = 0.024$
From Equation 12.20, $a_{CaCl_2\,(aq)} = (\gamma_{\pm} \times m_{\pm})^3 = (0.66 \times 0.03685)^3 = 1.44 \times 10^{-5}$
As expected the activity of the $CaCl_2$ has been increased by the addition of a solute with a common ion.

12.5   THE ACTIVITY OF IONS

Although the activity coefficient of ions in solution cannot be measured experimentally, mathematical models have been developed for estimating values, and the most widely used model for dilute solutions is that developed by Debye and Hückel in 1923. The deviation of the activity coefficient of electrolytes from Henrian behaviour is due to the inter-ionic attraction in solutions. In very dilute solutions the ions are far apart, separated by water molecules, and do not interact with each other. Hence Henrian behaviour is approached as solutions become more dilute. The partial molar Gibbs energy of an ion in solution is

$$G_i = G_i^0 + RT \ln a_i = RT \ln m_i + RT \ln \gamma_i$$

where the $RT \ln \gamma_i$ term is equal to the Gibbs energy needed to charge neutral particles to become ions. Debye and Hückel determined this Gibbs energy by first calculating the electrical potential of an ion due to its oppositely charged ionic atmosphere, then calculating the work done in charging the neutral particle to this potential. By equating this energy to the $RT \ln \gamma_i$ term they obtained the equation:

$$\log \gamma_i = -1.823 \times 10^6 \times \frac{z_i^2}{(\varepsilon T)^{2/3}} I_m^{1/2} \tag{12.21}$$

where ε is the relative permeability of the solvent and I_m is the *ionic strength* of the solution. Ionic strength is defined by the relation

$$I_m = 0.5 \sum_i m_i z_i^2 \tag{12.22}$$

where m_i is the molality of ion i and z_i is the charge on ion i. The unit of I_m is mol kg^{-1}. Ionic strength is a measure of the strength of the electrical field due to the ions in a solution. All the ions in a solution must be included in calculating I_m, not only the ions of the electrolyte of interest. For water at 25°C, $\varepsilon = 78.54$ and Equation 12.21 becomes

$$\log \gamma_i = -z_i^2 A I_m^{1/2} \tag{12.23}$$

where $A = 0.509$.

To relate the equation to an experimentally measurable property, Equation 12.23 can be expressed in terms of mean ionic activity coefficient:

$$\log \gamma_{\pm} = -|z_+ z_-| A I_m^{1/2} \tag{12.24}$$

Note the term $|z_+ z_-|$ indicates the product is positive; hence $\log \gamma_{\pm}$ will be negative and $\gamma_{\pm}$ will always be less than one. Equations 12.23 and 12.24 apply strictly only at infinite dilution and are referred to as the *Debye–Hückel limiting law*. However, in practice, they may be used for very dilute solutions (typically $I_m < 0.01$).

A variation of Equations 12.23 and 12.24, called the *extended Debye–Hückel law*, takes into account the difference in size of ions:

$$\log \gamma_i = \frac{-z_i^2 A I_m^{1/2}}{1 + B a \times 10^{-2} I_m^{1/2}} \tag{12.25}$$

$$\log \gamma_{\pm} = \frac{-|z_+ z_-| A I_m^{1/2}}{1 + B a \times 10^{-2} I_m^{1/2}} \tag{12.26}$$

where B is a dimensionless constant and has a value of 0.3281 at 25°C in water, and a is the effective diameter (pm)* of the ion in solution (or hydrated ionic radius, in the case of aqueous solutions). The effective diameter is an empirically fitted parameter; it is larger than the ionic radius because it includes the hydrated layer. Values of a for some common ions are listed in Table 12.2. Equations 12.25 and 12.26 are useful for more concentrated solutions (typically $I_m < 0.1$).

---

* 1 pm (or picometre) is 10^{-12} metres.

TABLE 12.2
Effective diameters (a) for some common ions in aqueous solutions

Ion	Effective diameter (pm)
H_3O^+	900
Li^+	600
Na^+, HSO_3^-, HCO_3^-, $H_2PO_4^-$	450
OH^-, F^-, HS^-, ClO_3^-	350
F^-, Cl^-, Br^-, I^-, CN^-, NO_3^-	300
Ag^+, NH_4^+	250
Mg^{2+}, Be^+	800
Ca^{2+}, Cu^{2+}, Zn^{2+}, Sn^{2+}, Mn^{2+}, Fe^{2+}, Ni^{2+}, Co^{2+}	600
Sr^{2+}, Ba^{2+}, Cd^{2+}, Hg^{2+}, S^{2-}	500
Pb^{2+}, SO_2^{2-}, SO_3^{2-}	450
SO_4^{2-}, $S_2O_3^{2-}$, HPO_4^{2-}	400
Al^{3+}, Fe^{3+}, Cr^{3+},	900
PO_4^{3-}, $Fe(CN_6^{3-})$	400

Source: Kielland, J. 1937. *J. Amer. Chem Soc.*, 59, 1675–1678.

EXAMPLE 12.3   Calculation of activity using the Debye–Hückel law for a strong electrolyte (complete dissociation)

i. Calculate the ionic strength of a 0.1 mol kg⁻¹ Na_2SO_4 solution at 25°C, the activity of the sodium and sulfate ions, the activity of Na_2SO_4 and the mean ionic activity of Na_2SO_4.

SOLUTION

Na_2SO_4 is a strong electrolyte so we assume complete dissociation:

$$Na_2SO_4(aq) = 2\,Na^+(aq) + SO_4^{2-}(aq)$$

$z_+ = 1$, $z_- = -2$; $v_+ = 1$, $v_- = 2$

Stoichiometric relations: $m_{Na^+} = 2\,m_{Na_2SO_4} = 0.2$ mol kg⁻¹; $m_{SO_4^{2-}} = m_{Na_2SO_4} = 0.1$ mol kg⁻¹

$$I_m = 0.5\sum_i m_i z_i^2 = 0.5\left(m_{Na^+} \times z_{Na^+}^2 + m_{SO_4^{2-}} \times z_{SO_4^{2-}}^2\right) = 0.5\times\left(0.2\times1^2 + 0.1\times2^2\right) = 0.30$$

CASE 1. USING THE DEBYE–HÜCKEL LIMITING LAW

$$\log\gamma_{Na^+} = -z_{Na^+}^2 A I_m^{1/2} = -0.509\times1^2\times0.3^{0.5} = -0.2788$$

Therefore, $\gamma_{Na^+} = 0.526$. Repeating the calculation for SO_4^{2-} ions gives $\gamma_{SO_4^{2-}} = 0.077$.

$$a_{Na^+} = \gamma_{Na^+}\times m_{Na^+} = 0.526\times0.2 = 0.105$$

$$a_{SO_4^{2-}} = \gamma_{SO_4^{2-}} \times m_{SO_4^{2-}} = 0.077 \times 0.1 = 0.0077$$

$$a_{Na_2SO_4} = a_{Na^+}^{\nu_+} \, a_{SO_4^{2-}}^{\nu_-} = 0.105^2 \times 0.0077^1 = 8.50 \times 10^{-5}$$

$$\gamma_\pm = \left(\gamma_+^{\nu_+} \gamma_-^{\nu_-} \right)^{1/\nu} = \left(0.526^2 \times 0.077^1 \right)^{1/3} = 0.277$$

Alternatively, $\gamma_\pm$ can be calculated directly:

$$\log \gamma_\pm = -\left| z_+ z_- \right| A I_m^{1/2} = -0.509 \left| z_{Na^+} \, z_{SO_4^{2-}} \right| I_m^{1/2} = -0.509 \times \left| 1 \times -2 \right| \times 0.3^{1/2} = -0.5576$$

Therefore, $\gamma_\pm = 0.277$.

CASE 2. USING THE EXTENDED DEBYE–HÜCKEL LAW

From Table 12.2, $a_{Na^+} = 450$ pm and $a_{Cl^-} = 300$ pm.

$$\log \gamma_{Na^+} = \frac{-0.509 \, z_{Na^+}^2 \, I_m^{1/2}}{1 + 0.3281 \times 450 \times 10^{-2} \times I_m^{1/2}} = \frac{-0.509 \times 1^2 \times 0.3^{1/2}}{1 + 0.3281 \times 450 \times 10^{-2} \times 0.3^{1/2}} = -0.07021$$

Therefore, $\gamma_{Na^+} = 0.7012$. By a similar calculation, $\gamma_{SO_4^{2-}} = 0.2418$.

$$a_{Na^+} = \gamma_{Na^+} \times m_{Na^+} = 0.7012 \times 0.2 = 0.1402$$

$$a_{SO_4^{2-}} = \gamma_{SO_4^{2-}} \times m_{SO_4^{2-}} = 0.2418 \times 0.1 = 0.0242$$

$$a_{Na_2SO_4} = a_{Na^+}^{\nu_+} \, a_{SO_4^{2-}}^{\nu_-} = 0.1402^2 \times 0.0242^1 = 4.76 \times 10^{-4}$$

$$\gamma_\pm = \left(\gamma_+^{\nu_+} \gamma_-^{\nu_-} \right)^{1/\nu} = \left(0.7012^2 \times 0.2418^1 \right)^{1/3} = 0.49$$

The experimentally determined value of $\gamma_\pm$ at 0.1 molality of Na_2SO_4 is 0.45 so the value calculated using the extended Debye–Hückel law is the more accurate.

COMMENT

If the calculation is repeated to cover ionic strength up to 1.0 mol kg^{-1} $\left(m_{Na_2SO_4} \approx 0.34 \text{ mol kg}^{-1} \right)$ the trends shown in Figure 12.4 are observed.

ii. Calculate the ionic strength of a solution at 25°C containing 0.1 mole kg^{-1} Na_2SO_4 and 0.2 mole kg^{-1} NaCl and the activity of the sodium, sulfate and chloride ions and the activities of NaCl and Na_2SO_4.

SOLUTION

As these are strong electrolytes we can assume they are completely dissociated in solution:

$$Na_2SO_4(aq) = 2\,Na^+(aq) + SO_4^{2-}(aq)$$

$$NaCl(aq) = Na^+(aq) + Cl^-(aq)$$

Dissociation of Na_2SO_4: $0.5 m_{Na^+} = m_{SO_4^{2-}} = m_{Na_2SO_4} = 0.1 \text{ mol kg}^{-1}$

Dissociation of NaCl: $m_{Na^+} = m_{Cl^-} = m_{NaCl} = 0.2 \text{ mol kg}^{-1}$

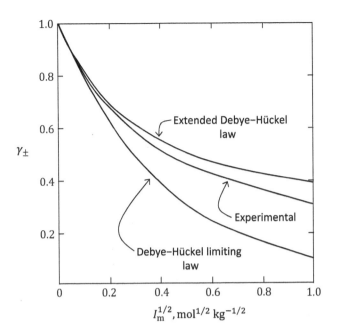

FIGURE 12.4 The variation of $\gamma_\pm \left(Na_2SO_4\right)$ with the square root of ionic strength. Values calculated using the Debye–Hückel limiting law and the extended law agree well with the experimental values at low ionic strength but deviate progressively as I_m increases. The extended law gives better agreement at all values of I_m.

$$m_{Na^+} = 0.2 + 0.2 = 0.4 \text{ mole kg}^{-1}$$

$$I_m = 0.5\sum_i m_i z_i^2 = 0.5 \times \left[\left(m_{Na^+} \times 1^2\right) + \left(m_{Cl^-} \times 1^2\right) + \left(m_{SO_4^{2-}} \times 2^2\right)\right]$$

$$= 0.5 \times \left[\left(0.4 \times 1^2\right) + \left(0.2 \times 1^2\right) + \left(0.1 \times 2^2\right)\right]$$

$$= 0.5 \times \left[0.4 + 0.2 + 0.4\right] = 0.5$$

Using the extended Debye–Hückel law:

From Table 12.2, $a_{Na^+} = 450$ pm, $a_{Cl^-} = 300$ pm, $a_{SO_4^{2-}} = 400$ pm.

$$\log\gamma_{Na^+} = \frac{-0.509\, z_{Na^+}^2\; I_m^{1/2}}{1 + 0.3281 \times 450 \times 10^{-2} \times I_m^{1/2}} = \frac{-0.509 \times 1^2 \times 0.5^{1/2}}{1 + 0.3281 \times 450 \times 10^{-2} \times 0.5^{1/2}} = -0.1761$$

Therefore, $\gamma_{Na^+} = 0.667$. By a similar calculation, $\gamma_{Cl^-} = 0.613$ and $\gamma_{SO_4^{2-}} = 0.179$.

$$a_{Na^+} = \gamma_{Na^+} \times m_{Na^+} = 0.667 \times 0.4 = 0.267$$

$$a_{Cl^-} = \gamma_{Cl^-} \times m_{Cl^-} = 0.613 \times 0.2 = 0.123$$

$$a_{SO_4^{2-}} = \gamma_{SO_4^{2-}} \times m_{SO_4^{2-}} = 0.179 \times 0.1 = 0.0179$$

$$a_{\text{Na}_2\text{SO}_4} = a_{\text{Na}^+}^{\nu_+} \, a_{\text{SO}_4^{2-}}^{\nu_-} = 0.267^2 \times 0.0179^1 = 1.28 \times 10^{-3}$$

$$a_{\text{NaCl}} = a_{\text{Na}^+}^{\nu_+} \, a_{\text{Cl}^-}^{\nu_-} = 0.267^1 \times 0.123^1 = 3.27 \times 10^{-2}$$

COMMENT

The activity of Na_2SO_4 in a solution containing 0.1 mole kg^{-1} is 4.76×10^{-4}. The activity of Na_2SO_4 at the same concentration but with 0.2 mole kg^{-1} NaCl is two orders of magnitude greater. The common ion effect increases the activity of Na_2SO_4 (as discussed in Section 12.4.4).

12.6   PARTIAL DISSOCIATION

In a solution of a weak electrolyte there is an equilibrium between free ions and any undissociated molecules,

$$A_{\nu_+}B_{\nu_-}(\text{aq}) = \nu_+ A^{z+}(\text{aq}) + \nu_- B^{z-}(\text{aq})$$

which is expressed by the equilibrium constant expression for the dissociation reaction,

$$K = \frac{a_{\text{A}^{z+}}^{\nu_+} \times a_{\text{B}^{z-}}^{\nu_-}}{a_{A_{\nu_+}B_{\nu_-}}} = \frac{a_+^{\nu_+} \times a_-^{\nu_-}}{a_e'} \tag{12.27}$$

where a_e' is the activity of the undissociated part of the dissolved electrolyte solute.

The activity terms can be expressed as the product of an activity coefficient and the molality:

$$K = \frac{\gamma_+^{\nu_+} \times \gamma_-^{\nu_-}}{\gamma_e'} \times \frac{m_+^{\nu_+} \times m_-^{\nu_-}}{m_e'} \tag{12.28}$$

where m_e' is the dimensionless molality of the undissociated portion of the electrolyte and γ_e' is the corresponding activity coefficient. Note, m_e' is less than m_e, the molality of the solution. For very weak electrolytes (K very small) m_e' will be almost equal to m_e, but in many cases of partial dissociation the difference between m_e' and m_e will be significant.

The self–ionisation of water illustrates the case for a very weak electrolyte:

$$H_2O(l) = H^+(\text{aq}) + OH^-(\text{aq})$$

$$K = \frac{a_{H^+} \, a_{OH^-}}{a_{H_2O}}$$

At 25°C, $K = 1.008 \times 10^{-14}$. Since K is very small, the degree of ionisation is also very small, and the activity of water will be almost one. Further, since $K = m_{H^+} = m_{OH^-}$ and activity can be approximated by molality at such low concentrations, $m_{H^+}^2 = 1.008 \times 10^{-14}$. Therefore,

$$m_{H^+} = m_{OH^-} = \left(1.008 \times 10^{-14}\right)^{1/2} = 1.004 \times 10^{-7} \text{ mol kg}^{-1}$$

This calculation involves the approximation that such a small fraction of water dissociates that its activity is effectively unchanged. This is reasonable in this case, but the assumption that the degree of dissociation is so small that the change in concentration (or activity) of the electrolyte solute can be neglected is not always justified. This is discussed in the following section.

12.6.1 DEGREE OF DISSOCIATION

The *degree of dissociation*, α, of an electrolyte solute is the fraction of 1 mole of the electrolyte that dissociates on forming a solution. Thus, if an electrolyte is added to water to form a solution of molality m_e the concentration of the electrolyte after dissociation m'_e is $m_e - \alpha m_e = m_e(1 - \alpha)$. From Equation 12.2, the corresponding concentration of cations is $v_+ \alpha m_e$ and of anions is $v_- \alpha m_e$. This can be summarised for the electrolyte $A_{v_+} B_{v_-}$ in the form of an Initial–Change–Equilibrium table as follows:

	$A_{v_+} B_{v_-}$	A^{z+}	B^{z-}
Initial	m_e	0	0
Change	$-m_e$	$+v_+ \alpha m_e$	$+v_- \alpha m_e$
Equilibrium	$m_e(1 - \alpha)$	$v_+ \alpha m_e$	$v_- \alpha m_e$

The Equilibrium line of the table is the sum of the Initial and Change lines. Equation 12.28 then becomes

$$K = \frac{\gamma_+^{v_+} \times \gamma_-^{v_-}}{\gamma'_e} \times \frac{\left(v_+ \alpha m_e\right)^{v_+} \times \left(v_- \alpha m_e\right)^{v_-}}{m_e(1 - \alpha)} \tag{12.29}$$

The undissociated electrolyte, being a neutral species, will obey Henry's law in dilute solutions and, further, if γ_+ and γ_- are assumed to be one, then

$$K = \frac{\left(v_+ \alpha m_e\right)^{v_+} \times \left(v_- \alpha m_e\right)^{v_-}}{m_e(1 - \alpha)} \tag{12.30}$$

If the value of K is known, then Equations 12.29 and 12.30 can be solved to obtain the value of α at any concentration of the electrolyte. This illustrated in Examples 12.4 and 12.5.

12.7 THE pH SCALE

Since the relation between H^+ and OH^- activities in water is $a_{H^+} \times a_{OH^-} = 1.008 \times 10^{-14}$, if an acid is added to water, the activity (and concentration) of hydrogen ions increases and the activity of OH^- ions decreases. The opposite effect occurs if a base is added. The activity of H^+ ions in 1 mol kg^{-1} HCl will be one, and in 1 mol kg^{-1} NaOH it will be 10^{-14} and may exceed these values for stronger acid or base solutions.

The hydrogen ion activity is conveniently expressed as the negative logarithm, which is called the pH of a solution. Thus, pH is defined as follows:

$$\text{pH} = -\log a_{H^+} = -\log\left(\gamma_{H^+} \times m_{H^+}\right) \tag{12.31}$$

In very dilute solutions, it may be assumed that $\gamma_{H^+} = 1$; then $\text{pH} = -\log m_{H^+}$. It follows that the pH of 1 mol kg^{-1} HCl will be 0 and the pH of 1 mol kg^{-1} NaOH will be 14. The pH of pure water is $-\log\left(10^{-7}\right) = 7$.

Since pH is defined in terms of a quantity that cannot be measured experimentally (the activity of hydrogen ions), Equation 12.31 is considered to be a notional definition, and in practice a primary standard is required against which calibration of pH is performed. This involves the use of an electrochemical cell known as the Harned cell.

EXAMPLE 12.4   Degree of dissociation and pH of a weak electrolyte

Calculate the degree of dissociation and pH of acetic acid in a solution containing 0.04 mol kg^{-1}.

SOLUTION

The dissociation equilibrium is:

$$CH_3COOH(aq) = CH_3COO^-(aq) + H^+(aq)$$

for which

$$K = \frac{a_{CH_3COO^-} \times a_{H^+}}{a_{CH_3COOH}} = \frac{\gamma_{CH_3COO^-} \times \gamma_{H^+}}{\gamma_{CH_3COOH}} \times \frac{m_{CH_3COO^-} \times m_{H^+}}{m_{CH_3COOH}}$$

The value of K at 25°C is 1.8×10^{-5}; $v_+ = 1$ and $v_- = 1$. The ICE table, therefore, is:

	CH$_3$COOH	H$^+$	CH$_3$COO$^-$
Initial	0.04	0	0
Change	$-0.04\,\alpha$	$+0.04\,\alpha$	$-0.04\,\alpha$
Equilibrium	$0.04(1-\alpha)$	$0.04\,\alpha$	$0.04\,\alpha$

Substituting,

$$K = \frac{\gamma_{CH_3COO^-} \times \gamma_{H^+}}{\gamma_{CH_3COOH}} \times \frac{0.04\alpha \times 0.04\alpha}{0.04(1-\alpha)} = \frac{0.04\alpha^2}{(1-\alpha)} = 1.8 \times 10^{-5}$$

Since γ_{CH_3COOH} is molecular in nature, undissociated acetic acid can be assumed to obey Henry's law $\left(\gamma_{CH_3COOH} = 1\right)$ at low concentration. Then,

$$\frac{\alpha^2}{(1-\alpha)} = \frac{1.8 \times 10^{-5}}{\gamma_{CH_3COO^-} \times \gamma_{H^+} \times 0.04} = z$$

Rearranging,

$$\alpha^2 + z\alpha - z = 0$$

If we assume $\gamma_{CH_3COO^-}$ and γ_{H^+} are equal to one, $z = 1.8 \times 10^{-5} / 0.04$, then solving (using the Goal Seek function in Excel™) gives $\alpha = 0.021$. Thus the equilibrium concentrations are:

$$m_{CH_3COOH} = 0.04(1-\alpha) = 0.0392 \text{ mol kg}^{-1}$$

$$m_{H^+} = 0.04\alpha = 8.40 \times 10^{-4} \text{ mol kg}^{-1}$$

$$m_{CH_3COO^-} = 0.04\alpha = 8.40 \times 10^{-4} \text{ mol kg}^{-1}$$

$$pH = -\log a_{H^+} = -\log\left(\gamma_{H^+} \times m_{H^+}\right) = 3.0759$$

More accurate values can be obtained by substituting the calculated ion concentrations into Equation 12.22 and substituting the calculated value of I_m in the Debye–Hückel limiting law to estimate the values of $\gamma_{CH_3COO^-}$ and γ_{H^+}. These values are then used to repeat the calculation until limiting values are attained. This can be performed manually, as in the table below (convergence is achieved after three iterations), or by using the Goal Seek function. The results are summarised in the table below which shows that convergence is achieved after only a few iterations.

Iteration		1	2	3
γ_{H^+}	9.666×10^{-1}	9.661×10^{-1}	9.660×10^{-1}	9.660×10^{-1}
$\gamma_{CH_3COO^-}$	9.666×10^{-1}	9.661×10^{-1}	9.660×10^{-1}	9.660×10^{-1}
$m_{CH_3COO^-}; m_{H^+}$	8.396×10^{-4}	8.683×10^{-4}	8.688×10^{-4}	8.688×10^{-4}
a_{H^+}	8.396×10^{-4}	8.393×10^{-4}	8.393×10^{-4}	8.393×10^{-4}
pH	3.0759	3.0761	3.0761	3.0761

The above approach is applicable even for moderately strong electrolytes such as $CuSO_4$ as the following example illustrates.

EXAMPLE 12.5 Activities in partially dissociated electrolyte solution

Estimate the activities of Cu^{2+}, SO_4^{2-} and undissociated $CuSO_4$ in a 0.5 mol kg^{-1} solution of copper sulfate at 25°C.

SOLUTION

$$CuSO_4(aq) = Cu^{2+}(aq) + SO_4^{2-}(aq)$$

$$K = \frac{a_{Cu^{2+}} \, a_{SO_4^{2-}}}{a_{CuSO_4}}$$

The value of K was calculated to be 4.987×10^{-3} in Example 12.1. The equilibrium constant is moderately small, so there will be an appreciable amount of undissociated $CuSO_4$ in the solution.

Let α be the fraction of $CuSO_4$ that dissociates. The Initial – Change–Equilibrium table is:

	$CuSO_4$	Cu^{2+}	SO_4^{2-}
Initial	0.5	0	0
Change	$-0.5\,\alpha$	$+0.5\,\alpha$	$+0.5\,\alpha$
Equilibrium	$0.5\,(1-\alpha)$	$0.5\,\alpha$	$0.5\,\alpha$

Therefore,

$$K = \frac{\gamma_{SO_4^{2-}} \, \gamma_{Cu^{2+}}}{\gamma_{CuSO_4}} \times \frac{0.5\alpha \times 0.5\alpha}{0.5(1-\alpha)}$$

As the solution is dilute we can assume for the undissociated $CuSO_4$ that $\gamma_{CuSO_4} = 1$. Therefore,

$$K = \left(\gamma_{SO_4^{2-}} \, \gamma_{Cu^{2+}}\right) \times \frac{0.5\alpha^2}{1-\alpha}$$

Let

$$z = \frac{K}{0.5\left(\gamma_{SO_4^{2-}} \gamma_{Cu^{2+}}\right)}$$

then, substituting and rearranging:

$$\alpha^2 + z\alpha - z = 0$$

If we assume $\gamma_{SO_4^{2-}}$ and $\gamma_{Cu^{2+}}$ are equal to one, $z = 5.7045 \times 10^{-3}/0.5$, then solving gives $\alpha = 0.0905$. Thus the equilibrium concentrations are

$$m_{CuSO_4} = 0.5(1-\alpha) = 0.4525 \text{ mol kg}^{-1}$$

$$m_{Cu^{2+}} = 0.5\alpha = 0.0475 \text{ mol kg}^{-1}$$

$$m_{SO_4^{2-}} = 0.5\alpha = 0.0475 \text{ mol kg}^{-1}$$

There is a significant amount of undissociated $CuSO_4$.

More accurate values can be obtained by substituting the calculated ion concentrations into Equation 12.22 and substituting the calculated value of I_m in the extended Debye–Hückel equation (Equation 12.25) to estimate the values of $\gamma_{Cu^{2+}}$ and $\gamma_{SO_4^{2-}}$. These values are then used to repeat the calculation until limiting values are attained. This can be performed manually, as in the table below (convergence is achieved after eight iterations), or by using the Goal Seek function. For $CuSO_4$, $z_+ = 2$, $z_- = 2$; $v_+ = 1$, $v_- = 1$. From Table 12.2, $a(Cu^{2+}) = 600$ pm and $a(SO_4^{2-}) = 450$ pm. The results are summarised in the table below which shows that convergence is achieved after eight iterations.

Iteration		1	2	3	4	5	6	7	8
α	0.0950	0.2745	0.3686	0.3978	0.4054	0.4073	0.4078	0.4079	0.4080
m_{CuSO_4}	0.4525	0.3627	0.3157	0.3011	0.2973	0.2963	0.2961	0.2960	0.2960
$m_{Cu^{2+}}$; $m_{SO_4^{2-}}$	0.0475	0.1373	0.1843	0.1989	0.2027	0.2037	0.2039	0.2040	0.2040
$\gamma_{Cu^{2+}}$	0.3329	0.2435	0.2240	0.2193	0.2182	0.2179	0.2178	0.2178	0.2178
$\gamma_{SO_4^{2-}}$	0.2884	0.1904	0.1695	0.1645	0.1633	0.1630	0.1629	0.1629	0.1629

The required activity values, therefore, are:

$$a_{Cu^{2+}} = \gamma_{Cu^{2+}} m_{Cu^{2+}} = 0.2178 \times 0.2040 = 0.0444$$

$$a_{SO_4^{2-}} = \gamma_{SO_4^{2-}} m_{SO_4^{2-}} = 0.1629 \times 0.2040 = 0.0332$$

$$a_{CuSO_4} = \gamma_{CuSO_4} m_{CuSO_4} = 1.0 \times 0.2960 = 0.2960$$

PROBLEMS

12.1 Calculate the approximate mass of H^+ and OH^- ions in 100 tonnes of water.

12.2 Using tabulated data (such as from the NBS compilation) calculate $\Delta_r H^0$, $\Delta_r S^0$, $\Delta_r G^0$ and K for the following reactions at 25°C:

$$Fe(s) + CuSO_4(aq) = Cu(s) + FeSO_4$$

$$2Ag^+(aq) + H_2S(g) = Ag_2S(s) + 2H^+(aq)$$

$$Ca^{2+}(aq) + 2NH_3(g) + 2H_2O(l) = Ca(OH)_2(s) + 2NH_4^+(aq)$$

12.3 Confirm that the following relations apply to a solution containing NaCl: $a_{NaCl} = (\gamma_-)^2 (m_{NaCl})^2 = (a_-)^2$

12.4 Confirm that the following relations apply to a solution containing $ZnCl_2$: $a_{ZnCl_2} = 1^1 \times 2^2 \times (\gamma_-)^3 (m_{ZnCl_2})^3 = 4(\gamma_-)^3 (m_{ZnCl_2})^3 = (a_-)^3$

12.5 Calculate the mean ionic activity and activity of each of the following salts in water if they are alone at a concentration of 0.2 mol kg⁻¹: NaCl, Na_2SO_4, $CaCl_2$ and $MgSO_4$. Data: At 0.2 mol kg⁻¹ and 25°C, $\gamma_\pm(NaCl) = 0.73$, $\gamma_\pm(Na_2SO_4) = 0.36$, $\gamma_\pm(CaCl_2) = 0.48$ and $\gamma_\pm(MgSO_4) = 0.13$.

12.6 Repeat the previous calculation for a solution of NaCl and Na_2SO_4 if the concentration of each is 0.2 mol kg⁻¹.

12.7 Estimate the solubility of CaF_2 in water at 25°C. Assume all activity coefficients are unity. Data: $\Delta_f G^0(CaF_2, s) = -1167.3$ kJ mol⁻¹; $\Delta_f G^0(Ca^{2+}, aq) = -553.58$ kJ mol⁻¹; $\Delta_f G^0(F^-, aq) = -278.79$ kJ mol⁻¹

12.8 Using the unit activity coefficient approximation, calculate the concentration of Ag^+ and S^{2-} ions in a solution saturated with Ag_2S at 25°C. If sodium cyanide is added to the solution the cyanide forms a complex with silver, $Ag^+(aq) + 2CN^-(aq) = Ag(CN)_2^-(aq)$. Calculate the concentration of $Ag(CN)_2^-$, Ag^+ and S^{2-} ions in a 0.15 mol kg⁻¹ sodium cyanide solution saturated with silver sulfide. Data: $\Delta_f G^0(Ag_2S, s) = -39.46$ kJ mol⁻¹, $\Delta_f G^0(Ag^+, aq) = 77.11$ kJ mol⁻¹; $\Delta_f G^0(S^{2-}, aq) = 85.8$ kJ mol⁻¹; $\Delta_f G^0(CN^-, aq) = 172.4$ kJ mol⁻¹; $\Delta_f G^0(Ag(CN)_2^-, aq) = 305.5$ kJ mol⁻¹

12.9 Estimate the solubility of barium sulfate in a 0.02 mol kg⁻¹ sodium sulfate solution. The solubility product for barium sulfate is 1.1×10^{-10}.

12.10 Estimate the solubility of CO_2 in water under a pressure of CO_2 of 1 bar. CO_2 dissolves in the molecular form in water. Assume unit activity coefficient in the solution. Data: $G^0(CO_2, aq) = -386$ kJ mol⁻¹; $G^0(CO_2, g) = -394$ kJ mol⁻¹.

12.11 Calculate the ionic strength of a solution containing 0.3 mol kg⁻¹ KCl and 0.5 mol kg⁻¹ $K_2Cr_2O_7$.

12.12 Estimate the activity coefficient and activity of SO_4^{2-} ions in a solution containing 0.02 mol kg⁻¹ Na_2SO_4 at 25°C using (i) the Debye–Hückel limiting law; (ii) the extended Debye–Hückel law. The Effective diameter of SO_4^{2-} ions is 400 pm. Assume complete dissociation of Na_2SO_4.

12.13 Calculate the ionic strength of a solution containing 0.02 mol kg⁻¹ $ZnCl_2$, then estimate the mean ionic activity coefficient of $ZnCl_2$ in the solution using the extended Debye–Hückel equation. Calculate the mean ionic molality, the mean ionic activity and activity of the $ZnCl_2$.

12.14 Estimate the pH of a solution containing 0.3 mol kg⁻¹ H_2SO_4. Assume unit activity coefficients.

12.15 What concentration of NaOH must be added to water to make the pH equal to 10.5? Assume unit activities.

12.16 Assuming unit activity coefficients, estimate the degree of dissociation of phosphoric acid, and the concentrations of H^+ and $H_2PO_4^-$ ions and undissociated H_3PO_4, in a solution containing 0.04 mol kg^{-1}. What is the pH of the solution? The dissociation reaction is: $H_3PO_4(aq) = H^+(aq) + H_2PO_4^-(aq)$; $K = 7.5 \times 10^{-3}$

12.17   Using the concentrations calculated in the previous question, estimate more accurate values of the ion activity coefficients by means of the Debye–Hückel limiting law, then recalculate the concentrations and repeat the calculation until convergence is obtained. What is the recalculated pH of the solution?

12.18   Repeat Problems 11.17 and 11.18 for the dissociation of hydrofluoric acid: $HF(aq) = H^+(aq) + F^-(aq)$; $K = 6.3 \times 10^{-4}$.

12.19   Estimate the activities of Cu^{2+}, Cl^- and $CuCl_2$ in a 0.2 mol kg^{-1} solution of cuprous chloride at 25°C if the dissociation reaction is: $CuCl_2(aq) = Cu^{2+}(aq) + 2\,Cl^-(aq)$. Use the unit activity coefficient approximation for the first estimate, then use the extended Debye–Hückel law and the iterative method to refine the estimates. Data: $\Delta_f G^0(CuCl_2, aq) = -197.9$ kJ mol^{-1} ; $\Delta_f G^0(Cu^{2+}, aq) = 65.49$ kJ mol^{-1}; $\Delta_f G^0(Cl^-, aq) = -131.23$ kJ mol^{-1}.

13 Phase equilibria: non-reactive systems

SCOPE

This chapter examines the equilibrium between phases, and the factors affecting the equilibrium state, for one-component (that is, single substances) and non-reactive two-component systems.

LEARNING OBJECTIVES

1. *Understand the phase rule and be able to apply it to one- and two-component non-reactive systems.*
2. *Be able to construct and interpret a p–T diagram for a pure substance.*
3. *Be able to apply the Clapyron and Clausius–Clapyron equations to simple problems.*
4. *Understand the main types of two-component phase diagrams and be able to interpret them.*
5. *Understand the thermodynamic basis of phase diagrams.*
6. *Understand the main ways in which phase diagrams can be determined.*
7. *Understand the equilibrium partitioning of a substance between phases.*

13.1 INTRODUCTION

By way of introduction, recall the following concepts introduced in Chapter 2. A *phase* is any part of a system which is physically homogeneous within itself and bounded by a surface so that it can be physically separated from the system. A phase may be composed of a single substance or it may be a solution of several substances. A phase need not be continuous; for example, a liquid dispersed as droplets throughout another immiscible liquid is a single phase. Multi-phase systems are said to be *heterogeneous* and single-phase systems are said to be *homogeneous*.

The chemical composition of non-reactive systems is expressed in terms of *components*, which are the chemically independent constituents comprising a system. It follows that the composition of phases comprising a system can also be expressed in terms of the components of the system as a whole. The *number of components C* in a system is the minimum number of independent species necessary to define the composition of all the phases of the system.

Though phases are mutually immiscible, a substance may occur in more than one phase in multi-phase systems; that is, a substance can be distributed between the phases of the system. For example, H_2O can occur as liquid and vapour in equilibrium (a one-component system), and alcohol added to water in a closed container (a two-component system) will become distributed between the liquid and vapour phases according to the conditions of temperature, pressure and concentration of alcohol. These are examples of non-reactive systems. CO_2 in the atmosphere is in equilibrium with CO_2 in the form of carbonic acid H_2CO_3 (or more accurately, as CO_3^{2-} ions) in the surface layers of the oceans. This is an example of distribution in a reactive multi-phase system. In this chapter we consider non-reactive systems; reactive systems are considered in Chapter 14.

13.2   EQUILIBRIUM IN MULTI-PHASE SYSTEMS

Consider a heterogeneous system composed of phases I, II, … in which components A, B, … are present. Let the total Gibbs energy of the system be G and the total Gibbs energy of each phase be (I), G(II) … . Now, consider the exchange of an infinitesimally small amount dn_A of component A between the phases. From the definition of partial molar quantity (Equation 9.3), the resulting Gibbs energy change of the phases at constant temperature and pressure is

$$dG(I) = G_A(I)\,dn_A(I)$$

$$dG(II) = G_A(II)\,dn_A(II)$$

$$\cdots$$

and the Gibbs energy change of the system is their sum:

$$dG = dG(I) + dG(II) + \cdots = G_A(I)dn_A(I) + G_A(II)dn_A(II) + \cdots$$

Since $dn_A(I) = -dn_A(II)$, that is, the amount of component A lost by one phase is equal to the amount gained by the other, then substituting into the previous equation,

$$dG = \left[G_A(I) - G_A(II)\right]dn_A(I)$$

At equilibrium, $dG = 0$ and, therefore,

$$G_A(I) = G_A(II) \tag{13.1}$$

Similarly,

$$G_B(I) = G_B(II), \text{ etc}$$

Equation 13.1 can be written in the form

$$\Delta G_A^0 = G_A(I) - G_A^0 = G_A(II) - G_A^0 = RT \ln a_A$$

Thus from the above, and Equation 9.48, we can make the following general (and equivalent) statements.

In a heterogeneous system at equilibrium:

- *The partial molar Gibbs energy of a component is the same in all phases;*
- *The chemical potential of a component is the same in all phases; and*
- *The activity of a component is the same in all phases.*

13.3   THE PHASE RULE

The phase rule is a tool which enables us to examine and greatly simplify the complex nature of multi-component, multi-phase systems. In examining such systems we are faced with the question: How many variables must be specified so that all the other variables of the system have unique values? The relevant variables are the state functions for the system (temperature, total pressure, partial pressures of the gaseous species) and the concentrations (or, alternatively, the activities) of the solid and liquid species. The answer to this question is important for at least two reasons. Firstly, in order to obtain relations (experimental or theoretical) between variables it is necessary to know how many variables must be specified (that is, measured or controlled) so that the relation is unique. Secondly, for controlling a process it is necessary to know how many variables must be controlled so that all other variables have unique values.

Consider a heterogeneous system containing C components. These will be distributed over a number of phases (even if in some cases the concentration of the component may be too small to be measured). Let the number of phases (I, II, III, …) be P. Assume that none of the species react with other species (the system is non-reactive) and that each species A, B, C, … is present in each phase. To specify the state of each phase we must specify the temperature, pressure and composition. Each phase contains C components and to specify its composition, therefore, requires $C-1$ variables (since $x_A + x_B + x_C + \cdots = 0$, the last variable is obtained by difference). The number of variables to specify for each phase then is $(C-1)$ plus 2 (temperature and pressure), that is, $C+1$. The number of variables required to completely specify P phases, therefore, is $P(C+1)$.

When the system is at equilibrium,

$$T(\mathrm{I}) = T(\mathrm{II}) = T(\mathrm{III}) = \cdots$$

$$p(\mathrm{I}) = p(\mathrm{II}) = p(\mathrm{III}) = \cdots$$

$$G_A(\mathrm{I}) = G_A(\mathrm{II}) = G_A(\mathrm{III}) = \cdots$$

For P phases there are $P-1$ equalities each for temperature and pressure, that is

$$T_1(\mathrm{I}) = T_2(\mathrm{II}),\ \ T_2(\mathrm{II}) = T_3(\mathrm{III}),\ \ T_3(\mathrm{III}) = T_4(\mathrm{IV}),\ldots T(P-1) = T$$

$$p_1(\mathrm{I}) = p_2(\mathrm{II}),\ \ p_2(\mathrm{II}) = p_3(\mathrm{III}),\ \ p_3(\mathrm{III}) = p(\mathrm{IV}),\ldots p(P-1) = p$$

Similarly, there are $P-1$ equalities for chemical potential for each of the C species:

$$G_A(\mathrm{I}) = G_A(\mathrm{II}),\ \ G_A(\mathrm{II}) = G_A(\mathrm{III}),\ \ G_A(\mathrm{III}) = G_A(\mathrm{IV}),\ldots G_{P-1}(P-1) = G_A$$

Therefore, the number of independent equations relating $C+2$ variables (C species plus temperature and pressure) is $(P-1)(C+2)$. Now, let F be the difference between the number of variables and the number of equations, then

$$F = P(C+1) - (P-1)(C+2) = PC + P - PC - 2P + C + 2$$

Therefore,

T and p are both variables: $F = C + 2 - P$ (13.2)

This is the *phase rule*. F is called the *variance* or the *degrees of freedom* of the system. It is the number of variables whose value may be chosen freely, and which must be chosen, before the system is in a unique, or *determinant*, state.

In the derivation of Equation 13.2 we assumed that each species A, B, C, … is present in each phase. Now, suppose that one of the components is absent from one of the phases; then the number of variables and the number of equations are both reduced by one and Equation 13.2 is unchanged. If pressure is not considered a variable, that is, it is held constant, then the number of variables required to completely specify P phases is PC, and the number of independent equations is $(P-1)(C+1)$. In this case the difference is:

T is a variable; p is not a variable: $F = C + 1 - P$ (13.3)

13.4 ONE-COMPONENT SYSTEMS

A one-component system consists of one pure substance. Pure substances can exist in three states – solid, liquid and gas. For a pure substance, these states each represent a phase. It is the relationship between these that is of interest here. Applying the phase rule for this case, $C = 1$, and the degrees of freedom are

$$F = 1 + 2 - P = 3 - P$$

When only one phase is present, $P = 1$ and $F = 2$. This means that both temperature and pressure can be varied independently over a certain range and the substance will remain in that particular state. Liquid water, for example, remains liquid over a range of temperatures and pressures. Therefore, a single phase is represented by an area on a pressure–temperature graph, as shown in Figure 13.1.

When two phases are in equilibrium (solid–liquid, solid–gas or liquid–gas), $P = 2$ and $F = 1$, and only one of temperature and pressure can be independently varied. The system is said to be *univariant*. We can choose the temperature of the system but not the pressure, or *vice versa*. For example, at 100°C the pressure of the single-component two-phase water–steam system is 1 atm. In this case, the equilibrium of two phases is represented by a line in the phase diagram (Figure 13.1). There are three lines, representing the three equilibria: Solid–liquid, solid–gas, liquid–gas.

When the three phases (solid, liquid and gas) co-exist, $P = 3$ and $F = 0$, and neither temperature nor pressure can be varied independently. This condition can only be achieved at a unique temperature and pressure, and the equilibrium of the three phases is therefore represented by a point on the phase diagram, called the *triple point*. The triple point of water occurs at 273.16 K (0.01°C) and a pressure of 611.7 Pa. The system is said to be *invariant*. There is an upper limit to the vapour pressure curve in Figure 13.1, the *critical point*, the point at which the distinction between the vapour and liquid disappears and only one phase is present. At temperatures and pressures greater than those of the critical point, *supercritical fluid* fills the container (Section 3.4.1).

The curves in Figure 13.1 could in principle be determined experimentally by applying pressure to a substance in a closed container and determining the state of the substance over a range of temperatures. This could then be repeated at various pressures. In practice, there are other, often

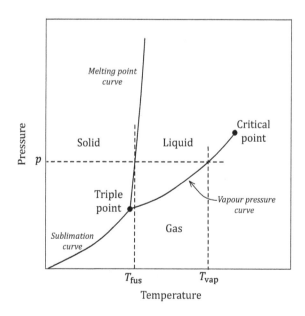

FIGURE 13.1 Typical p–T diagram for a pure substance. T_{fus} and T_{vap} are the melting/freezing point and boiling/condensation point, respectively, of the substance at pressure p.

simpler, ways to construct such a diagram, and the principles for doing this are discussed later in the chapter.

In Section 2.2 we saw that the pressure of the vapour in equilibrium with the liquid or solid form of a substance, the vapour pressure of the substance, increases with increasing temperature. In an open system, molecules escaping from the surface are not contained within the system but diffuse away (as, for example evaporation of water or gasoline from an open container). In the case where the substance is liquid when the vapour pressure reaches that of the external pressure (for example that of the atmosphere) vapourisation begins to occur throughout the bulk of the liquid, and the vapour can expand into the surroundings. This is the phenomenon of *boiling*, and the temperature at which this occurs at a particular external pressure is the *boiling point*, T_{vap}. If the external pressure is 1 atm the boiling point is called the *normal boiling point*, and if it is 1 bar it is called the *standard boiling point*. In the case where the substance is still solid when the vapour pressure reaches that of the external pressure (for example dry ice, or solid carbon dioxide, in air) bulk vapourisation occurs from the solid. This phenomenon is called *sublimation*, and the temperature at which it commences is the *sublimation temperature*, T_{sub}.

13.4.1 AN EXAMPLE OF A *P–T* DIAGRAM: CARBON

The diagram in Figure 13.2 represents the present state of knowledge of the pressure–temperature relationship for elemental carbon. The diagram is limited to the well-known solid forms of carbon (graphite and diamond) and liquid carbon. Less well-known forms such as amorphous glassy carbon and carbon black and the crystalline fullerenes, such as C_{60} buckyballs and C_{70} buckyfootballs, are not considered here.

The graphite – diamond – liquid triple point is approximately 5000 K, 12 GPa. The binding energy between atoms of carbon is very large, and this is the reason for the very high melting point of its solid forms (~5000 K). Also, it is the reason carbon atoms in a stable configuration require a large activation energy to produce a transformation to a different stable phase. Thus, very high

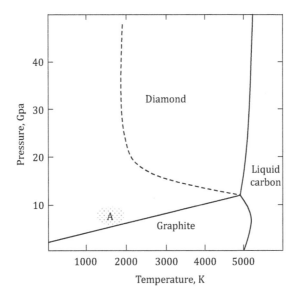

FIGURE 13.2 The *p–T* diagram for carbon. The dashed curve is the *p–T* threshold of very fast (< 1 ms) solid–solid transformation of graphite to diamond. The shaded area A represents the region for commercial production of diamonds from graphite using catalysts to affect the transformation. As a result, the diamonds are slightly contaminated by the catalyst material. Source of data: F. Bundy, P. *et al.* 1996. *Carbon* 34(2), 141–153.

temperatures are often required to initiate spontaneous transformations from one solid phase to another.

Because of the high binding and activation energies, carbon polymorphs exist as metastable phases well into a T, p region where a different solid phase is thermodynamically stable. For example, diamond survives indefinitely at ambient temperature under conditions where graphite is the thermodynamically stable form. Conversely, except at very high temperatures, graphite persists at pressures far into the diamond stability field. Catalysts can be used to lower the activation energy (and, hence, the temperature of transformation), and this has made it possible to produce diamonds economically from graphite on a commercial scale.

13.4.2 Stability of Phases – The Clapyron and Clausius–Clapyron Equations

Consider two phases, I and II, of pure substance B. These could be solid, liquid or gas. B(I) can be transformed into B(II) by the transition reaction:

$$B(I) = B(II)$$

There are three possible states for a closed system containing B in which the temperature and pressure are held constant:

B(I) and B(II) co-exist in equilibrium; in this case, $\Delta G = 0$, and $G(I) = G(II)$.
Only B(I) is stable; in this case, $\Delta G > 0$, and $G(II) > G(I)$.
Only B(II) is stable; in this case, $\Delta G < 0$, and $G(I) > G(II)$.

Thus the stable phase at any temperature and pressure is the phase with the smallest Gibbs energy.

At constant pressure, $dG / dT = -S$ (Equation 8.13), and, since S can have only positive values, G will always decrease as temperature is increased. The effect is greatest for gases and least for solids since $S(g) \gg S(l) > S(s)$. Thus, the typical change in the Gibbs energy of a substance on heating the pure solid substance at constant pressure will have the general form shown in Figure 13.3. Note, the relative slopes of the lines in the figure reflect the effect of temperature on G. The stable phase at any temperature is the phase with the smallest Gibbs energy. The points of intersection of the

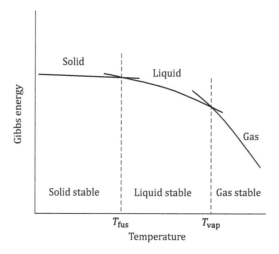

FIGURE 13.3 The variation of G with temperature for a pure substance which undergoes the transformation s → l → g at constant pressure.

curves give the transition temperatures – melting point, boiling point, etc. For the case where the solid transforms directly to gas (that is, it sublimes), the curve for the liquid state will lie above the intersection of the curves for the solid and gaseous states.

At constant temperature, $dG/dp = V$ (Equation 8.12). Since V is always positive, G will increase with increase in pressure at constant temperature. The effect is greatest for gases and, usually, least for solids since $V(g) \gg V(l) > V(s)$. H_2O, As, Sb and Bi are exceptions since for these substances $V(s) > V(l)$. Figure 13.4 shows the variation of G at two pressures as a pure substance is heated. The result of an increase in pressure is an elevation in both the boiling point (ΔT_{vap}) and the melting point (ΔT_{fus}) with the exception that the melting point is depressed for those substances for which $V(s) > V(l)$.

To summarise:

- *The state of a substance with the smallest Gibbs energy at a particular temperature and pressure is the state that is stable at that temperature and pressure.*
- *Alternatively, the magnitude of ΔG for the transformation reaction between two states determines which of the states is stable at a particular tyemperature and pressure.*

Since the Gibbs energy of a substance is the same in each phase at equilibrium, then from Equation 8.11,

$$dG(I) = V(I)dp - S(I)dT$$

$$dG(II) = V(II)dp - S(II)dT$$

and, subtracting,

$$d\left[G(II) - G(I)\right] = \left[V(II) - V(I)\right]dp - \left[S(II) - S(I)\right]dT$$

or

$$d(\Delta G) = \Delta V \, dp - \Delta S \, dT \tag{13.4}$$

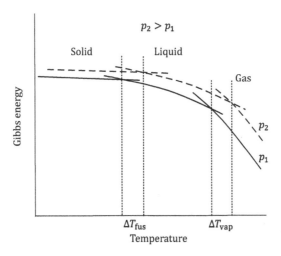

FIGURE 13.4 The effect of pressure on the variation of G with temperature for a pure substance which undergoes the transformation s→l→g.

At any point on the any of the three lines in Figure 13.1, two phases are in equilibrium, and since $d(\Delta G)$ will then be zero, Equation 13.4 becomes:

$$\Delta V\,dp - \Delta S\,dT = 0$$

Therefore, for 1 mole of substance

$$\frac{dp}{dT} = \frac{\Delta S_m}{\Delta V_m} \tag{13.5}$$

Since $\Delta G_m = \Delta H_m - T\Delta S_m$ and $\Delta G_m = 0$, $\Delta H_m = T\Delta S_m$. Substituting into Equation 13.5 yields

$$\frac{dp}{dT} = \frac{\Delta H_m}{T\Delta V_m} \tag{13.6}$$

Equations 13.5 and 13.6 are forms of the *Clapyron equation*. These equations describe the slope of any two-phase equilibrium line for a pure substance and enable calculation of the boundary line.

The solid–liquid boundary: melting point

The slope of the solid–liquid boundary line is given by

$$\frac{dp}{dT} = \frac{\Delta_{fus}H_m}{T_{fus}\,\Delta_{fus}V_m} \tag{13.7}$$

where T_{fus} is the melting point of the substance, $\Delta_{fus}H_m$ is the enthalpy of fusion and $\Delta_{fus}V_m$ the change in volume of the substance on melting. $\Delta_{fus}V_m$ does not change greatly with temperature. Furthermore, $\Delta_{fus}H_m\,/\,T_{fus}$ (which is equal to the entropy change from the solid to the liquid form, $\Delta_{fus}S_m$) also does not change greatly with temperature. Therefore, dp/dT is approximately constant; that is, the relationship between temperature and pressure is approximately linear (see Figure 13.5).

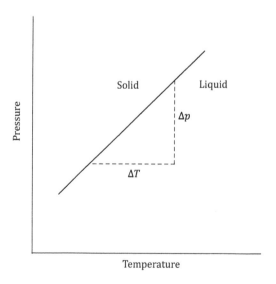

FIGURE 13.5 The solid–liquid boundary. The relationship is approximately linear with a slope of $\Delta p/\Delta T$.

EXAMPLE 13.1   The effect of pressure on melting poin

The melting point of ice at 1 bar is 0°C for which $\Delta_{fus}H^0 = 6.007$ kJ. Estimate the temperature at which ice will melt at 1000 bar.

SOLUTION

Data: $V_m(H_2O, s, 298.15) = 19.65$ mL mol^{-1}; $V_m(H_2O, 1, 298.15) = 18.01$ mL mol^{-1}
The transformation is:

$$H_2O(s) = H_2O(l); \quad \Delta_{fus}H^0 = 6.007 \text{ kJ mol}^{-1}$$

From Equation 13.7

$$\Delta T = \frac{\Delta p \, T_{fus} \, \Delta_{fus}V_m}{\Delta_{fus}H_m}$$

and

$$T - 273.15 = \frac{(1000-1) \times 10^5 \times 273.15 \times (18.01 - 19.65) \times 10^{-6}}{6007} = -7.45 \text{ K}$$

The temperature at which ice melts at 1000 bar, therefore, is $273.15 - 7.45 = 265.70$ K.

The liquid–gas boundary: vapour pressure and boiling point

The slope of the liquid–vapour boundary line is given by

$$\frac{dp}{dT} = \frac{\Delta_{vap}H_m}{T\Delta_{vap}V_m}$$

Since $V(vap) \gg V(l)$, then $\Delta V = V(vap)$ and if the gas behaves ideally, $V_m = \dfrac{RT}{p}$ and, therefore

$$\frac{dp}{dT} = \frac{p\,\Delta_{vap}H_m}{RT^2}$$

or

$$\frac{dp}{p} = \frac{\Delta_{vap}H_m}{RT^2}\, dT$$

$$\frac{d\ln p}{dT} = \frac{\Delta_{vap}H_m}{RT^2} \tag{13.8}$$

which is called the *Clausius–Clapyron* equation.

If it is assumed that $\Delta_{vap}H_m$ is constant (independent of temperature),

$$\int d\ln p = \frac{\Delta_{vap}H_m}{R}\int \frac{dT}{T^2}$$

Therefore,

$$\ln p = \frac{-\Delta_{vap}H_m}{RT} + c \tag{13.9}$$

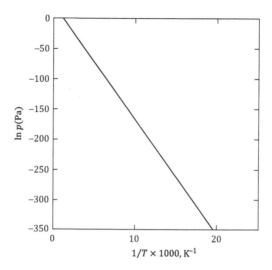

FIGURE 13.6 The variation of the vapour pressure of lithium with temperature. The slope of the liquid–vapour line is −19.134; therefore the enthalpy of vapourisation of lithium over the liquid temperature range is approximately $8.314 \times 19.134 \times 1000 = 161$ kJ mol^{-1}.

where c is the integration constant. Alternatively, integration over the range (p_1, T_1) to (p_2, T_2) yields

$$\ln p_2 - \ln p_1 = \left(\frac{-\Delta_{vap}H_m}{R}\right)\left(\frac{1}{T_2} - \frac{1}{T_1}\right) \tag{13.10}$$

Equations 13.9 and 13.10 are alternative forms of the *Clausius–Clapyron* equation. Equation 13.9 implies there is a linear relation between the natural logarithm of the vapour pressure of a pure liquid and $1/T$, with slope equal to $-\Delta_{vap}H_m / R$ (see Figure 13.6). Equation 13.10 is useful for estimating the vapour pressure of a pure liquid if the value is known at some other temperature or for determining the enthalpy of vapourisation if the variation of vapour pressure with temperature is known.

EXAMPLE 13.2   The effect of temperature on vapour pressure

The boiling point of water at 1 bar is 99.63°C at which $\Delta_{vap}H^0 = 40.893$ kJ mol^{-1}. Estimate the vapour pressure of water at 70°C.

SOLUTION
The transformation is:

$$H_2O(l) = H_2O(g); \quad \Delta_{vap}H^0 = 40.893 \text{ kJ}$$

From Equation 13.10,

$$\ln p_2 = \ln p_1 + \left(\frac{-\Delta_{vap}H_m}{R}\right)\left(\frac{1}{T_2} - \frac{1}{T_1}\right)$$

$$= \ln(10^5) + \left(\frac{-40893}{8.314}\right) \times \left(\frac{1}{(70 + 273.15)} - \frac{1}{(99.63 + 273.15)}\right) = -1.13929$$

Therefore, $p_2 = 0.32$ bar.

The solid–gas boundary: sublimation temperature

Similar equations apply for the solid–gas boundary as for the liquid–gas boundary with the exception that $\Delta_{sub}H_m$ replaces $\Delta_{vap}H_m$ in Equations 13.8 to 13.10:

$$\frac{d \ln p}{dT} = \frac{\Delta_{sub}H_m}{RT_{sub}^2} \tag{13.11}$$

$$\ln p = \frac{-\Delta_{sub}H_m}{RT_{sub}} + c \tag{13.12}$$

$$\ln p_2 - \ln p_1 = \left(\frac{-\Delta_{sub}H_m}{R}\right)\left(\frac{1}{T_2} - \frac{1}{T_1}\right) \tag{13.13}$$

Since $\Delta_{sub}H_m > \Delta_{vap}H_m$, Equation 13.11 predicts a steeper slope for the sublimation curve than for the vapourisation curve, as shown in Figure 13.1.

13.4.3 THE EFFECT OF EXTERNAL PRESSURE ON VAPOUR PRESSURE

When a substance is heated in a vessel that is open to the atmosphere it is subject to an external pressure. The vapour pressure of a substance (Section 2.2) is the pressure exerted by the vapour of the substance in a closed system containing only that substance. Thus the question arises: Is the vapour pressure of a substance at a particular temperature the same when it is heated in an open vessel as the vapour pressure measured as described in Section 2.2? To answer this question, we now derive an equation for the effect of an external pressure on the vapour pressure of a substance.

Since the Gibbs energies of two phases in equilibrium are equal (Equation 13.1),

$$G_m(g) = G_m(l)$$

where $G_m(g)$ is the Gibbs energy of 1 mole of the gaseous substance in equilibrium with 1 mole of the liquid substance. Therefore,

$$dG_m(g) = dG_m(l)$$

and, from Equation 8.11, at constant temperature

$$V_m(g)\,dp(g) = V_m(l)\,dp(l)$$

Rearranging, and assuming the vapour behaves ideally ($pV=RT$),

$$\frac{dp(g)}{dp(l)} = \frac{V_m(l)}{V_m(g)} = \frac{V_m(l)\,p(g)}{RT}$$

or

$$\frac{dp(g)}{p(g)} = \frac{V_m(l)}{RT}\,dp(l)$$

This equation relates the pressure applied to a liquid to its vapour pressure. A similar equation can be written for a solid substance and its vapour. If we assume $V(l)$ is constant (that is, the external pressure has a negligible effect on the volume of a liquid or solid), then integrating between $p_1(g)$ and $p_2(g)$,

$$\ln\left(\frac{p_2(g)}{p_1(g)}\right) = \frac{V_m(l)\left[p_2(l) - p_1(l)\right]}{RT} \tag{13.14}$$

where $p(g)$ is the vapour pressure of the liquid and $p(l)$ is the pressure applied to the liquid. A similar equation applies for solids.

EXAMPLE 13.3   The effect of external pressure on the vapour pressure of a liquid

The vapour pressure of water at 50°C is 12.344 kPa in a one-component system. Calculate the change in vapour pressure of water when the pressure applied to the water is increased from 12.344 kPa to 101325 Pa (1.0 atm), that is, when the water is in a container open to the atmosphere. Data: $V_m = (H_2O) = 18.3 \times 10^{-5}$ m^3.

SOLUTION

Using Equation 13.14,

$$\ln\left(\frac{p_2(g)}{p_1(g)}\right) = \frac{18.3 \times 10^{-5} \times (101\ 325 - 12.344)}{8.314 \times (50 + 273.15)} = 0.00069$$

Therefore,

$$\frac{p_2(g)}{p_1(g)} = 1.00069$$

Thus, the vapour pressure of water has increased by about 0.07% due to the external pressure of 1 atm.

COMMENT

Repeating the calculation for an external pressure of 10 atm gives a 0.7% increase in vapour pressure.

The change in equilibrium vapour pressure with external pressure calculated using Equation 13.14 is found to be small in most cases, and the Clausius–Clapyron equation for a one-component system allows quite accurate prediction of vapour pressures under normal open atmospheric conditions.

13.5   TWO-COMPONENT SYSTEMS

In one-component systems the variables of interest are temperature and pressure. In two-component or binary systems, the variables of interest are temperature, pressure and composition. From the phase rule, since $C = 2$, then $F = 4 - P$. Thus, at an invariant point (at which T, P and x have unique values), $F = 0$, and four phases co-exist in equilibrium. Along a univariant curve (along which only one of T, P and x can be varied independently), $F = 1$, and three phases co-exist.

The effects of temperature, pressure and composition on the phases which exist at equilibrium can be represented graphically, but while three-dimensional phase diagrams (T, p, x) can be constructed, particularly with the aid of computer graphics, they are often difficult to visualise and interpret. To represent a system with three variables in two dimensions it is necessary to hold one variable constant, and for this purpose pressure is usually chosen, and often it is held at 1 bar (or 1 atm). Such a two-dimensional representation can be thought of as a section through the three-dimensional diagram at a particular pressure. This makes it possible to present temperature–composition phase diagrams which show which phases are stable in the various regions of the diagram. When pressure is held constant, or the state of the system is independent of pressure, then according to Equation 13.3 three phases co-exist at univariant points and two phases co-exist along univariant curves.

For ease of interpretation of phase diagrams, composition is often expressed in mass percent, but mole fraction is sometimes used, and temperature is usually expressed in Celsius. Phase diagrams are constructed from direct experimental measurements on the system of interest or calculated from thermodynamic data. These techniques are discussed in Sections 13.6 and 13.7.

13.5.1 SOLID–LIQUID, SOLID–SOLID AND LIQUID–LIQUID SYSTEMS

Solid–liquid and liquid–liquid systems are frequently encountered in materials, metallurgical and geological applications, for example alloy systems (such as Fe–C, Cu–Sn, Pd–Sn, etc.) and mineral-forming systems (such as $CaO–SiO_2$, $SiO_2–Al_2O_3$, etc.). We consider first the case where the components A and B of a binary system are mutually soluble over the entire composition range. The simplest phase diagram of this type is shown in Figure 13.7(a) which is typical of systems in which the solid and liquid solutions are ideal or in which the degree of deviation from ideality of the solid and liquid is similar. The upper curve is known as the *liquidus* and represents the temperatures above which only liquid phase is present. The area above the curve is labelled L to indicate liquid phase. The lower curve is known as the *solidus* and represents the temperatures below which only solid is present, in this case a solid solution of A and B of variable composition. The area below the curve is labelled α to indicate the solid solution phase. Between the liquidus and solidus curves solid and liquid coexist in a state of equilibrium, and the area between the curves is labelled L + α. The lowest point on the temperature axis, at which the solidus and liquidus lines intersect, is the melting point of pure A, and the highest point is the melting point of pure B.

If there is a small deviation from ideality the central portion of both curves will be displaced slightly upwards (negative deviations) or downwards (positive deviation) and will pass through a maximum or minimum, and at lower temperature the solid solution will separate into two phases, α and β. Such a feature is called a *miscibility gap*. The downward displaced curve illustrated in Figure 13.7(b) is the more common since the solid solution usually has greater positive deviation from ideality than the liquid phase. With even greater positive deviation in the solid solution a *eutectic* type of diagram is obtained as illustrated in Figure 13.7(c). This can be considered as being derived from Figure 13.7(b) by further lowering of the central portions of the liquidus and solidus curves and increasing the size of the miscibility gap (that is, limiting the extent of solid solution). Three phases co-exist at the *eutectic point* (α, β and liquid) and hence F = 0. The eutectic point therefore is invariant, meaning for a given system the eutectic composition and temperature are unique. Further intrusion of the miscibility gap into the solidus curve of Figure 13.7(c) results in a *peritectic* type of phase diagram, illustrated in Figure 13.7(d), the *peritectic point* being the temperature and composition at which the three phases α, β and liquid co-exist in equilibrium.

Some systems have a miscibility gap in the liquid state; that is, the components are not completely miscible. An example is shown in Figure 13.7(e). It is common for this gap to become narrower as the temperature increases and eventually disappear at some temperature, called the *critical temperature*. The point at which two liquid phases and the solid solution phase are in equilibrium is called the *monotectic point*. There are, however, many systems in which complete miscibility can never be reached (at practically attainable temperatures) at atmospheric pressure, for example, the Fe–Pb system (see Figure 13.8).

If components A and B have a tendency to form a compound AB_x (where x is the stoichiometric coefficient), then diagrams of the type in Figure 13.7(f) are common. This diagram can be visualised as two binary eutectic systems $A–AB_x$ and $AB_x–B$. If AB_x is a stoichiometric compound the vertical (dashed) line at composition AB_x is the boundary between the two systems. If AB_x is a non-stoichiometric compound the variable composition is represented by the area of phase β.

It is possible in some systems that there are no molten phases within the temperature range of interest. In these cases, diagrams of the same form as those in Figure 13.7 would exhibit a solid solution phase at the higher temperatures instead of a liquid phase. In these cases the terms *eutectoid*,

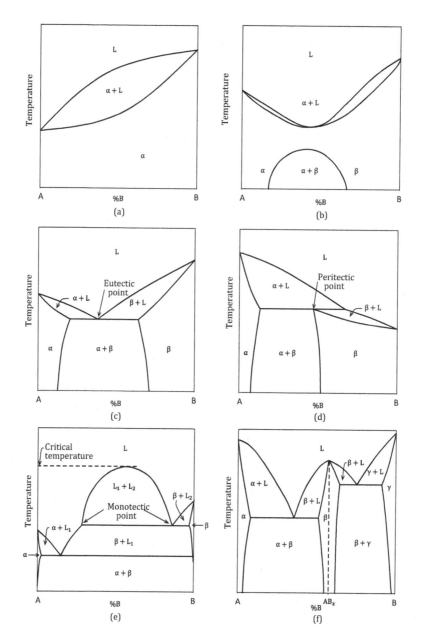

FIGURE 13.7 Common types of binary phase diagrams: (a) system with complete solid solubility, (b) system with a minimum in the solidus and liquidus curves and partial miscibility in the solid state, (c) eutectic system with partial solid solubility, (d) peritectic system with partial solid solubility, (e) system with a liquid miscibility gap and (f) eutectic system with compound formation.

peritectoid and *monotectoid* are used rather than the corresponding liquid–solid terms of eutectic, pertitectic and monotectic.

13.5.2 INTERPRETING PHASE DIAGRAMS

Phase diagrams contain much useful information. Figure 13.9 will serve to demonstrate features common to all phase diagrams.

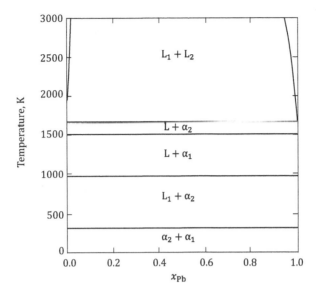

FIGURE 13.8   The Pb–Fe phase diagram exhibiting immiscibility even at temperatures greater than 3000 K. (Source of data: http://resource.npl.co.uk/mtdata/phdiagrams/fepb.htm)

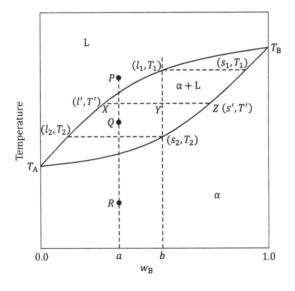

FIGURE 13.9   Binary phase diagram illustrating the principles of phase diagram interpretation.

1. *If the temperature and composition of the system are known, then the number of phases, the types of phases present and their composition are known.*

This is clear from the foregoing discussion. For example, at point P one phase, liquid, is present; at point Q two phases, liquid and solid solution, are present; and at point R one phase, solid solution, is present. The bulk composition in each case is given by point a on the composition axis. Lines drawn horizontally as shown in the figure are called *tie lines*. Their significance is that they join phases that coexist at a particular temperature; their end-points give the composition of the phases in equilibrium. Tie lines are horizontal because they join coexisting phases at the same temperature. Thus at bulk composition b and temperature T', the system will consist of a mixture of liquid of composition l' and a solid solution of composition s'.

If a melt of A and B of composition b and temperature above the liquidus temperature is slowly cooled to T_1, solid of composition s_1 will be begin to be precipitated. If cooling is continued more solid will precipitate, and the bulk composition of the solid will move downwards along the solidus curve towards A. The liquid will become progressively depleted in B (since the solid phase is richer in B than the solution), and the composition of the liquid will correspondingly move downwards along the liquidus line towards A. This will continue until temperature T_2 is reached at which the last remaining liquid, of composition l_2, solidifies and the composition of the solid solution reaches that of b. Note that during the cooling process the composition of the solid solution is continually changing and becoming enriched in component A. For this to occur in practice requires the cooling process to be very slow so diffusion of A and B in the solid solution can maintain the equilibrium composition at all times. This illustrates another important feature of phase diagrams: They are equilibrium diagrams. If the cooling is too fast non-equilibrium, or metastable, solid phases form.

  2. *If the temperature and composition of the system are known, then the relative amount of each phase is known.*

Consider the system of composition b and temperature T' in Figure 13.9. The mass of component B is equal to the sum of the mass of B in the two phases:

$$w_B m_T = l' m_l + s' m_s$$

where m_T is the total mass of A and B in the system, w_B is the mass fraction of B in the system, m_l is the mass of liquid of composition l' (mass fraction of B) and m_s is the mass of solid of composition s' (mass fraction of B). The mass of B is also given by:

$$w_B m_T = w_B (m_l + m_s) = w_B m_l + w_B m_s$$

Equating these two expressions

$$l' m_l + s' m_s = w_B m_l + w_B m_s$$

Rearranging,

$$\frac{m_s}{m_l} = \frac{w_B - l'}{s' - w_B}$$

or, in terms of the line XYZ in Figure 13.9,

$$\frac{m_s}{m_l} = \frac{XY}{YZ} \tag{13.15}$$

Equation 13.15 is known as the *lever rule*. The name derives from the analogy to a mechanical lever; the two-phase region is analogous to a lever balanced on a fulcrum.

  If we consider one unit of total mass, then

$$m_T = m_l + m_s = 1$$

and

$$m_l = w_l; \quad m_s = w_s$$

where w_l is the mass fraction of liquid and w_s is the mass fraction of solid. Therefore, from Equation 13.15

$$\frac{w_s}{1-w_s} = \frac{XY}{YZ}$$

Rearranging results in

$$w_s = \frac{XY}{XY+YZ}$$

Similarly,

$$w_l = \frac{YZ}{XY+YZ}$$

3. *If the composition of the system is known, then the equilibrium internal structure (micro-structure) of the solid formed by cooling of the liquid can be predicted.*

Cooling from the liquid state to the solid state results in the development of an internal structure of the solid, called the *microstructure*. If the cooling is sufficiently slow that equilibrium is maintained throughout the cooling process, the microstructure of the solid can be predicted. For a phase diagram of the type illustrated in Figure 13.7(a), the evolution of the microstructure of the solid as the liquid is cooled is illustrated in Figure 13.10. When temperature T_1 is reached, solid crystals of α begin to form within the liquid. As cooling continues the amount of crystals increases and liquid decreases. Their compositions change also as discussed previously. By the time temperature T_2 is reached all liquid has solidified and all the crystals have grown to occupy the volume of the system. The progressive development of the microstructure, points W, X, Y and Z, is illustrated in the figure.

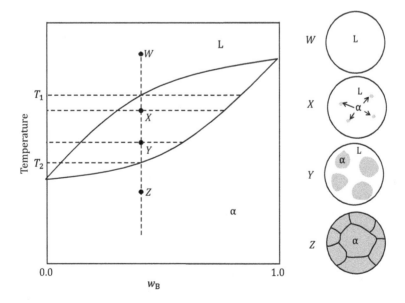

FIGURE 13.10 The equilibrium development of the microstructure in a binary system exhibiting complete solid solubility as it is cooled from the molten state. In cases where there is no initial liquid phase (only solid solutions), the sequence of microstructure development is the same.

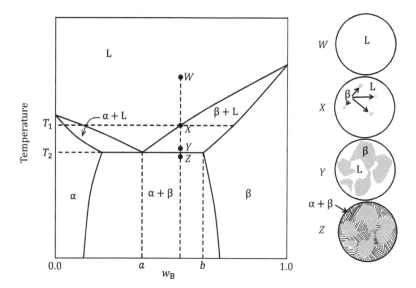

FIGURE 13.11 The equilibrium development of the microstructure in a binary eutectic system as it is cooled from the molten state. In the cases where there is no initial liquid phase, only solid solutions, the sequence of microstructure development is the same.

A more complex case is illustrated in Figure 13.11, for a eutectic system. As the liquid cools, the state of the system crosses into the $\alpha+L$ or $\beta+L$ field, depending on the bulk composition. At the composition indicated, at temperature T_1 solid β begins to crystallise. The amount of β increases as the system continues to cool, and the composition of β and liquid change as described under Point 1, until temperature T_2 is reached. At this temperature, the liquid has the composition a (called the eutectic composition) and the solid has the composition b. According to the phase diagram, below T_2 only solid is present. As the system cools through T_2, the remaining liquid solidifies. It does this isothermally, that is, at constant temperature T_2. This requires both solid α and solid β to crystallise simultaneously, and in doing so they form a mixture of α and β. This mixture often has a lamella (or layered) structure, and the microstructure is referred to as a eutectic structure. The progressive development of the microstructure, points W, X, Y and Z, is illustrated in the figure.

13.5.3 LIQUID–VAPOUR SYSTEMS

Thus far, while considering two-component systems, the gas phase has been ignored. This is acceptable when the vapour pressure of both components is sufficiently small to be neglected. When the vapour pressure of one or both components is significant in the temperature range of interest, liquid–vapour phase diagrams can be constructed which show the composition of the liquid and vapour phases in equilibrium at a particular pressure (often 1 atm or 1 bar).

Miscible liquids

For ideal systems, or those not deviating greatly from ideality, the form of the diagram will be similar to that of Figure 13.9 but with the liquid phase region in the lower part and vapour phase in the upper part. An example is shown in Figure 13.12. The lower curve represents the boiling point of the liquid (at the particular pressure p) and the upper curve the temperature at which vapour begins to condense (at pressure p). The principles of interpretation of the diagram are the same as for solid–liquid, solid–solid and liquid–liquid diagrams.

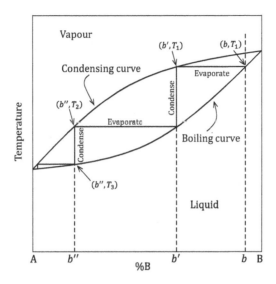

FIGURE 13.12 Temperature–composition phase diagram at constant pressure for an ideal, or nearly ideal, solution in which component A is more volatile than B.

Distillation

Consider what happens when a system of composition b in Figure 13.12 is cooled from the vapour state. When the temperature crosses the condensing curve, liquid richer in A than the bulk composition will begin to condense. The composition of the vapour and liquid will move along the condensing and boiling curves, respectively, towards A as cooling continues. At temperature T_1, the liquid ceases to boil, and at temperatures below T_1 the liquid composition remains constant. The total vapour pressure of A and B at temperatures less than T_1 will be less than p and will vary with temperature as shown in Figure 9.6.

Now consider what happens when liquid of composition b is heated. It begins to boil at T_1, and the initial vapour has composition b'. In simple distillation, the vapour is drawn off as it is formed and then condensed. The first trace amount of condensate has composition b'. As the remaining liquid becomes depleted in component A, the condensate becomes progressively enriched in A. In *fractional distillation*, the boiling and condensation cycle is repeated successively. If the condensate of composition b' is collected and reheated it boils at T_2 and yields vapour of composition b'' which if condensed will yield a liquid of composition b''. The cycle can be repeated until the desired levels of purities of A and B are obtained. Thus it is possible to obtain almost complete separation of components A and B.

In practice, fractional distillation is usually performed in a continuous manner in a fractionating column rather than in the batch manner described. A fractionating column contains a series of plates or stages where the liquid and vapour are in equilibrium (see Figure 13.13). These correspond to the horizontal steps in Figure 13.12. Liquid feed is introduced at some intermediate position, liquid moves downwards into a boiler at the base, and vapour moves in counter-current flow upwards to a condenser at the top of the column. Part of the condensed liquid at the top is drawn off as separated component B, and liquid from the boiler at the base is extracted as separated component A. Part of the condensed liquid at the top is returned as reflux to the top of the column.

Azeotropes

Many systems have temperature–composition diagrams similar to that in Figure 13.12, but systems which exhibit strong deviation from ideality have forms as shown in Figure 13.14. The presence of

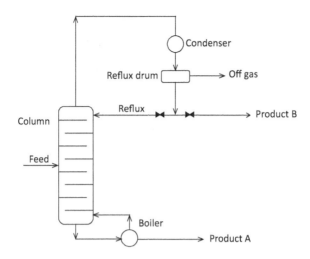

FIGURE 13.13   Typical layout of an industrial-scale system for fractional distillation.

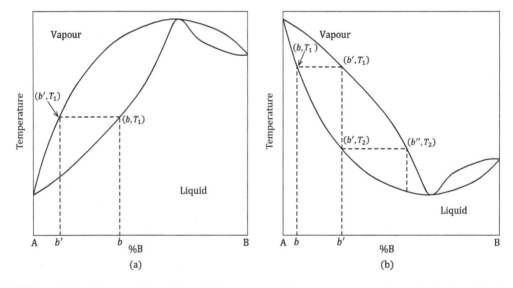

FIGURE. 13.14   Solutions of A and B exhibiting azeotropic behaviour: (a) negative deviation from ideality, (b) positive deviation from ideality.

a maximum or minimum in the phase boundary has an important implication. At the composition at the maximum or minimum, the liquid and vapour curves touch, and the composition of the liquid and vapour in equilibrium are the same. If a solution of this composition is heated to the boiling point evaporation then occurs without any change in composition of the liquid. Such solutions are said to be *azeotropic*, or *constant boiling mixtures*, and A and B are said to form an *azeotrope*. Hydrochloric acid and water, for example, form an azeotrope at 80 mass percent water which boils at 108.6°C. Both figures can be divided into two parts by a vertical line through the maximum or minimum, and it is evident that each part has the appearance of the simple system in which the boiling point changes systematically with composition. Each part may be interpreted in the same manner as for the simple system.

Consider a solution of composition b in Figure 13.14(a). If the liquid is heated to the boiling point T_1, vapour of composition b' will be formed. If the vapour is continuously removed, the composition of the remaining liquid will move upwards along the boiling curve (and the boiling point will increase), and the composition of the vapour will move upwards along the condensing curve. When the azeotropic composition is reached, the vapour has the same composition as the liquid and any further vapourisation then occurs without any change in composition. If the system is fractionally distilled, the composition of the vapour, and hence the condensate, will move towards A if the original mixture lies to the left of the maximum and towards B if the original mixture lies to the right, while the composition of the residual liquid will move towards the azeotropic composition.

A similar situation occurs for systems of the type in Figure 13.14(b). In this case, the composition of the condensate will move towards the azeotropic composition while the composition of the residual liquid will move towards A if the original mixture lies to the left of the minimum and towards B if the original mixture lies to the right. It will be apparent that, in both types of systems, it is possible to obtain in the nearly pure state by fractional distillation only the component present in excess of the amount required to form the constant boiling point solution.

Immiscible liquids

Now consider the case where the system consists of liquids A and B which are immiscible or nearly immiscible (for example, oil and water). In this case we can consider the system to be a mixture of A and B, rather than a solution of A and B. Assuming the vapours of A and B are ideal, the total vapour pressure of the system is $p = p_A + p_B$. If the temperature is increased until the total vapour pressure equals 1 atm, boiling will commence. This will occur at a temperature lower than either pure A or pure B would boil because boiling commences when the total pressure equals 1 atm.

This is the basis of *steam distillation* (or *steam stripping*), which enables some heat-sensitive volatile organic compounds to be distilled at a lower temperature than their normal boiling point, for example, the extraction of oil from the leaves of eucalyptus trees. The condensed vapour consists of both water and the organic compound in proportion to the vapour pressure of the components; oils of low volatility thus distil at low abundance relative to water. If A and B are completely immiscible the condensed vapour would form two layers which can be separated.

13.6   THERMODYNAMIC BASIS OF PHASE DIAGRAMS

Since the relative stability of phases depends on their Gibbs energies, it is to be expected that this provides the thermodynamic basis for phase diagrams. In this section, we show this relationship in a qualitative way; however, it is possible to calculate the phase diagram of systems from thermodynamic data – experimental or derived from thermodynamic models – using the general approach described here. The procedure involves determining which phase is stable or which phases are in equilibrium at particular temperatures and compositions. If, at constant pressure, values of Gibbs energy of the possible phases are known over the composition range at a series of temperatures, the corresponding temperature–composition phase diagram can be constructed. Similarly, a pressure–composition diagram can be constructed if the values at different pressures and constant temperature are known.

When two phases are in equilibrium, the chemical potential (or activity) of a component is the same in both phases. Furthermore, the intercepts of the tangent to a Gibbs energy curve at the two extremities give the values of the chemical potential of the components (Section 9.3). Therefore, when two phases are in equilibrium the tangent to the curves for the phases must be common to both since the chemical potential of each component is the same in each phase. The points where the common tangent touches the Gibbs energy curves (determined numerically or graphically) give the compositions of the two phases in equilibrium. This technique is called the *common tangent construction*.

The relationship between the Gibbs energy of mixing and a binary temperature–composition diagram is shown schematically in Figure 13.15 for a system with a miscibility gap. The $\Delta_{mix}G_m$

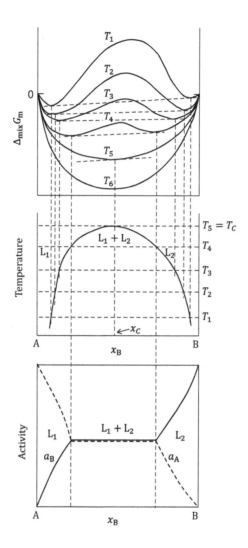

FIGURE 13.15 Relationship between the Gibbs energy of mixing and a temperature–composition phase diagram containing a miscibility gap. T_c is the critical temperature, and x_c is the critical composition.

curves are shown for temperatures T_1 to T_6, where the increasing numbers designate increasing temperatures. The points where the common tangents touch the curves give the compositions of the equilibrium phase(s) at the corresponding temperatures. The curve joining these points defines the phase boundary of the miscibility gap as a function of temperature. Temperature T_5 is the critical temperature of the system. At temperatures greater than T_5 there is complete miscibility, and the $\Delta_{mix}G_m$ curve has the familiar shape of a mixing curve. The variation of the activity of A and B at T_4 is shown in the bottom graph. For both components the standard state is the pure liquid at 1 bar and temperature T_4. The activities of A and B are constant across the miscibility gap because the chemical potentials of A and B are constant.

In this example, components A and B are both liquids in the temperature range considered (that is, at temperatures above their melting points). The most convenient standard state for A and B is, therefore, liquid A and liquid B at 1 bar pressure and the relevant temperature, respectively. The situation would be similar if both A and B were solids in the temperature range of interest (that is, at temperatures below their melting points). Then the phases would be α and β (solid solutions of A and B), and the miscibility gap would occur in the solid state. The standard states of A and B would be solid A and solid B, respectively, at 1 bar and the relevant temperature.

The situation is more complex when one of the components is solid and the other is liquid at the temperature of interest. Consider the phase diagram shown in Figure 13.16(a). Temperature T is above the melting point of component A and below the melting point of component B. Figure 13.16(b) shows the Gibbs energy of mixing curves. The 'liquid' curve is for the mixing of A and B assuming both are liquids; that is, the activity of A is relative to liquid A at temperature T, and the activity of B is relative to pure supercooled liquid B at T. The 'solid' curve is for the mixing assuming A and B

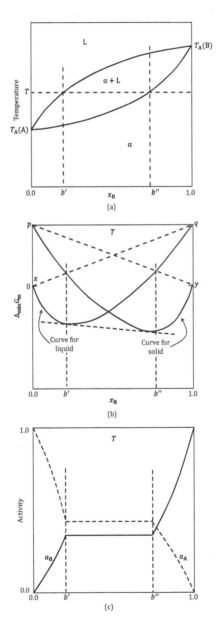

FIGURE 13.16 Relationships between the thermodynamic properties of the system AB and its phase diagram: (a) the phase diagram of AB, (b) the Gibbs energies of mixing at temperature T, (c) activity of A relative to pure solid A; activity of B relative to pure liquid B.

are both solids; that is, the activity of A is relative to pure superheated solid A at T, and the activity of B is relative to pure solid B at T.

At T, the stable states of pure A and B are located at $\Delta_{mix}G_m = 0$, at $x_B = 0$ for liquid A and $x_B = 1$ for solid B. Since

$$\Delta_{mix}G_m = x_A \Delta G_A + x_B \Delta G_B$$

at $x_B = 0$, $x_A = 1$ and $\Delta_{mix}G_m = \Delta G_A^0$. Therefore, point p is equal to $G_A^0(s) - G_A^0(l)$ and, because T is less than $T_{fus}(B)$, $G_A^0(s) - G_A^0(l)$ is greater than zero. The Gibbs energy of fusion of A at T_2, therefore, is given by:

$$\Delta_{fus}G_A^0 = -\left[G_A^0(s) - G_A^0(l) \right]$$

Similarly, point q is equal to $G_B^0(l) - G_B^0(s)$ and, because T is greater than $T_{fus}(A)$, $G_B^0(l) - G_B^0(s)$ is greater than zero. The Gibbs energy of fusion of B at T then is given by:

$$\Delta_{fus}G_B^0 = G_B^0(l) - G_B^0(s)$$

The line xq represents the Gibbs energy of the unmixed liquid forms of A and B, and the line py represents the Gibbs energy of the unmixed solid forms of A and B. The equation of the line xq is:

$$\Delta_{mix}G_m = x_B \Delta_{fus}G_B^0$$

and for the line py is:

$$\Delta_{mix}G_m = -x_A \Delta_{fus}G_A^0$$

The value of $\Delta_{fus}G_A^0$ at T can be determined as follows. Since

$$\Delta_{fus}G_A^0 = \Delta_{fus}H_A^0 - T\Delta_{fus}S_A^0$$

and, if we assume $\Delta_{fus}H_A^0$ and $\Delta_{fus}S_A^0$ are independent of temperature (this is equivalent to assuming $C_p(A,s) = C_p(A,l)$), then at T

$$\Delta_{fus}G_A^0 = \frac{\Delta_{fus}H_A^0 \times T_{fus}(A)}{T_{fus}(A)} - \frac{\Delta_{fus}H_A^0 \times T}{T_{fus}(A)}$$

$$= \Delta_{fus}H_A^0 \left[\frac{T_{fus}(A) - T}{T_{fus}(A)} \right]$$

A similar equation applies to component B.

At any particular composition, the formation of a liquid solution from pure liquid A and solid B can be considered as consisting of two steps: The melting of B, for which $\Delta G = x_B \Delta_{fus}G_B^0$, and the mixing of liquid A and liquid B to form a solution, for which

$$\Delta_{mix}G_m = RT\left(x_A \ln a_A + x_B \ln a_B \right)$$

Therefore, the Gibbs energy change for the formation of 1 mole of liquid from liquid A and solid B is

$$\Delta_{mix}G_m(s) = RT\left(x_A \ln a_A + x_B \ln a_B \right) + x_B \Delta_{fus}G_B^0$$

which is the equation of the 'liquid' curve in Figure 13.16(b). Similarly, the formation of a solid solution from pure liquid A and solid B consists of the steps: Solidification of liquid B and the mixing of liquid A and solid B to form a solid solution. Thus, the 'solid' curve is given by:

$$\Delta_{mix}G_m(l) = RT\left(x_A \ln a_A + x_B \ln a_B\right) - x_A\Delta_{fus}G_A^0$$

In both cases, the activity of A is relative to pure liquid A and the activity of B is relative to pure solid B.

The common tangent to the two curves touches the 'liquid' curve at composition b' and the 'solid' curve at composition b''. Hence at temperature T, liquid of composition b' is in equilibrium with solid of composition b''. As the temperature is decreased the magnitude of px will decrease and qy will increase shifting the 'solid' and 'liquid' curves relative to one another such that the positions of b' and b'' move to the left. Similarly, as the temperature is increased the curves move relative to one another such that b' and b'' move to the right. The paths of b' and b'' with change of temperature trace out the liquidus and solidus curves, respectively.

At temperatures less than $T_{fus}(A)$, $\Delta_{fus}G_A^0$ and $\Delta_{fus}G_B^0$ are both positive, and the 'liquid' curve will lie above the 'solid' curve at all compositions indicating that the solid solution phase is stable at all compositions. Similarly, at temperatures greater than $T_{fus}(B)$, $\Delta_{fus}G_A^0$ and $\Delta_{fus}G_B^0$ are both negative, and the 'solid' curve will lie above the 'liquid' curve at all compositions indicating that the liquid phase is stable at all compositions.

Figure 13.16(c) illustrates how the activities of A and B vary with composition at temperature T.

13.7 DETERMINATION OF PHASE DIAGRAMS

Historically, the method used for determining phase boundaries in systems has been the use of cooling curves. The temperature of a sample is measured as it is allowed to cool from an elevated temperature, and the shape of the curve of temperature *versus* time is then examined. When no phase change occurs, the curve is smooth as illustrated in Figure 13.17(a). When a pure substance is cooled through a transition temperature (for example, a melting point), its temperature is maintained at the transition temperature until the transition is complete, as illustrated in Figure 13.17(b). In practice, the true transition temperature can be difficult to determine from a cooling curve because of the non-equilibrium conditions during a dynamic test. This is illustrated in the cooling curve shown in Figure 13.17(c), which shows the effect of supercooling; the dip in the curve at the start of freezing is caused by a delay in the start of crystallisation as a result of the need to first nucleate some solid phase.

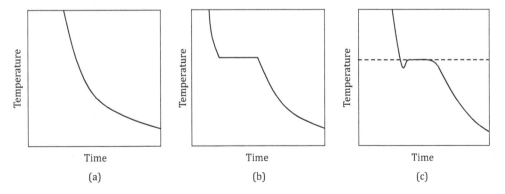

FIGURE 13.17 Cooling curves of a pure substance: (a) cooling with no phase change, (b) ideal cooling with a phase change, (c) actual cooling with a phase change.

The transition over a temperature range that occurs during cooling through a two-phase liquid + solid field results in a reduced slope to the curve between the liquidus and solidus temperatures, as shown in Figure 13.18. By preparing samples covering the range of compositions, the shape of the liquidus and solidus curves can be determined. Cooling curves can be similarly used to investigate all other types of phase boundaries. Figure 13.19 illustrates the approach for the case of a eutectic system.

Another method is to prepare a sample of the required components of known bulk composition, heat it to a known temperature, allow equilibrium to be established, then identify the phases (usually after quenching the sample quickly so as to retain the equilibrium structure). This is repeated at different temperatures until the phase boundary temperature is established at that composition. The procedure is then repeated at different compositions across the composition range. In this way, liquidus and solidus lines, and other phase transition lines, can be determined.

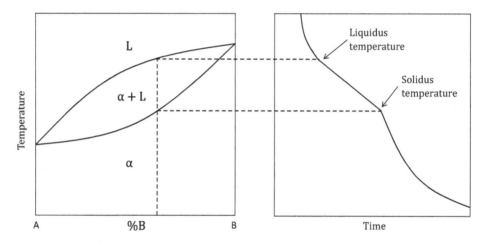

FIGURE 13.18   Cooling curve for a mixture of A and B for the case where the phase change occurs over a range of temperature.

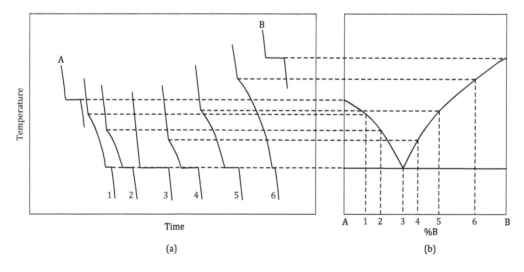

FIGURE 13.19   Construction of a eutectic phase diagram by the cooling curve method: (a) cooling curves of pure A and B and mixtures of composition 1–6, (b) the phase boundary lines.

The experimental determination of phase diagrams is a slow and repetitive exercise, and today experimental measurements are supplemented with mathematical modelling. This reduces the amount of experimentation needed and improves the overall accuracy. The approach has been systematised in recent decades and is known as the CALculation of PHAse Diagrams (CAPHAD) approach (www.calphad.org/). The CALPHAD approach* is based on the fact that a phase diagram is a manifestation of the equilibrium thermodynamic properties of the system, which are the sum of the properties of the individual phases. In the CALPHAD approach, experimental information on phase equilibria in a system and thermodynamic information obtained from thermochemical and thermophysical studies are collected and assessed. The thermodynamic properties of each phase are then described with a mathematical model containing adjustable parameters. The parameters are evaluated by optimising the fit of the model to the available information. It is then possible to calculate the phase diagram as well as the thermodynamic properties of the phases. The aim is to obtain a consistent description of the phase diagram and the thermodynamic properties so as to reliably predict the set of stable phases and their thermodynamic properties in regions for which experimental information is not available and for metastable states during simulations of phase transformations.

13.8 PARTITIONING OF COMPONENTS BETWEEN PHASES

If a substance that is soluble (or partially soluble) in several phases of a system is added to the system, it will become distributed or partitioned between the phases (Figure 13.20). At equilibrium of the system the chemical potential (and activity) of the substance will be the same in each phase. The constancy of the activity, however, does not mean that its concentration is necessarily the same in all the phases since its activity coefficient can, and most likely will, be different in each phase.

From Equation 9.47,

$$x_B = \frac{a_B}{\gamma_B}$$

and, therefore, component B will be present at highest concentration in the phase in which its activity coefficient is least and at lowest concentration in the phase in which its activity coefficient is greatest. Since at equilibrium,

$$a_B(I) = a_B(II)$$

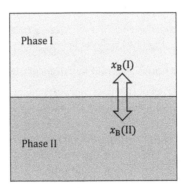

FIGURE 13.20 The partitioning of substance B between two phases.

* Spencer, P. J. 2008. A brief history of CALPHAD. *Computer Coupling of Phase Diagrams and Thermochemistry*, 32, 1–8.

then

$$\gamma_B(I) \times x_B(I) = \gamma_B(II) \times x_B(II)$$

and, therefore,

$$L_B(I / II) = \frac{x_B(I)}{x_B(II)} = \frac{\gamma_B(II)}{\gamma_B(I)} \qquad (13.16)$$

where $L_B(I/II)$, the ratio of the mole fractions, is the *distribution coefficient* or *partition ratio* of substance B in the system. It follows that:

- Substance B will partition preferentially to Phase I if $\gamma_B(I) < \gamma_B(II)$.
- Substance B will partition preferentially to Phase II if $\gamma_B(I) > \gamma_B(II)$.

This makes it possible to design systems to collect a substance preferentially in one phase, either because it is a valuable/useful substance or because we wish to remove it from the system.

It is sometimes found that the distribution coefficient is constant, or approximately constant, irrespective of the amount of the substance added to the system, that is

$$L_B(I / II) = \frac{x_B(I)}{x_B(II)} = \text{constant}$$

It will be seen from Equation 13.16 that for this relationship to hold, the ratio $\gamma_B(II) / \gamma_B(I)$ must be independent of the concentration of B in both phases. This will be so in ideal or nearly ideal solutions and also within the Henry's law region in both phases. In the latter case, the relationship will normally hold true for sparingly soluble substances or for more soluble substances added in small amounts. For example, when elemental iodine I_2 is shaken with the immiscible liquids water and carbon tetrachloride it dissolves to a small extent in the molecular form in both solvents with a constant distribution coefficient,

$$\frac{m_{I_2}(\text{org})}{m_{I_2}(\text{aq})} \approx 90 \left(\text{at } 25°C\right)$$

PROBLEMS

13.1   The vapour pressure of solid uranium hexafluoride (UF_6) is given by the relation

$$\ln p(\text{Pa}) = 29.411 - \frac{5893.5}{T}$$

and the vapour pressure of liquid uranium hexafluoride by the relation

$$\ln p(\text{Pa}) = 22.254 - \frac{3479.9}{T}$$

Calculate the temperature and pressure of the triple point.

13.2   The standard Gibbs energy change for the transition of graphite to diamond is 5.956 kJ mol^{-1} at 1000 K. The molar volume of graphite and diamond are 3.417 and 5.298 mL mol^{-1}, respectively. Calculate the minimum pressure required to convert graphite to diamond at 1000 K. Assume the change in volume on transition is independent of temperature and pressure. Calculate the minimum pressure at 1800 K if the standard Gibbs energy change is 9.435 kJ mol^{-1}. Compare your answers to values estimated from Figure 13.2.

13.3   The vapour pressure of solid Cl_2 is 352 Pa at –112°C and 35 Pa at –126.5°C. The vapour pressure of liquid Cl_2 is 1590 Pa at –100°C and 7830 Pa at –80°C. Estimate the molar enthalpies of sublimation, vapourisation and fusion.

13.4   Construct the p–T diagram for benzene in the vicinity of its triple point using the following data: Triple point: 4.83 kPa and 5.4°C; $\Delta_{fus}H = 9.8$ kJ mol^{-1}; $\Delta_{vap}H = 30.8$ kJ mol^{-1} (near the triple point); $V_m(s) = 85.71$ mL; $V_m(l) = 86.76$ mL.

13.5   Determine the enthalpy of vapourisation of carbon disulphide (CS_2) given the variation of its vapour pressure with temperature.

T (°C)	–73.8	–44.7	–22.5	–5.1	28.0	46.5	69.1	104.8	136.3	175.5	222.8	256.0
p (kPa)	0.13	1.33	5.33	13.33	53.3	101.3	202.7	506.6	1013	2027	4053	6080

13.6   Calculate the change in the melting point of tin when the pressure is increased from 1 to 3 atm. Data: $\rho_{Sn(s)} = 7310$ kg m^{-3}; $\rho_{Sn(l)} = 6990$ kg m^{-3}; $T_{fus}(Sn) = 231.9$°C; $\Delta_{fus}H^0 = 7.029$ kJ mol^{-1}.

13.7   Estimate the relative change in the vapour pressure of mercury when the external pressure is increased from 1 to 1000 atm given $V_m(Hg) = 15.025 \times 10^{-6}$ m^3.

13.8   The vapour pressure of water increases from 0.031 atm at 25°C to 1 atm at 100°C. Calculate the average enthalpy of vapourisation of water over the range 25°C to 100°C.

13.9   With reference to the accompanying phase diagram, describe the evolution of the microstructures formed when melts of composition a, b and c are slowly cooled to room temperature.

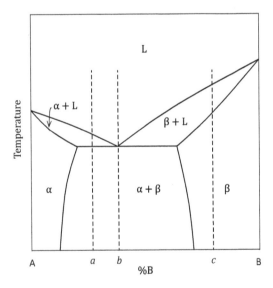

13.10   When mixtures of finely powdered silicon and germanium are heated melting occurs as follows:

x_{Si}	0	0.25	0.40	0.62	0.80	0.90	1.00
Start of melting (°C)	940	1010	1070	1170	1275	1340	1412
Completion of melting (°C)	940	1160	1235	1310	1370	1395	1412

    i.   Plot the points and construct the phase diagram of the Si–Ge system.

    ii.   Describe the sequence of events and the development of the microstructure as a melt of composition $x_{Si} = 0.3$ is cooled slowly from 1500 to 900°C.

    iii. What is the mass fraction of each phase?

13.11   The results of cooling curve tests for the gold–antimony system are as follows:

Mass% Sb	0	10	20	30	40	50	55	60	70	80	90	100
Start of freezing (°C)	1063	730	470	400	445	455	460	495	545	580	610	631
Completion of freezing (°C)	1063	360	360	360	360	360	460	460	460	460	460	631

    i.   Plot the points and construct the phase diagram for the Au–Sb system, and identify the phases present.

    ii.   Describe the sequence of events and the development of the microstructure as melts of 20, 50 and 80 mass% Sb are cooled slowly from 800 to 300°C.

    iii. In each case what is the mass fraction of each phase?

13.12   High purity silicon is required for electronic applications. It has been proposed to lower the concentration of impurity elements in metallurgical grade silicon by vacuum refining which involves applying a low pressure to molten silicon to allow the impurity elements to volatilise. Show that in a vapour–molten metal system at equilibrium

$$L_{M/Si} = \frac{p_M^0 \, \gamma_M \, x_M}{p_{Si}^0}$$

where M is an impurity element in silicon. Assume M is monatomic in the alloy and in the vapour state. Using the data in the table, determine which elements could be amenable to vacuum refining. Assume the impurity content of each element is $x_M = 0.01$.

Impurity	Al	Mg	Zn	Pb	Ni	Mn
p_M^0, bar	1.66×10^{-4}	7.75	30.1	0.11	2.45×10^{-6}	8.61×10^{-3}
γ_M^0	0.37	0.02	1.87	51.3	0.010	0.0016

FURTHER READING

Campbell, F. C. (ed). 2012. *Phase Diagrams: Understanding the Basics.* Materials Park, OH: ASM International.

Lukas, H., S. G. Fries and B. Sundman. 2007. *Computational Thermodynamics – The Calphad Method.* Cambridge: Cambridge University Press.

14 Phase equilibria: reactive systems

SCOPE

This chapter examines in detail the equilibrium in multi-phase systems in which chemical reactions may occur.

LEARNING OBJECTIVES

1. *Understand the meaning of the terms component and species, and be able to distinguish between them.*
2. *Be able to determine the number of independent reactions in a reactive system.*
3. *Understand the implications of the phase rule for reactive systems, and be able to apply it to simple systems.*
4. *Be able to construct the predominance area diagram for reactive ternary systems.*
5. *Understand the theory of the distribution of reactive elements between two or more phases.*

14.1 INTRODUCTION

Chapter 13 examined phase equilibria in non-reactive systems. We now examine the case of multi-phase equilibria which involve chemical reactions. We encountered this previously with gas–solid reactions in Chapter 10 but now broaden the discussion to include solid–solid, solid–liquid, liquid–liquid and gas–liquid equilibria in which reactions may occur between species in different phases.

14.2 THE PHASE RULE FOR REACTIVE SYSTEMS

In Chapter 13, components were defined as chemically independent constituents of a system. We now define another, related term, species. *Species** are atoms, molecules, molecular fragments, ions, etc. that are subject to a chemical process (or to a measurement) and that are sufficiently stable to exist over the duration of the process (or measurement). *The composition of reactive systems is expressed in terms of the amounts or relative amounts (mass percent, mole fraction, etc.) of the species considered to be present.*

In a non-reactive system, the number of components is equal to the number of species in the system. However, if some species enter into reaction with one another this is not the case. The equilibrium of the system will involve one or more reaction equilibrium in addition to the temperature, pressure and phase equilibria. In this case, the number of independent equations (that relate the variables) to be specified is increased by r, the *number of independent reactions* possible between the species. Suppose there are s chemical species, some of which may be inert; then, following the approach in Section 13.3, the number of independent equations is

$$(P-1)(s+2)+r$$

---

* The noun species is both singular and plural and can, therefore, refer to a single substance (for example, the species Fe_2O_3) or a number of substances (for example, the species Fe, Fe_2O_3, O_2, …). The word shares a Latin origin with the singular noun specie, but their meanings have diverged over time and the words species and specie are now unrelated. The noun specie refers primarily to coin money.

Now assume that those species that are not inert are in chemical equilibrium with each other via appropriate chemical reactions. Then, again following the approach of Section 13.3, the number of variables is

$$P(s+1)$$

The variance or degrees of freedom of the system is the difference:

$$F = P(s+1) - ((s+2)(P-1)+r) = s - r + 2 - P \tag{14.1}$$

The variance, it will be recalled, is the minimum number of variables whose value must be chosen so that the state of the system is unique (or determinant). Comparing Equation 14.1 with Equation 13.2, it is apparent that the number of components in the system is given by:

$$C = s - r \tag{14.2}$$

In some applications there may also be other restrictions on a system, and the effect of each of these is to reduce the number of components by one since each restriction adds another independent equation relating the variables. Some common examples are:

- The concentration or activity of a particular species is constant (for example, the activity of a species will be one if the system is saturated with respect to that species).
- The concentrations of two species in a particular phase are in a fixed ratio. This could occur either because the system is designed so that the ratio remains constant during reaction or because of a stoichiometric restriction due to the nature of a reaction (for example, in the decomposition reaction $NH_4Cl(s) = NH_3(g) + HCl(g)$, there is the restriction that $n_{NH_3} = n_{HCl}$ in the gas phase).
- In ionic solutions, the charge balance* must be such that the solution as a whole is electrically neutral.
- The reactions are independent of pressure.
- Though a reaction between species is thermodynamically feasible it does not occur due to a kinetic barrier (for example, the temperature is too low).

A rule for determining whether a stoichiometric and charge balance equation reduces the number of components is:

Stoichiometric relationships that reduce the number of components are those in which every species in the relationship appears in an equation for the equilibrium constant of an independent reaction.

Why this is so will be made clear by the particular cases examined in Example 14.2.

Thus, the number of components in a chemically reactive system is given by the relation:

$$C = s - r - z \tag{14.3}$$

where z is the number of restrictions. The *phase rule for reactive systems* then becomes:

$$F = C + 2 - P = s - r - z + 2 - P \tag{14.4}$$

---

* Solutions are electrically neutral, so for solutions containing ions the number of positive charges equals the number of negative charges. Charge balance equations express this relationship by equating the molar concentrations of the positive and negative charges (Equation 12.3).

For simple systems the number of independent reactions can be obtained by simple inspection of the possible species. For more complex systems the following procedure is useful.

- Select the species that are regarded as being present in the system. For any particular system it is necessary to decide which of the actual possible species can be neglected for the purpose of the application in mind.
- Write balanced chemical equations for the formation of all the species from their constituent atoms. (Sometimes, a species will be an element, and, clearly, an equation is then unnecessary.)
- Combine these equations in pairs in such a way as to eliminate all free atoms not considered to be present in that state in the system so that the final set of equations includes all the selected species.

The resulting number of equations gives the number of independent reactions. Depending on how the equations are combined it is possible to arrive at different reactions in the set of independent reactions, and it is immaterial which set is selected, the decision being based on the application in mind, for example, which composition variables can be measured or controlled.

EXAMPLE 14.1   Determine the number of independent reactions in a system

i. Consider the system M–C–O, where M is a divalent metal. At high temperatures, the following species may be present: M(s), C(s), MO(s), MC(s), CO(g), CO$_2$(g). Determine the number of independent reactions in this system.

SOLUTION

First, write the chemical equations to form the species from their constituent elements:

$$\begin{aligned}
\text{M} + \text{O} &= \text{MO} & a \\
\text{M} + \text{C} &= \text{MC} & b \\
\text{C} + \text{O} &= \text{CO} & c \\
\text{C} + 2\,\text{O} &= \text{CO}_2 & d
\end{aligned}$$

Now combine the equations in pairs until all non-species are removed, ensuring each of the above equations is used at least once.

$$\begin{aligned}
c - a: \quad & \text{MO} + \text{C} = \text{M} + \text{CO} & e \\
b - a: \quad & \text{MO} + \text{C} = \text{MC} + \text{O} & f \\
F + c: \quad & \text{MO} + 2\,\text{C} = \text{MC} + \text{CO} & g \\
2c - d: \quad & \text{CO}_2 + \text{C} = 2\,\text{CO} & h
\end{aligned}$$

Therefore, e, g and h are independent equations, and there are three independent equations.

Other combinations are possible, and in each case there will be three independent equations. For example,

$$\begin{aligned}
-(a+c): \quad & \text{MO} + \text{CO} = \text{M} + \text{C} + \text{O} & i \\
i + d: \quad & \text{MO} + \text{MC} = 2\,\text{M} + \text{CO}_2 & j \\
-(a+b): \quad & \text{MO} + \text{MC} = 2\,\text{M} + \text{C} + \text{O} & k \\
k + c: \quad & \text{MO} + \text{MC} = 2\,\text{M} + \text{CO} & l
\end{aligned}$$

Therefore, equations j and l, together with equation b (which contains only species), make another three independent reactions.

ii. Now consider $O_2(g)$ as an additional species. What effect will this have on the number of independent reactions?

SOLUTION

There will be one additional reaction forming a compound from its constituent elements:

$$2\,O = O_2 \qquad m$$

Combine this to eliminate the elemental oxygen:

$$d-m: \quad C + O_2 = 2\,CO \quad n$$

There are now four independent reactions.

EXAMPLE 14.2   Determine the degrees of freedom in a system

i. The system $Fe - FeO - Fe_2O_3$.

SOLUTION

The system has one independent reaction, $4\,FeO = Fe_3O_4 + Fe$.
    There are three species (Fe, FeO, Fe_2O_3) and three phases.

$$C = s - r - z = 3 - 1 - 0 = 2$$

(for example, Fe, FeO or Fe, Fe_2O_3 or FeO, Fe_2O_3)

$$F = C + 2 - P = 2 + 2 - 3 = 1$$

If either T or p is fixed, all other intensive variables have unique values.

ii. The decomposition of calcium carbonate: $CaCO_3(s) = CaO(s) + CO_2(g)$.

SOLUTION

$$s = 3; \quad r = 1; \quad P = 3$$
$$C = s - r - z = 3 - 1 - 0 = 2$$
$$F = C + 2 - P = 2 + 2 - 3 = 1$$

If T is fixed, the pressure of CO_2 is unique. The relationship between temperature and pressure of CO_2 was calculated in Example 10.4.

iii. The dissociation of ammonium chloride: $NH_4Cl(s) = NH_3(g) + HCl(g)$.

SOLUTION

$$s = 3; \quad r = 1; \quad P = 2; \quad z = 1$$

since $n_{NH_3} = n_{HCl}$ (in the gas phase)

$$C = s - r - z = 3 - 1 - 1 = 1 \text{ (namely, NH}_4\text{Cl)}$$
$$F = C + 2 - P = 1 + 2 - 2 = 1$$

If the degree of freedom is taken to be temperature, then specifying T determines the values of p_{HCl} and p_{NH_3}.

COMMENT

That the system is univariant can be demonstrated as follows. Since the activity of solid NH_4Cl is one,

$$K = p_{HCl} \times p_{NH_3}$$

where

$$p_{HCl} = \frac{n_{HCl}}{n_{HCl} + n_{NH_3}} p$$

and

$$p_{NH_3} = \frac{n_{NH_3}}{n_{HCl} + n_{NH_3}} p.$$

But since $n_{NH_3} = n_{HCl}$, then

$$p_{HCl} = \frac{n_{NH_3}}{2\, n_{NH_3}} p = 0.5\, p = p_{NH_3}$$

and,

$$K = 0.5\, p \times 0.5\, p = 0.25\, p^2$$

Thus, $p_{HCl} = p_{NH_3} = \left(\dfrac{K}{0.25} \right)^{0.5} = 2\, K^{0.5}$ and is a function of temperature only.

Note the difference between this example and the decomposition of limestone. Even though there is a stoichiometric relationship in the former case, $n_{CaO} = n_{CO_2}$, this does not affect the number of components because the addition or removal of CaO will not alter the equilibrium of the reaction (since the activity of CaO is always one irrespective of the amount present):

$$K = p_{CO_2}$$

In the case of ammonium chloride dissociation, addition or removal of some NH_3 will alter the equilibrium concentration of HCl since $K = p_{HCl} \times p_{NH_3}$ and K is constant. This illustrates the rule previously given that the only stoichiometric relationships that reduce the number of components are those in which every species in the stoichiometric relationship appears in an equation for the equilibrium constant of an independent reaction.

   iv. Dissociation of water: $H_2O(l) = H^+(aq) + OH^-(aq)$.

SOLUTION

$$s = 3; \quad r = 1; \quad P = 1; \quad z = 1 \ (\text{since } n_{H^+} = n_{OH^-})$$

The stoichiometric restriction applies because changing the concentration of either H^+ or OH^- will affect the concentration of the other since $K = a_{H^+} \times a_{OH^-}$, assuming $a_{H_2O} = 1$. Then

$$C = s - r - z = 3 - 1 - 1 = 1$$

and

$$F = C + 2 - P = 1 + 2 - 1 = 2$$

Thus, we can vary temperature and pressure independently; at any fixed temperature and pressure of the system, the concentrations (or activities) of H^+ and OH^- ions have unique values, both being equal to 1.004×10^{-7} mol kg^{-1} at 25°C and 1 bar (Section 12.6).

v. Solution of a soluble salt in water.

SOLUTION

When aluminium chloride is added to water, it may dissociate, hydrolyse and partially precipitate as aluminium hydroxide. The independent reactions are:

$$H_2O(l) = H^+(aq) + OH^-(aq) \quad K_{H_2O} = a_{H^+} \times a_{OH^-}$$

$$AlCl_3(s) = Al^{3+}(aq) + 3\,Cl^-(aq) \quad K_{AlCl_3} = \frac{a_{Al^{3+}} \times a_{Cl^-}^3}{AlCl_3}$$

$$Al(OH)_3(s) = Al^{3+} + 3\,OH^- \quad K_{Al(OH)_3} = a_{Al^{3+}} \times a_{OH^-}^3$$

There are seven species (H_2O, H^+, OH^-, $AlCl_3$, Al^{3+}, Cl^-, and $Al(OH)_3$) and two phases (water and $Al(OH)_3$; $AlCl_3$ is not a phase because it completely dissolves). There is one restriction, namely, the charge balance equation.

$$3m_{Al^{3+}} + m_{H^+} = m_{Cl^-} + m_{OH^-}$$

Therefore,

$$C = s - r - z = 7 - 3 - 1 = 3; \quad \text{for example, } H_2O, AlCl_3 \text{ and } Al(OH)_3$$

and

$$F = C + 2 - P = 3 + 2 - 2 = 3 \ \left(T, p \text{ and one of } m_{Al^{3+}}, m_{H^+}, m_{Cl^-} \text{ or } m_{OH^-} \right).$$

COMMENT

If $AlCl_3$ was added to the solution until it was saturated, then solid $AlCl_3$ would have to be included as a phase, and the degrees of freedom would then be two.

14.3 PHASE STABILITY DIAGRAMS

Consider the system M–O where M represents a divalent metal. Suppose there are three possible species: M, MO and O_2. The activities of M and MO (when they are present) are one and each is a phase. Thus, there are three phases. Only one independent reaction is possible:

$$M(s) + 1/2 \, O_2(g) = MO(s)$$

The number of components is $C = s - r - z = 3 - 1 - 0 = 2$; for example, M and O_2; M and MO; or O_2 and MO. The degree of freedom in the system is:

$$F = C + 2 - P = 2 + 2 - 3 = 1$$

The system is univariant. If the temperature is fixed, for example, the partial pressure of oxygen is determinant. The equilibrium relationship between temperature and pressure, therefore, can be represented by a line, as illustrated for the case of Pb–O–PbO in Figure 14.1.

Now assume the metal can form both an oxide and a carbide at sufficiently high temperatures. The system is now M–O–C, and there are six possible species: M, MO, MC, C, CO and CO_2. Oxygen will be present also but in very small amounts at equilibrium and will be neglected. CO and CO_2 occur in the gas phase; the solids are mutually immiscible, and each, therefore, is a phase ($P = 5$). There are three independent reactions (Example 14.1), for example,

$$MO + C = M + CO$$

$$MO + 2\,C = MC + CO$$

$$CO_2 + C = 2\,CO$$

Therefore,

$$C = s - r - z = 6 - 3 - 0 = 3; \quad \text{for example, M, C and MO}$$

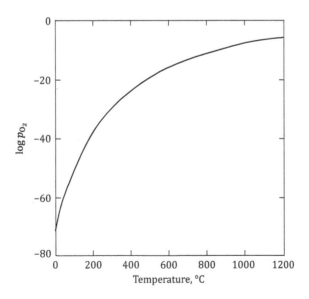

FIGURE 14.1 The variation with temperature of the equilibrium partial pressure of oxygen in the system of Pb–O–PbO, calculated from $\Delta_r G^0$ for the reaction $2\,PbO = 2\,Pb + O_2$; $K = p_{O_2}$.

and

$$F = C + 2 - P = 3 + 2 - P = 5 - P$$

If the temperature is fixed, then $F = 4 - P$. Thus, in addition to the gas phase, one, two or three condensed phases can coexist in equilibrium but not all four solid phases. This is illustrated in Figure 14.2 for the system Ca–C–O. The lines which describe the equilibrium of the gas with two condensed phases are defined by the equilibria:

$$CaO(s) + CO(g) = Ca(s) + CO_2(g) \tag{1}$$

$$Ca(s) + 2\,CO(g) = CaC(s) + CO_2(g) \tag{2}$$

$$CaO(s) + 3\,CO(g) = CaC(s) + 2\,CO_2(g) \tag{3}$$

$$CO_2(g) + C(s) = 2\,CO(g) \tag{4}$$

Along each line two solid phases are in equilibrium with each other and with the gas phase (that is, $P = 3$) and, at a fixed temperature, $F = 4 - P = 4 - 3 = 1$. Thus, if one of p_{CO_2} or p_{CO} is fixed then so is the other. At points P and Q three condensed phases and the gas phase co-exist and the system is invariant ($F = 0$). In the areas between the lines a single phase co-exists with the gas phase and $F = 2$; that is, at a fixed temperature both p_{CO_2} and p_{CO} can be varied independently.

The equations for the lines in Figure 14.2 are derived from the equilibrium constant expressions for reactions (1) to (4). Thus, for reaction (1)

$$K_1 = \frac{p_{CO_2}}{p_{CO}}$$

and, therefore

$$\log p_{CO} = \log p_{CO_2} - \log K_1$$

Similarly, for reactions (2), (3) and (4),

$$\log p_{CO} = 1/2 \log p_{CO_2} - 1/2 \log K_2$$

$$\log p_{CO} = 2/3 \log p_{CO_2} - 1/3 \log K_3$$

$$\log p_{CO} = 1/2 \log p_{CO_2} + 1/2 \log K_4$$

Thus, the slopes of the lines are 1, 1/2, 2/3 and 1/2, respectively.

Figure 14.2 is called a *phase stability diagram* or *predominance area diagram*. Similar diagrams can be constructed for most reactive ternary systems, and they are useful for gaining an understanding of the thermodynamics of systems. Predominance area diagrams can also be constructed for aqueous systems. In these it is usual to plot the redox potential (a measure of the oxygen potential) of the system against the pH (a measure of the activity of hydrogen ions). Except for solid phases, for which the activity is one, calculation of the diagrams must be done at an assumed molality of the relevant species. This is discussed further in Chapter 16.

14.4   THE DISTRIBUTION OF ELEMENTS BETWEEN PHASES

Section 13.8 examined the equilibrium distribution of a substance between several phases for the case when no reactions occur in the system. The possibility that an element may be distributed between several phases in different chemical forms (species) is an important concept in systems in

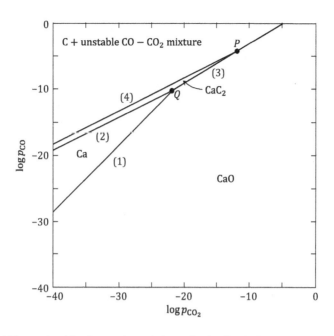

FIGURE 14.2 Stability regions for the condensed phases in the Ca–C–O system at 1200°C. The line numbers correspond to Reactions (1) to (4).

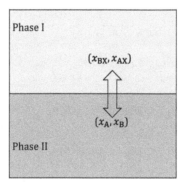

FIGURE 14.3 The equilibrium distribution of substances A and B between two phases in the reactive system A(I) + BX(II) = AX(II) + B(I).

which chemical reaction is possible. This can be illustrated using the two-phase system shown in Figure 14.3 in which there is one independent reaction:

$$A(II) + BX(I) = AX(I) + B(II)$$

In this system, elements A and B are soluble in phase II but not in phase I, and substances AX and BX are soluble in phase I but not phase II. Since

$$K = \frac{a_{AX} \, a_B}{a_A \, a_{BX}} = \frac{x_{AX} \, \gamma_{AX} \, x_B \, \gamma_B}{x_A \, \gamma_A \, x_{BX} \, \gamma_{BX}} \tag{14.5}$$

then

$$L_{I/II} = \frac{x_{AX}}{x_A} = K \times \frac{\gamma_A}{\gamma_{AX}} \times \frac{\gamma_{BX}}{\gamma_B} \times \frac{x_{BX}}{x_B} \tag{14.6}$$

where $L_{I/II}$, the ratio of the concentration of AX in phase I to the concentration of A in phase II, is the *partition ratio* or *distribution coefficient* of A in the system.

A number of factors will influence the magnitude of the distribution coefficient:

- The equilibrium constant is unique for each reaction and is a function of temperature and Gibbs energy of the reaction. A large value of K will favour the presence of A as AX in phase I; a low value of K will favour the presence of A in phase II.
- The activity coefficients of species of solutions are functions of temperature and composition of the solution, and the relative values of γ_A, γ_B, γ_{BX} and γ_{AX} will influence the magnitude of the distribution coefficient.
- If the concentration of BX is increased the reaction will be displaced to the right, and *vice versa* if the concentration is decreased.

Equilibria of this type between phases sometimes make it possible to collect a substance preferentially in one phase using an appropriate reaction, with a view to either recovering it or removing it from the other phase.

The degree of separation depends on the above intensive properties, but it also depends on the ratio of the extensive properties of volume (or mass) of the two phases. If V_I is the volume of phase I, V_{II} the volume of phase II and c is concentration, with V and c in appropriate units (such as mL and mol mL^{-1}, or L and x), then the fraction of A, F_A, in phase I at equilibrium (the *fractional recovery*) is given by

$$F_A = \frac{V_I\, c_{A,I}}{V_{II}\, c_{A,I} + V_I\, c_{A,II}} = \frac{(V_I/V_{II})c_{A,I}}{(V_{II}/V_{II})c_{A,I} + (V_I/V_{II})c_{A,II}}$$

$$= \frac{(V_I/V_{II})(c_{A,I}/c_{A,II})}{(V_{II}/V_{II})(c_{A,I}/c_{A,II}) + (V_I/V_{II})(c_{A,I}/c_{A,II})} = \frac{(V_I/V_{II})L_{I/II}}{(V_I/V_{II})L_{I/II} + 1}$$

where, in this case, $L_{I/II} = c_{A,I} / c_{A,II}$. Rearranging leads to:

$$\frac{1}{F_A} = 1 + \frac{V_{II}}{L_{I/II}\, V_I}$$

The recovery of A in phase I varies directly with the value of $L_{II/I}$ and with the ratio of V_I and V_{II} – the larger these are, the greater is the recovery of A.

14.4.1 SOLVENT EXTRACTION

Solvent extraction is a technique used to separate compounds or metal complexes based on their relative solubilities in two immiscible liquids, usually water (a polar compound) and an organic solvent (a non-polar compound). Solvent extraction technology is used on a commercial scale to recover uranium, vanadium, molybdenum, copper, nickel and rare earth elements from solutions obtained by leaching their ores. The metal ion or ion complex in the aqueous solution is mixed with an immiscible organic phase containing an extractant, and the desired component transfers from the aqueous phase to the organic phase. The *loaded* organic phase is then separated, and the extracted metal is *stripped* from the organic phase to form a purified, concentrated aqueous solution from which the element can be recovered in a further operation. The stripped organic is recycled back to the extraction step.

The reaction can be represented as follows:

$$M^{2+}(aq) + 2\, RH(org) = MR_2(org) + 2\, H^+(aq)$$

where M is a metal* and RH is the organic extractant.

$$K = \frac{a_{H^+}^2 \, a_{MR_2}}{a_{M^+} \, a_{RH}^2}$$

If Henry's law is obeyed,

$$K = \frac{a_{H^+}^2 \, c_{MR_2}}{c_{M^+} \, c_{RH}^2} = \frac{a_{H^+}^2 \, L_{org/aq}}{c_{RH}^2}$$

where $L_{org/aq} = c_{MR_2\,(org)} / c_{M^+\,(aq)}$. Therefore,

$$\log L_{org/aq} = \log K + 2\,\text{pH} + 2\log c_{RH}$$

The magnitude of $L_{org/aq}$ depends on the magnitude of K, the pH of the aqueous solution and the concentration of the extractant in the organic phase. The value of K depends on the Gibbs energy of the reaction, which in turn will depend on the nature of the extractant used.

An important parameter for control purposes is the pH of the aqueous phase. Some typical relations are shown in Figure 14.4 for common metals for the extractant C9 Aldoxime (a phenolic oxime molecule with a branched hydrocarbon R group of nine carbon atoms). The curves predict, for example, that copper and molybdenum are strongly extracted at pH = 2.0 whereas ferric iron is extracted only slightly and nickel, cobalt, zinc and manganese are not extracted. Thus, copper and molybdenum could be recovered selectively from an aqueous solution containing, in addition, ferric iron, nickel, cobalt, zinc and manganese.

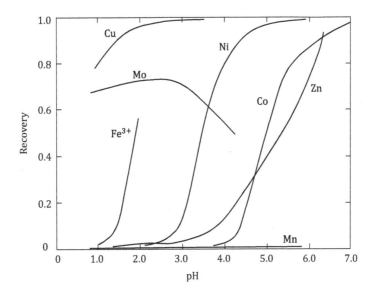

FIGURE 14.4   Solvent extraction curve for some common metals using the extractant C9 Aldoxime (25°C). Source of data: BHSF Redbook – Mining solutions <www.mining-solutions.basf.com>.

---

* M is assumed to be divalent, but similar equations apply for other valence states.

14.4.2 Distribution of elements in gas–slag–metal systems

In the high temperature production of metals, molten metal is frequently in contact with molten slag, and minor elements present in the metal will become partitioned between the two phases. This effect is made use of to refine impure metals by preferentially collecting unwanted elements in the slag or gas phase by reacting them with a suitable oxidising agent, usually oxygen. Alternatively, in other processes, the desired elements are preferentially collected in the slag for further recovery in a separate process.

Consider the case of impure molten copper in contact with a molten slag consisting of SiO_2, CaO and FeO. In this system, under the conditions at which copper is refined (typically, at around 1200°C and $p_{O_2} = 10^{-6}$ bar), SiO_2 and CaO can be considered to be inert (because they are much more stable than copper oxide); their presence in the slag is to lower the melting point of the slag and help collect unwanted impurities. Let M represent the elements present in the metal phase. Then any of M present in the slag at equilibrium will be in the form MO*. The system is illustrated in Figure 14.5.

The partition of element M between the metallic and slag phases is governed by the equilibrium of the heterogeneous reaction

$$M + 0.5\, O_2(g) = MO$$

for which the equilibrium constant is:

$$K = \frac{a_{MO}}{a_M\, p_{O_2}^{0.5}} = \frac{x_{MO}\, \gamma_{MO}}{x_M\, \gamma_M\, p_{O_2}^{0.5}}$$

Rearranging leads to

$$L_{MO/M} = \frac{x_{MO}}{x_M} = \frac{K \gamma_M\, p_{O_2}^{0.5}}{\gamma_{MO}}$$

A number of factors will influence the magnitude of the distribution coefficient:

- The equilibrium constant is unique for each reaction and is a function of temperature and the Gibbs energy of formation of MO. A large value of K will favour the presence of M as MO in the slag; a low value will favour the presence of M in the alloy phase.

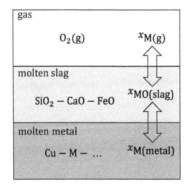

FIGURE 14.5 The equilibrium distribution of the element M between gas, slag and metal phases in the reactive system M–O.

---

* For simplicity, it is assumed all the elements represented by M are divalent. Other valencies can be easily accommodated by modifying the subsequent equations.

- The activity coefficients of species in solutions are functions of temperature and composition of the solution, and the relative values of γ_M and γ_{MO} will influence the magnitude of the distribution coefficient.
- The larger the partial pressure of oxygen, the greater will be the magnitude of $L_{MO/M}$.

The distribution of a minor element between a slag and metal at equilibrium, therefore, is a function of the Gibbs energy of the reaction (that is, the stability of the oxide), temperature, slag composition, metal composition and the oxygen potential of the system. The equilibrium concentration of M present in the slag (as MO), for example, could be increased by

- Raising the activity coefficient of M in the metal
- Raising the oxygen potential of the system or
- Lowering the activity coefficient of MO in the slag

Of these, oxygen has the greatest effect because in practice it can be varied over many orders of magnitude whereas activity coefficients might be varied at most by an order of magnitude.

Elements with significant vapour pressure at the relevant temperature will also be present in the gas phase. For a closed system, the distribution of element M between the metal and gas phases is controlled by the equilibrium of the reaction:

$$M(l) = M(g)$$

The *volatility* of element M relative to that of the host metal Cu is given by:

$$L_{M/Cu} = \frac{p_{M(g)}}{p_{Cu}^0}$$

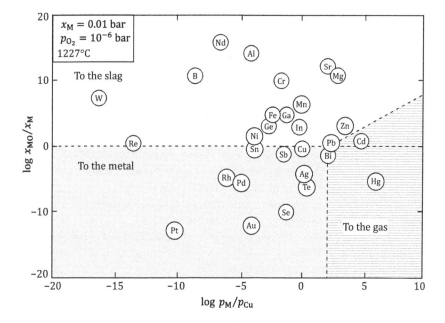

FIGURE 14.6 Relative distributions of minor elements between the metal, slag and gas phases under copper refining conditions. Source of data: Nakajima, K. *et al.* 2011. *Environmental Science & Technology*, 45, pp. 4929–4936.

where p_{Cu}^0 is the vapour pressure of pure copper and $p_{M(g)}$ is the vapour pressure of M dissolved in copper. A high value of $L_{M/Cu}$ means a high proportion of M will be volatilised. Since $a_M = p_M / p_M^0$ (Equation 9.52) and $a_M = \gamma_M x_M$,

$$L_{M/Cu} = \frac{a_M p_M^0}{p_{Cu}^0} = \frac{p_M^0 \gamma_M x_M}{p_{Cu}^0}$$

Figure 14.6 shows values of $\log L_{MO/M}$ plotted against values of $\log L_{M/Cu}$ for minor elements commonly present in copper smelting systems calculated at 1227°C and $p_{O_2} = 10^{-6}$ bar assuming a concentration of minor elements in the copper of $x_M = 0.01$. Appropriate values of activity coefficients were taken from the literature. As can be seen, stable elements such as Pt, Au, Ag and Te preferentially remain in the metal phase whereas reactive elements such as Al, Mg and Cr preferentially remain in the slag. Elements such as Sn, Sb and Pb will have significant concentrations in both slag and metal phases while volatile elements such as Cd and Hg will concentrate preferentially in the gas phase. Elements that cannot be preferentially collected in the slag or gas phase would have to be removed from the copper (to produce pure copper) in a separate process. In practice, these elements are removed (and recovered) by electrorefining (Section 17.1).

PROBLEMS

14.1 For the following reactions determine the number of phases, the number of components and the degrees of freedom in the system.

$$ZnS(s) = Zn(g) + 0.5\, S_2(g)$$

$$MgCO_3(s) = MgO(s) + CO_2(g)$$

$$CuSO_4.5H_2O(s) = CuSO_4(s) + 5\, H_2O(g)$$

$$2\, HCl(g) = H_2(g) + Cl_2(g)$$

14.2 The species ZnS, ZnO, $ZnSO_4$, SO_2, SO_3 and O_2 are possible in the Zn–S–O system which is industrially important in the oxidation roasting of zinc sulphide concentrates.
  i.   How many phases are possible?
  ii.   Show that no more than three independent chemical reactions are needed to describe the chemistry of the system.
  iii. Does the stoichiometry of the system impose any additional constraints on the system?
  iv.   Determine the degrees of freedom of the system. Explain the significance of the value.

14.3 When phosphoric acid (H_3PO_4) is added to water it dissociates. Assume the species in the solution are: H_3PO_4, H^+, OH^-, H_2O, $H_2PO_4^-$, HPO_4^{2-} and PO_4^{3-}.
  i.   What is the minimum number of independent equations required to define the chemistry of the system? Write a set of independent reactions.
  ii.   How many additional constraints are imposed by conditions of either electroneutrality or stoichiometry?
  iii. Determine the number of components.
  iv.   Determine the degrees of freedom of the system. Explain the significance of the value.

14.4 When the metallic nitrate MNO_3 is heated to a certain temperature in a sealed vessel the possible species at equilibrium in significant quantities are found to be: MNO_2, MNO_3, M_2O, $N_2(g)$, $O_2(g)$, $NO(g)$ and $NO_2(g)$. Derive a set of independent reactions assuming MNO_2, MNO_3 and M_2O are mutually immiscible, and determine the degrees of freedom in the system.

14.5 Construct the predominance area diagram for the Fe–Cl–O system at 800 K assuming the possible species are Fe, Fe_2O_3, Fe_3O_4, $FeCl_2$ and $FeCl_3$. $FeCl_2$ and $FeCl_3$ are volatile species. What is the vapour pressure of each at 800 K? Let the x-axis be $\log p_{O_2}$ and the y-axis be $\log p_{Cl_2}$.

14.6 Construct the predominance area diagram for the Cu–S–O system at 873 K assuming the possible species are Cu, Cu_2S, CuO, Cu_2O, $CuSO_4$, $CuO.CuSO_4$, O_2 and SO_2. Let the x–axis be $\log p_{O_2}$ (from −2 to −14) and the y–axis be $\log p_{SO_2}$ (from 10 to −4). Is it possible to oxidise Cu_2S directly to metallic copper at 873 K? If so, under what conditions? Under what conditions could copper sulphide present in a copper ore be converted to a water-soluble form for extraction by leaching? Data:

Species	Cu_2S	CuO	Cu_2O	$CuSO_4$	$CuO.CuSo_4$	SO_2
$\Delta_f G^0(873)$ kJ mol^{-1}	−104.82	−78.50	−104.77	−449.91	−532.06	−298.10

14.7 Molten solutions of MnO and FeO and of Mn and Fe are both approximately ideal and immiscible. Determine the mole fraction of manganese in the Mn–Fe alloy which is in equilibrium with a MnO–FeO melt at 1900 K which contains 0.30 mole fraction MnO. Determine the equilibrium oxygen pressure of the system. If a partial pressure of oxygen of 10^{-10} bar was imposed on the system, what would be the equilibrium mole fraction of Mn in the alloy? Data: $\Delta_f G^0(MnO,1900) = -239.08$ kJ mol^{-1}; $\Delta_f G^0(FeO,1900) = -145.16$ kJ mol^{-1}.

14.8 Aluminium scrap often consists of scrap from various sources, and when remelted the resulting metal may contain undesired impurities. Some of these can be removed by oxidation, and they enter a molten slag (or flux) phase, and others may volatilise and are removed in the gas phase. Determine which of the following impurity elements can be removed from aluminium either as oxides in a slag phase or as gaseous elements in a gas phase: Ca, Fe, Zn, Mg, Si, Cd. Assume the aluminium is melted and refined at 1100 K and the impurity elements are present at mole fraction of 0.01. Assume also that the activity coefficients of the impurity elements in aluminium and the activity coefficients of the impurity oxides in the slag are both equal to one (that they behave ideally).* The refining reactions occur at $p_{O_2} = 10^{-42}$ bar, which is approximately the value at which Al begins to oxidise to Al_2O_3. The vapour pressure of Al at 1100 K is 1.5×10^{-9} bar.

---

* This assumption is grossly inaccurate; however, the assumption does not invalidate the conclusions as the effect of the value of activity coefficient is outweighed by other variables.

15 Complex equilibria

SCOPE

This chapter examines the problem of how the equilibrium composition can be calculated for systems in which several independent reactions can occur.

LEARNING OBJECTIVES

1. *Understand the difference between reactive systems in which a single reaction may occur and systems in which more than one independent reaction may occur and the computational approaches that can be used to find the equilibrium composition of these systems.*
2. *Understand the principle of Gibbs energy minimisation, and be able to apply it to systems with one independent reaction.*
3. *Be aware of the software packages available for performing Gibbs energy minimisation calculations.*

15.1 INTRODUCTION

So far we have been concerned only with the equilibrium of single reactions, but often more than one reaction is possible in a system. Sometimes, all except the main reaction can be ignored if the amounts of products formed by the other reactions are small or unimportant. However, in other situations small amounts of products may be important, or there may be several reactions all of which occur to a significant extent. This chapter examines how the equilibrium state of a system can be calculated given the initial quantities of reactants, the temperature and pressure. Only closed systems are considered, that is, those in which no material is added or subtracted during the reaction. This corresponds to batch processing in practice.

There are two broad approaches to solving this type of problem. The first is based on the possible independent reactions and may be called the stoichiometric approach. It leads to a series of equations to be solved and requires a new set of equations to be derived for each system studied. The second approach is based on the possible species in the system, the Gibbs energy of which, when minimised, gives the composition of the system at equilibrium (Section 10.2). This approach, called Gibbs energy minimisation, is easier to systematise and thus solve using generic computer algorithms.

15.2 THE STOICHIOMETRIC APPROACH

In this approach, two types of equations are written to describe a reacting system:

- Those that describe the state of equilibrium of the system. These are based on expressions for the equilibrium constants for the independent reactions possible within the system as defined.
- Those that describe arbitrary (such as temperature and pressure of the system) and stoichiometric restrictions.

When the number of equations that can be written equals the number of variables they may be solved.

To illustrate the approach consider a gas mixture which contains the elements carbon, hydrogen and oxygen. At high temperatures the predominant stable species are CO, CO_2, H_2, H_2O and CH_4.

There are five species and two independent reactions. Therefore, there are three components, which can be taken as C, H and O. The independent reactions may be chosen arbitrarily, for example

$$CO(g) + H_2O(g) = CO_2(g) + H_2(g) \tag{15.1}$$

$$CO(g) + 3\,H_2(g) = CH_4(g) + H_2O(g) \tag{15.2}$$

The equilibrium constant expressions (assuming ideal gas behaviour) are

$$K_{15.1} = \frac{p_{CO_2}\,p_{H_2}}{p_{CO}\,p_{H_2O}} \tag{15.3}$$

and

$$K_{15.2} = \frac{p_{CH_4}\,p_{H_2O}}{p_{CO}\,p_{H_2}^3} \tag{15.4}$$

The stoichiometry relations are expressed in terms of moles of species. To solve the complete set of equations it is necessary to use a consistent set of variables, and, therefore, the partial pressures of species should be also expressed in terms of moles. Since (Equation 3.7),

$$p_B = p \times \frac{n_B}{n}$$

where, in this case

$$n = n_{CO} + n_{CO_2} + n_{H_2} + n_{H_2O} + n_{CH_4} \tag{15.5}$$

then, substituting for the partial pressure terms in Equations 15.3 and 15.4 gives

$$K_{15.1} = \frac{n_{CO_2}\,n_{H_2}}{n_{CO}\,n_{H_2O}} \tag{15.6}$$

and

$$K_{15.2} = \frac{n_{CH_4}\,n_{H_2O}}{n_{CO}\,n_{H_2}^3} \times \left(\frac{n}{p}\right)^3 \tag{15.7}$$

The numbers of moles of C, H and O are conserved during the reaction. Therefore,*

$$n_C = n_{CO} + n_{CO_2} + n_{CH_4} \tag{15.8}$$

$$n_H = 2n_{H_2} + 2n_{H_2O} + 4n_{CH_4} \tag{15.9}$$

$$n_O = n_{CO} + 2n_{CO_2} + n_{H_2O} \tag{15.10}$$

If we know the number of moles of C, H and O in the mixture, the total pressure and temperature (which fixes the values of the equilibrium constants of Equations 15.1 and 15.2) then the six independent Equations 15.5 to 15.10 contain six unknowns, namely n_{CO_2}, n_{CO}, n_{H_2}, n_{H_2O}, n_{CH_4} and n. Therefore, these equations can be solved to determine the equilibrium number of moles of each species.

---

* For example, each mole of H_2, H_2O and CH_4 contains 2, 2 and 4 moles of H atoms, respectively; therefore $n_H = 2n_{H_2} + 2n_{H_0} + 4n_{CH_4}$. Similarly for C and O.

To illustrate, suppose we start with a gas mixture of 1 mole of CO and 2 moles of H_2O and heat this to 800°C at a total pressure of 1 bar. We want to find the equilibrium composition of the gas mixture. In this case, Equations 15.6 and 15.7 become:

$$1.406 = \frac{n_{CO_2}\, n_{H_2}}{n_{CO}\, n_{H_2O}} \tag{15.11}$$

and

$$0.0400 = \frac{n_{CH_4}\, n_{H_2O}}{n_{CO}\, n_{H_2}^{3}} \times n^3 \tag{15.12}$$

Since the number of moles of carbon, hydrogen and oxygen are conserved, using the initial composition of the gas mixture,

$$n_C = n_{CO} + n_{CO_2} + n_{CH_4} = 1 + 0 + 0 = 1$$

$$n_H = 2n_{H_2} + 2n_{H_2O} + 4n_{CH_4} = 0 + 2 \times 2 + 0 = 4$$

$$n_O = n_{CO} + 2n_{CO_2} + n_{H_2O} = 1 + 0 + 2 = 3$$

and Equations 15.8 to 15.10, therefore, become

$$n_{CO} + n_{CO_2} + n_{CH_4} = 1 \tag{15.13}$$

$$2n_{H_2} + 2n_{H_2O} + 4n_{CH_4} = 4 \tag{15.14}$$

$$n_{CO} + 2n_{CO_2} + n_{H_2O} = 3 \tag{15.15}$$

The six equations 15.5 and 15.11 to 15.15 contain the six unknown values and, therefore, can be solved.

In practice solving the equations is not easy since they are usually non-linear. In relatively simple systems, such as the above example, the successive approximation method is suitable. In this approach, as many unknowns as possible are eliminated by solving those equations which are linear; in the above example, four unknowns can be eliminated by solving Equations 15.13 to 15.15 for four unknowns in terms of the other two and substituting these in Equations 15.11 and 15.12. The remaining two non-linear equations in two unknowns can then be solved by successive approximation. There are other more advanced numerical methods for solving non-linear equations which are described in mathematical texts. There are also commercial software products (for example, Polymath) and the Solver add-in in Microsoft Excel™ which solve non-linear equations. The solution for the above problem is summarised in Table 15.1.

TABLE 15.1

The solution of equations 15.5, 15.11 – 15.15

Species	Initial no. moles	Initial composition (vol%)	Equilibrium no. moles at 800°C	Equilibrium composition (vol%)
$H_2O(g)$	2.0	66.7	1.325	44.2
$CO_2(g)$	0		0.675	22.5
$H_2(g)$	0		0.675	22.5
$CO(g)$	1.0	33.3	0.325	10.8
$CH_4(g)$	0		5.32×10^{-5}	0.0018
Total	3.0	100%	3.0	100%

15.3 GIBBS ENERGY MINIMISATION

Other than for very simple cases this method requires the use of a computer, and a number of commercial packages to perform the calculations are available. The method is based on the principle that the Gibbs energy of a system at constant temperature and pressure is a minimum when the system is at equilibrium.

The approach can be illustrated using a simple example in which the computational procedures do not overly confuse things. Consider the reaction:

$$CO(g) + H_2O(g) = CO_2(g) + H_2(g) \tag{15.16}$$

Suppose the reaction occurs at 800°C and 1 bar and initially the system consists of a mixture of 1 mole of CO and 2 moles of H_2O. This is the same problem as solved in Section 15.2 except methane has been omitted for simplicity since we know that under the conditions specified there will be a negligible amount of methane formed (Table 15.1).

Let α be the number of moles of CO that react as the mixture comes to equilibrium. The stoichiometry of Equation 15.16 indicates that α moles of H_2O will react with α moles of CO to form α moles of CO_2 and α moles of H_2. Therefore, the equilibrium composition will be:

$$n_{CO} = 1 - \alpha; \quad n_{H_2O} = 2 - \alpha; \quad n_{CO_2} = 0 + \alpha = \alpha; \quad n_{H_2} = 0 + \alpha = \alpha$$

The total number of moles is:

$$n = n_{CO} + n_{H_2O} + n_{CO_2} + n_{H_2} = (1 - \alpha) + (2 - \alpha) + \alpha + \alpha = 3$$

Therefore, the equilibrium composition in terms of mole fractions is:

$$x_{CO} = \frac{1 - \alpha}{3}; \quad x_{H_2O} = \frac{2 - \alpha}{3}; \quad x_{CO_2} = \frac{\alpha}{3}; \quad x_{H_2} = \frac{\alpha}{3}$$

This can be summarised in the form of an Input–Change–Equilibrium table (Section 10.6):

	CO	H₂O	CO₂	H₂O	Total
Initial	1	2	0	0	3 moles
Change	$-\alpha$	$-\alpha$	$+\alpha$	$+\alpha$	0 moles
Equilibrium	$1 - \alpha$	$2 - \alpha$	α	α	3 moles
Eqlm. pressure	$(1 - \alpha)p/3$	$(2 - \alpha)p/3$	$\alpha p/3$	$\alpha p/3$	p bar

The total Gibbs energy of the system is the sum of the Gibbs energies of all the species in the system,

$$G = \sum_{B=1}^{s} n_B \, G_B$$

where s is the number of species in the system. For an ideal gas,

$$G_B = G_B^0 + RT \ln p_B$$

For systems involving liquid phases the appropriate equation would be:

$$G_B = G_B^0 + RT \ln a_B$$

For an ideal gas mixture, $p_B = x_B\, p$ (where p is the total pressure of the system) and, therefore,

$$G = \sum_{B=1}^{s} n_B\, G_B^0 + \sum_{B=1}^{s} RT\, n_B \ln(x_B\, p) \tag{15.17}$$

For the reaction under consideration, $p = 1$, and Equation 15.17 becomes

$$G = n_{CO}\, G_{CO}^0 + n_{H_2O}\, G_{H_2O}^0 + n_{CO_2}\, G_{CO_2}^0 + n_{H_2}\, G_{H_2}^0$$

$$+ RT\, n_{CO} \ln x_{CO} + RT\, n_{H_2O} \ln x_{H_2O} + RT\, n_{CO_2} \ln x_{CO_2} + RT\, n_{H_2} \ln x_{H_2}$$

$$= (1-\alpha)G_{CO}^0 + (2-\alpha)G_{H_2O}^0 + \alpha\, G_{CO_2}^0 + \alpha\, G_{H_2}^0$$

$$+ RT\left[(1-\alpha)\ln\frac{(1-\alpha)}{3} + (2-\alpha)\ln\frac{(2-\alpha)}{3} + \alpha\ln\frac{\alpha}{3} + \alpha\ln\frac{\alpha}{3} \right]$$

Dividing throughout by RT and collecting like terms,

$$\frac{G}{RT} = \frac{(1-\alpha)}{RT}G_{CO}^0 + \frac{(2-\alpha)}{RT}G_{H_2O}^0 + \frac{\alpha}{RT}\left(G_{CO_2}^0 + G_{H_2}^0\right)$$

$$+ (1-\alpha)\ln\frac{(1-\alpha)}{3} + (2-\alpha)\ln\frac{(2-\alpha)}{3} + 2\alpha\ln\frac{\alpha}{3} \tag{15.18}$$

The relevant values of the Gibbs energies of the reactants and products at 800°C are:

CO_2	−649.24 kJ
CO	−340.67 kJ
H_2O	−465.71 kJ
H_2	−157.63 kJ

(Alternatively, $\Delta_f G^0$ values for the reactants and products could be used.) These are substituted into Equation 15.18 and values of G/RT calculated for particular values of α. The value of α at the minimum value of G/RT can then be found graphically as illustrated in Figure 15.1, from which it is evident that the minimum occurs at around 0.65–0.70, as expected from the earlier calculation.

To find a more accurate value of α, the right side of Equation 15.18 can be differentiated with respect to α and the derivative set equal to zero:

$$\frac{-G_{CO}^0 - G_{H_2O}^0 + G_{CO_2}^0 + G_{H_2}^0}{RT} + 2\ln\alpha - \ln(1-\alpha)(2-\alpha) = 0$$

that is,

$$\ln\left[\frac{(1-\alpha)(2-\alpha)}{\alpha^2}\right] = \frac{G_{CO_2}^0 + G_{H_2}^0 - G_{CO}^0 - G_{H_2O}^0}{RT}$$

or

$$\ln\left[\frac{(1-\alpha)(2-\alpha)}{\alpha^2}\right] = \frac{\left(-649.24 - 157.63 + 340.64 + 465.71\right)\times 1000}{8.314\times\left(800+273.15\right)}$$

$$= -0.0583$$

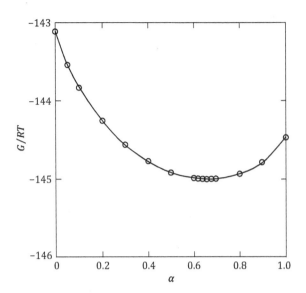

FIGURE 15.1 G/RT as a function of α calculated using Equation 15.18.

Solving algebraically (or numerically) gives $\alpha = 0.675$. Since

$$n_{CO} = 1 - \alpha; \quad n_{H_2O} = 2 - \alpha; \quad n_{CO_2} = \alpha; \quad n_{H_2} = \alpha$$

therefore,

$$n_{CO} = 0.325; \quad n_{H_2O} = 1.325; \quad n_{CO_2} = 0.675; \quad n_{H_2} = 0.675$$

As expected, this corresponds to the equilibrium composition in Table 15.1.

The above example is rather trivial, but it demonstrates the concept of Gibbs energy minimisation. Single-reaction systems, as in the above example, are relatively easy to analyse, but more complex systems with several possible reactions and one or more condensed phases in addition to the gas phase are much more difficult and require a different solution approach. It is not necessary to specify the independent reactions, just the species that may be present at equilibrium. The total Gibbs energy for a system of s species consisting of a solid, liquid and gas phase is:

$$\frac{G}{RT} = \sum_{B=1}^{s} n_B G_B = \frac{1}{RT} \sum_{B=1}^{s} n_B(g) G_B^0(g) + \sum_{B=1}^{s} n_B(g) \ln f_B(g)$$

$$+ \frac{1}{RT} \sum_{B=1}^{s} n_B(l) G_B^0(l) + \sum_{B=1}^{s} n_B(l) \ln a_B(l)$$

$$+ \frac{1}{RT} \sum_{B=1}^{s} n_B(s) G_B^0(s) + \sum_{B=1}^{s} n_B(s) \ln a_B(s)$$

$$+ \sum_{B=1}^{s} n_B(g) \ln p$$

(15.19)

where, for the gas phase (assuming ideal behaviour),

$$G_B = G_B^0 + RT \ln p_B$$

and for the solid and liquid phases,

$$G_B = G_B^0 + RT \ln a_B$$

where $a_B = 1$ if the solid and liquid phases are pure substances and $a_B < 1$ if species B is present in a solid or liquid solution.

The problem is to find the values of n_B which correspond to the minimum in G/RT. Not every combination of n_B is acceptable since there are two constraints on the system:

- The law of conservation of mass must be obeyed (that is, the number of moles of each element is fixed according to the initial conditions).
- The values of n_B must be positive at equilibrium.

The equations comprising the function to be minimised and the constraints were traditionally solved using the Lagrange multiplier method,* but this method can be difficult due to the resulting set of non-linear equations. Computer algorithms, based on sequential quadratic programming and generalised reduced gradient approaches, have been developed to linearise equations, and these have enabled Gibbs energy minimisation to become the most widely used technique for computational equilibrium calculations. Commercial software products are available for performing the calculations. The Microsoft Excel™ spreadsheet Solver feature, which uses the generalised reduced gradient method, can also be used, and examples of its application are available in the literature.† Discussion of the mathematics is beyond the scope of this book.‡

The above example involved a single reaction, and the Gibbs energy minimisation method was used to obtain the solution merely to demonstrate the methodology. It could have been more easily solved using the method for single reactions described in Section 10.6. In that case,

$$K = \frac{p_{CO_2} \, p_{H_2}}{p_{CO} \, p_{H_2O}}$$

$$= \frac{\alpha/3 \times \alpha/3}{(1-\alpha)/3 \times (2-\alpha)/3} \quad \text{(from the ICE table)}$$

$$= \frac{\alpha^2}{(1-\alpha)(2-\alpha)}$$

Knowing the value of K at 800°C ($K = 1.05489$), the equation can be solved for α. Using the 'Goal Seek' function in Excel™ a value of $\alpha = 0.675$ is obtained, as expected.

---

* Eriksson, G. 1971. *Acta Chem. Scand.*, 25, 2651–2658.Eriksson, G. and E. Rosen. 1973. *Chemica Scripta*, 4, 193–194. Eriksson, G. 1975. *Chemica Scripta*, 8, 100–103.

† See for example: Lwin, Y. 2000. Chemical equilibrium by Gibbs energy minimization on spreadsheets, *Internat. J. Engineering Education*, 16(4), pp. 335–339. da Silva, A. L. and N. C. Heck. 2008. A procedure to compute equilibrium concentrations in multicomponent systems by Gibbs energy minimization on spreadsheets. CONAMET/SAM–2008. Congreso Internacional de Metalurgia y Materiales. Univerdidad Tecnológica Metropolitana, Santiago, Chile, 40–48.

‡ It is the author's contention that the mathematics is best left to mathematicians. Once the basic concept of Gibbs energy minimisation is understood it is better for students of chemical thermodynamics to become proficient in an appropriate computer package and to use it to solve a variety of complex probems, rather than spend time setting up spreadsheets to solve the problems.

15.4   COMMERCIAL SOFTWARE TO PERFORM GIBBS ENERGY MINIMISATION

Some of the better known commercial software products for performing Gibbs energy minimisation calculations are:

- HSC Chemistry™; <www.outotec.com/products/digital–solutions/hsc–chemistry>; developed by Outotec.
- FactSage™; <www.factsage.com>; an integrated database/computing system for chemical thermodynamics introduced in 2001; it is the fusion of the earlier FACT Win/F*A*C*T and the ChemSage/SOLGASMIX thermochemical packages.
- MTDAT™; <www.npl.co.uk/science–technology/mathematics–modelling–and–simulation>; developed by the UK National Physical Laboratory.
- Thermo–Calc™; <www.thermocalc.com>; developed at the Division of Physical Metallurgy at KTH, Stockholm, Sweden.
- Stream Analyser™; <www.olisystems.com>; developed by OLI Systems Inc; useful for aqueous systems for its ability to simulate non-ideal behaviour in concentrated solutions of aqueous species.

To illustrate how a problem is set up using commercial software, the first six columns of the input table for the HSC Chemistry software for the problem posed at the beginning of the chapter are shown in Table 15.2. In this system there is only one phase, which has been designated as GAS. If there were other phases they would be given appropriate names. The species selected for the system being studied are entered into column 1. A few extra species have been added in this example to illustrate the point that as many species may be specified as deemed necessary. The second column is the temperature at which the reactants are added, the third is the activity coefficient of each species in the gas phase. Ideal behaviour has been assumed. The amounts of input species (in kmol in this example) are entered into column 4. As only CO and H_2O are present initially, all other species have zero input. The remaining input columns are not needed for this problem.

The pressure of the system (1 bar) and temperature at which the reactions will occur (800°C) are then entered in a second table. The 'Calculate' button is then pressed, and a 'Results' table is generated. The data from this table may be copied and pasted into a spreadsheet and processed as required, or, alternatively, the software has the option of plotting two-dimensional graphs of any selected set of variables. The equilibrium composition values extracted from the 'Results' table are summarised in Table 15.3. These show only trace amounts of O_2, CH_4, C_2H_6 and C_3H_8 are formed. The very low oxygen pressure reflects the highly reducing nature of the gas mixture.

TABLE 15.2

Input table for HSC Chemistry to calculate the composition of a gas phase in which multiple reactions are possible

Species	Temperature, °C	Activity coefficient	Amount unit	Initial amount	Initial amount, %
Phase: Gas				3	100
CO(g)	25	1	kmol	1	33.33333
CO_2(g)	25	1	kmol		0
H_2(g)	25	1	kmol		0
H_2O(g)	25	1	kmol	2	66.66667
O_2(g)	25	1	kmol		0
CH_4(g)	25	1	kmol		0
C_2H_6(g)	25	1	kmol		0
C_3H_8(g)	25	1	kmol		0

TABLE 15.3

Equilibrium composition at 800°C and 1 bar of the gas mixture specified in Table 15.2

Species	Amount (kmol)	Volume %
$CO(g)$	0.325	10.8
$CO_2(g)$	0.675	22.5
$H_2(g)$	0.675	22.5
$H_2O(g)$	1.325	44.2
$O_2(g)$	4.96E−18	1.652E−16
$CH_4(g)$	5.268E−05	1.756E−03
$C_2H_6(g)$	1.33E−12	4.446E−11
$C_3H_8(g)$	1.19E−20	3.955E−18
Total	3.000	100.0

TABLE 15.4

Input table for HSC Chemistry to calculate the reaction products when chalcopyrite is reacted with graphite and calcium oxide

Species	Feed temperature, °C	Activity coefficient	Amount unit	Initial amount
Phase 1: Gas				0
O2(g)	25	1	kmol	
S2(g)	25	1	kmol	
SO2(g)	25	1	kmol	
CO(g)	25	1	kmol	
CO2(g)	25	1	kmol	
Phase 2: Chalcopyrite				1
$CuFeS_2$	25	1	kmol	1
Phase 3: Graphite				2
C	25	1	kmol	2
Phase 4: Lime				4
CaO	25	1	kmol	4
Phase 5: Ca sulfide				0
CaS	25	1	kmol	
Phase 6: Copper				0
Cu	25	1	kmol	
Phase 7: Iron				0
Fe	25	1	kmol	
Phase 8: Wustite				0
FeO	25	1	kmol	
Phase 9: Magnetite				0
Fe_3O_4	25	1	kmol	
Phase 10: Ca ferrite				0
$*2CaO*Fe_2O_3$	25	1	kmol	0

A more complex problem is the reduction of the mineral chalcopyrite $CuFeS_2$ using carbon as the reductant and calcium oxide as the collector for sulfur. The system is Cu–Fe–S–C–Ca–O, and numerous reaction products are possible. A preliminary assessment would limit these to the most likely ones: Cu, Fe, FeO, Fe_3O_4, Fe_2O_3, CaS, $2CaO.Fe_2O_3$, $CO_2(g)$, $CO(g)$, $O_2(g)$, $S_2(g)$, $SO_2(g)$. Table 15.4 is an

TABLE 15.5

Equilibrium composition at 400–800°c and 1 bar of the system specified in Table 15.4

Species	Input, kmol	Equilibrium amounts, kmol				
		400°C	500°C	600°C	700°C	800°C
O2(g)	0.00	5.26E-32	3.92E-28	5.80E-25	8.54E-23	1.06E-20
S2(g)	0.00	4.75E-19	6.48E-17	3.45E-15	4.46E-14	7.56E-13
SO2(g)	0.00	1.33E-16	3.09E-15	3.42E-14	1.11E-13	4.36E-13
CO(g)	0.00	0.00	0.02	0.08	0.23	1.60
CO2(g)	0.00	0.25	0.24	0.21	0.14	0.20
CuFeS$_2$	1.00	0.00	0.00	0.00	0.00	0.00
C	2.00	1.75	1.74	1.71	1.64	0.20
CaO	4.00	1.00	1.00	1.00	1.00	2.00
CaS	0.00	2.00	2.00	2.00	2.00	2.00
Cu	0.00	1.00	1.00	1.00	1.00	1.00
Fe	0.00	0.00	0.00	0.00	0.00	1.00
FeO	0.00	0.00	0.00	0.00	0.00	0.00
Fe$_3$O$_4$	0.00	0.00	0.00	0.00	0.00	0.00
*2CaO*Fe$_2$O$_3$	0.00	0.50	0.50	0.50	0.50	0.00

extract of the first five columns of the input file. The reactants are 1 kmol CuFeS$_2$, 2 kmol C and 4 kmol CaO. It is assumed the gas species are ideal and that there is no solid solution between the condensed phases; that is, each of the species Cu, Fe, FeO, Fe$_2$O$_3$, Fe$_2$O$_3$ and CaS, 2CaO.Fe$_2$O$_3$ are a single phase. The result of Gibbs energy minimisation calculations at 400–800°C, extracted from the 'Results' file, are listed in Table 15.5. The following points of interest can be noted:

- All the chalcopyrite is consumed.
- The main gaseous products are CO and CO$_2$.
- Metallic copper is produced at all temperatures.
- The sulfur is almost entirely captured as CaS.
- Metallic iron is produced at 800°C.
- Iron is captured as 2CaO.Fe$_2$O$_3$ at the lower temperatures.

The dominant reactions at $T < {\sim}750°C$ are, therefore

$$CuFeS_2 + 0.5\,C + 3\,CaO = Cu + 0.5\,2CaO.Fe_2O_3 + 2\,CaS + 0.5\,CO$$

$$CuFeS_2 + 0.25\,C + 3\,CaO = Cu + 0.5\,2CaO.Fe_2O_3 + 2\,CaS + 0.25\,CO_2$$

while the dominant reactions at $T > {\sim}750°C$ are:

$$CuFeS_2 + 2\,C + 2\,CaO = Cu + Fe + 2\,CaS + 2\,CO$$

$$CuFeS_2 + C + 2\,CaO = Cu + Fe + 2\,CaS + CO_2$$

PROBLEMS

15.1 Consider the dissociation of 1 mole of N_2O_4: $N_2O_4(g) = 2\,NO_2(g)$.
    i. Set up the ICE table for this reaction, and derive an equation of the form of Equation 15.18 for the reactive mixture.
    ii. Calculate values of $\Sigma G/RT$ at a total pressure of 1 bar and 100°C, and plot them as a function of α. Data: $G^0(N_2O_4$, g, $373.15) = -105.116$ kJ mol⁻¹; $G^0(NO_2$, g, $373.15) = -56.896$ kJ mol⁻¹.
    iii. Fit a curve, and find the minimum point.
    iv. What are the equilibrium concentrations of N_2O_4 and NO_2 in the gas mixture?
    v. Repeat the calculation at total pressure of 10 bar.

15.2 Repeat the above calculation of equilibrium concentrations of N_2O_4 and NO_2 in the gas mixture using the method described in Section 10.6 for single-reaction equilibria.

15.3 Consider the reaction of hydrogen with nitrogen to form ammonia: $N_2(g) + 3\,H_2(g) = 2\,NH_3(g)$. Assume the starting mixture is 5 moles each of H_2 and N_2 and zero moles of NH_3.
    i. Set up the ICE table for this reaction, and derive an equation of the form of Equation 15.18 for the reactive mixture.
    ii. Calculate values of $\Sigma G/RT$ at 200 bar pressure and 800 K, and plot them as a function of α. The relevant values of the Gibbs energies at 800 K are: $G^0(N_2$, g$) = -161.77$ kJ mol⁻¹; $G^0(H_2$, g$) = -112.94$ kJ mol⁻¹; $G^0(NH_3$, g$) = -211.67$ kJ mol⁻¹.
    iii. Fit a curve, and find the minimum point.
    iv. What are the equilibrium concentrations of H_2, N_2 and NH_3 in the gas mixture?
    v. Repeat the above calculation for $p = 100$ bar.
    vi. Repeat the calculation for an initial composition of 5 moles H_2 and 2 moles N_2 at 200 bar.

15.4 The gas from a pyrite (FeS_2) burner of a sulfuric acid plant has the following average composition: 8 vol% SO_2, 13 vol% O_2 and 79 vol% N_2. This gas is passed through a reactor at 800 K and 1 atm in the presence of a catalyst to convert the SO_2 to SO_3. Calculate the equilibrium composition of the gas from the reactor.

15.5 A mixture of 15 g iron sulfide and 10 g zinc metal powders is sealed in an inert container and heated in a furnace to 1000°C. The following reaction occurs: $FeS(s) + Zn(s) = ZnS(s) + Fe(s)$.
    i. Assuming no solid solutions form between the reactants and products, in what direction will the reaction proceed, and what will be the final composition of the system?
    ii. In fact, Zn and Fe do form a solid solution. Assuming ZnS and FeS do not, what will be the equilibrium composition of the system? Assume the alloy solution is ideal.
    iii. ZnS and FeS actually do form a solid solution. Assuming the solution is ideal,[*] calculate its equilibrium composition of the system at 1000°C.
    Data: At 1000°C, $G^0(Fe) = -62.31$ kJ, $G^0(Zn) = -83.53$ kJ, $G^0(FeS) = -250.00$ kJ, $G^0(ZnS) = -321.19$ kJ.

15.6 20 g Fe and 100 g NiO powders are mixed and heated in a sealed container to 600°C. The following reaction occurs: $NiO(s) + Fe(s) = Ni(s) + FeO(s)$.
    i. Assuming no solid solutions form between the reactants and products, in what direction will the reaction proceed, and what will be the final composition of the system?

---

[*] It is actually not ideal, but, as shown in Problem 9.9, it is not far from ideal so this assumption will give an answer that is reasonably accurate.

ii. In fact, Ni and Fe form a solid solution. Assuming NiO and FeO do not, what will be the equilibrium composition of the system? Assume the alloy solution is ideal.

iii. NiO and FeO actually also form a solid solution. What will be the equilibrium composition of the system for the case where both alloy and oxide solid solutions form? Assume both solutions are ideal.

iv. The assumption of ideality is reasonable for the NiO–FeO system but not for the Ni–Fe system. In the relevant composition range, γ_{Ni} will be 1 but γ_{Fe} will much less than one. What effect will this have on the answers to part iii? Assume a value of $\gamma_{Fe} = 0.05$.

v. The reaction is exothermic. Would the metallic solid solution have a higher or lower concentration of Ni at 1000°C than at 600°C?

vi. What effect would pressure have on this reaction?

Data: At 600°C, $G^0(Ni) = -36.84$ kJ, $G^0(NiO) = -291.62$ kJ, $G^0(Fe) = -34.20$ kJ, $G^0(FeO) = -336.38$ kJ. Ni and Fe have negligible vapour pressures at 600°C.

FURTHER READING

Liu, Z. and Y. Wang. 2016. *Computational Thermodynamics of Materials*. Cambridge: Cambridge University Press.

Lukas, H., S. G. Fries and B. Sundman. 2007. *Computational Thermodynamics: The Calphad Method*. Cambridge: Cambridge University Press.

Smith, W. R. and R. W. Missen. 1982. *Chemical Reaction Equilibrium Analysis: Theory and Algorithms*. New York: John Wiley & Sons.

16 Electrochemistry

SCOPE

This chapter examines the nature and thermodynamics of electrochemical reactions.

LEARNING OBJECTIVES

1. *Understand the nature of electronic and ionic conductors and the nature of electrochemical reactions.*
2. *Be familiar with the meaning of the following terms, and be able to give examples: Electrode, electrolyte, oxidation, reduction, redox reaction, cathode, anode, conductivity, electrolytic cell, galvanic cell, half-cell reaction, cell reaction, standard electrode potential, contact potential, polarisation, overvoltage.*
3. *Be able to identify and describe the main types of galvanic and electrolytic cells in terms of half-cell and cell reactions, and determine the cell potential from standard electrode potential data.*
4. *Be able to apply the Nernst equation and interpret the results.*
5. *Be able to calculate the total potential of an electrochemical cell, the heat generated and mass of matter transferred during operation.*

16.1 INTRODUCTION

In electronic conductors such as metals an electric current is the result of the movement of electrons in the structure of the conductor when an electrical potential difference is applied. The flow of electrons is not accompanied by any significant movement of matter since electrons have negligible mass. Aqueous solutions of ionic compounds and molten salts, on the other hand, contain mobile ions which have an electrical charge. When a potential difference is applied the ions will move in a direction depending on their charge. As for electronic conduction the electric current is a movement of electric charge, but in these cases it is carried by ions having significant mass. Ionic conduction, therefore, is accompanied by transfer of mass from one place to another. Ionic conduction plays a major role in some important phenomena and processes – the generation and storage of electrical energy, corrosion and the oxidation of metals, electroplating and electrolytic extraction and refining of metals, for example.

16.2 DEFINITIONS OF AMPERE, COULOMB AND VOLT

The base unit of electric current I is the ampere A, defined as the current which if maintained in two straight parallel conductors of infinite length and negligible circular cross-section placed 1 metre apart in vacuum would produce between these conductors a force of 2×10^{-7} Nm^2. The charge transported by a current of 1 ampere in 1 second is the Coulomb C (1 C = 1 A s).

An electrical potential E (also called the electric field potential, potential difference or electrostatic potential) is the amount of work needed to move a unit of positive charge from a reference point to a specific point inside an electrical field without producing an acceleration. The unit of electrical potential is the Volt V, defined as the difference in electrical potential between two points of a conducting wire when an electric current of 1 ampere dissipates 1 watt of power between those points (1 V = 1 J C^{-1}).

16.3 ELECTROCHEMICAL REACTIONS

Electrochemistry is the branch of chemistry that studies chemical reactions which take place at the interface of an electron conductor (the *electrode*, usually a metal or a semi-conductor) and an ionic conductor (the *electrolyte*) and which involve electron transfer between the electrode and the electrolyte or species in solution. If a chemical reaction is driven by an externally applied potential difference, as in electrolysis, or if a potential difference is created by a chemical reaction, as in a battery, it is an electrochemical reaction.

Chemical reactions in which electrons are transferred directly between molecules are called oxidation/reduction reactions, or *redox reactions*. The *reducing agent* (or reductant) is the electron donor, and the *oxidising agent* (or oxidant) is the electron acceptor. In contrast, electrochemical reactions are reactions in which oxidation and reduction reactions are separated in space or time and connected by an external electrical circuit. The concept is illustrated in Figure 16.1 which represents the electrochemical reaction

$$YA + Z = ZA + Y$$

where Y and Z are elements and A is an anion (such as O^{2-}, S^{2-}, SO_4^{2-}, etc.). A mixture of $Y + YA$ and of $Z + ZA$ are joined by two conductors – an electronic conductor and an ionic conductor through which only A^{2-} anions can pass. For simplicity, we assume Y and Z are both divalent. The redox reaction is:

$$Y^{2+} + Z = Z^{2+} + Y$$

However, this reaction could be carried out as two physically separate reactions:

$$\text{Reduction:} \quad Y^{2+} + 2\,e^- = Y$$

$$\text{Oxidation:} \quad Z = Z^{2+} + 2\,e^-$$

Note, by definition, *oxidation* is the loss of electrons by a molecule, atom or ion and *reduction* is the gain of electrons by a molecule, atom or ion.* When summed, the oxidation and reduction reactions give the redox reaction.

If the reaction $Y^{2+} + Z = Z^{2+} + Y$ is spontaneous ($\Delta G < 0$), then the reaction proceeds as follows. Two electrons leave a Z atom to form a Z^{2+} ion and travel along the electronic conductor

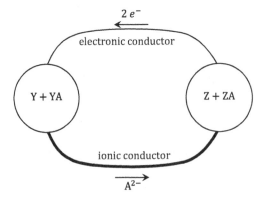

FIGURE 16.1   Schematic representation of an electrochemical reaction.

---

* A simple way to remember this is think of the phrase Oil Rig. OIL: Oxidation is Loss (of electrons). RIG: Reduction is Gain (of electrons).

from right to left. On arrival at the Y + YA mixture the electrons convert a Y^{2+} ion to a Y atom. Simultaneously, an A^{2-} ion leaves the Y + YA mixture and travels along the ionic conductor from left to right to the Z + ZA mixture. The movement of the electrons along the electronic conductor creates an electrical potential difference which can be measured by placing an externally opposing potential difference in the circuit (a potentiometer) which is adjusted until no current flows, and the reaction ceases.* If the voltage is increased further, the reaction will be driven in the reverse direction. Thus, the voltage required to just stop the reaction is the voltage generated by the spontaneous reaction (operating as a battery) or the voltage that must be exceeded in order to reverse the reaction (for reducing ZA to Z).

16.3.1   AN EXAMPLE OF AN ELECTROCHEMICAL REACTION

When silver tarnishes, it combines with sulfur, and a thin layer of black silver sulfide forms on the surface. One way to clean the silver is to physically remove the silver sulfide by polishing with an abrasive or chemical that reacts with the silver sulfide. In both cases, some silver is lost. Another method is to reverse the chemical reaction and turn the silver sulfide back into silver. In this case, the silver remains in place and is not lost. To do this, the item is placed on a sheet of aluminium foil on the bottom of a container and a strong solution of sodium bicarbonate and sodium chloride is added to cover the item. The tarnish will soon begin to disappear. The reaction is faster if the solution is warm.

Aluminium has a greater affinity for sulfur than has silver, and silver sulfide reacts spontaneously with aluminium since the Gibbs energy of reaction is negative:

$$3\,Ag_2S(s) + 2\,Al(s) = 6\,Ag(s) + Al_2S_3(s); \Delta G^0 = -481\,kJ$$

The reaction between silver sulfide and aluminium takes place when the two are in contact while immersed in solution. In this case, the solution is the ionic conductor (which carries sulfur ions from the silver to the aluminium), and the aluminium foil is the ionic conductor. This is illustrated in Figure 16.2. The reaction is electrochemical:

$$\text{Reduction:} \quad 3\,Ag^+ + 3\,e^- = 3\,Ag$$

$$\text{Oxidation:} \quad Al = Al^{3+} + 3\,e^-$$

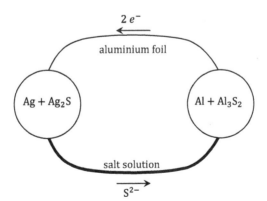

FIGURE 16.2   Schematic representation of the electrochemical mechanism by which tarnish can be removed from silver.

---

* In practice, a high impedance (electronic) voltmeter drawing negligible current is used to measure cell potentials.

Combining the two half reactions gives the redox equation for the reaction:

$$Al + 3\,Ag^+ = Al^{3+} + 3\,Ag$$

Since the reaction

$$Al_2S_3(s) + 6\,H_2O(aq) = 2\,Al(OH)_3(aq) + 3\,H_2S(g)$$

can also occur, the sodium bicarbonate is added to prevent the formation of a layer of aluminium hydroxide which would act as a barrier and prevent further reaction. The sodium chloride increases the ionic conductivity of the solution since it dissociates into Na^+ and Cl^- ions.

16.4   CONDUCTORS AND CONDUCTION

In *electronic conductors* current is carried by electrons which are free to move though the conductor under an applied potential difference. There are two types, metallic conductors and semi-conductors. Metallic conductors have conductivities many times greater than semi-conductors. Semi-conductor materials have electrical conductivity values between that of conductors – such as copper and gold – and insulators, such as glass. Semi-conductor materials are generally covalent compounds. Depending on temperature some electrons can break free and leave holes in the structure which are positively charged. Application of an electric field will cause electrons to move in one direction (called N type conductivity) and holes to move in the opposite direction (called P type conductivity). The total current is the sum of the two. Increasing the temperature allows more electrons to break free and hence increases conductivity whereas for metals the opposite is true. The conducting properties of semi-conductors may be altered in useful ways by the controlled introduction of impurities (doping) into the crystal structure. Solid silicon and arsenic are semi-conductors as are many of the sulfides and oxides of heavy metals.

In *ionic conductors* (also called *electrolytic conductors*) the current is carried by anions and cations, and the passage of current through ionic conductors, therefore, is associated with the transport of chemical species. This makes the separation of ions possible, primarily on the basis of their charge. Compounds which dissociate into ions when they are dissolved in solutions, heated above a certain temperature or when they melt are called *electrolytes* (Section 12.1).

Ohm's law applies to all conductors, namely,

$$E = I\,R \tag{16.1}$$

where E is the potential difference in Volts (V) across the conductor, I the current in Amperes (A) and R the resistance in ohms (Ω) of the conductor. While this relation is quite familiar for electronic conductors, it applies equally to ionic conductors because these also offer a resistance to the movement of ions carrying the charge.

The *conductivity* κ_B of a substance is defined as

$$\kappa_B = \frac{I\,l}{\Delta V\,A} \tag{16.2}$$

where A is the cross-sectional area of the substance (m^2) and ΔV is the potential difference (V) across length l (m). The unit of conductivity is $ohm^{-1}\,m^{-1}$ or, since by definition 1 Siemens $= 1\,ohm^{-1}$, the unit is $S\,m^{-1}$. The conductivity of metals is typically of the order $10^7\,S\,m^{-1}$, that of ionic melts in the range 10^2 to $10^3\,S\,m^{-1}$, that of semi-conductors 10^{-7} to $10^4\,S\,m^{-1}$. Good electrical insulators have a conductivity of the order $10^{-20}\,S\,m^{-1}$. Distilled water has a conductivity of around 10^{-6} to $10^{-5}\,S\,m^{-1}$. The value for water increases greatly when ionic salts are dissolved in it as may be seen, for the case of KCl additions, in Figure 16.3. Solid electrolytes such as yttria-stabilised zirconia and sodium β–alumina at elevated temperatures can have ionic conductivities of oxygen and sodium

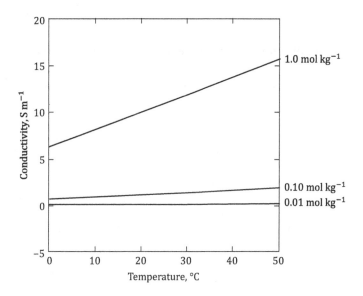

FIGURE 16.3 Conductivity of water containing dissolved potassium chloride at three molalities. Source of data: Haynes, W. M. (ed.). 2016. *CRC Handbook of Chemistry and Physics*, 97th ed. Boca Baton: CRC Press.

ions, respectively, comparable in magnitude to those of ionic compounds dissolved in water.* These structures contain vacancies through which the oxygen or sodium ions can move at high temperatures. In the case of yttria-stabilised zirconia, Y^{3+} ions replace Zr^{4+} ions and create vacancies in the anion lattice.

For comparing the conductivity of electrolytes in a solution, rather than of the solution itself, use is made of the concept of *molar conductivity*.[†] This is defined as

$$\Lambda_B = \frac{\kappa}{c} \tag{16.3}$$

where c is the concentration of the electrolyte expressed in equivalents[‡] per cubic metre of solution. The unit of molar conductivity is $ohm^{-1}\ m^2\ mol^{-1}$ or $S\ m^2\ mol^{-1}$. The values of conductivity of some common electrolytes in water are listed in Table 16.1 and of some ionic salts at their melting points in Table 16.2. The molar conductivities of many salts in aqueous solutions and salt melts are of comparable magnitude.

16.5   ELECTROCHEMICAL CELLS

A device in which an electrochemical reaction is carried out is called an *electrochemical cell*. This consists of two electrodes (electronic conductors) in contact with an electrolyte (ionic conductor). The electrode at which the reduction reaction takes place (gain of electrons) is called the *cathode*, and the electrode at which the oxidation reaction (loss of electrons) takes place is called the *anode*. An electrode and its electrolyte form an *electrode compartment*. Both electrodes may share the same compartment or may have their own compartment in which case they are connected by an *ionic bridge* to complete the circuit.

---

* A useful review of solid electrolytes is: Goodenough, J. B. and P. Singh. 2015. Review Solid electrolytes in rechargeable electrochemical cells. *J. Electrochem. Soc.* 162(14), A2387A2392.

† In older literature, the term *equivalent conductance* is used rather than molar conductivity.

‡ The *equivalent* of a salt is the number of moles of the salt divided by the number of charges on the ions it forms in the solution.

TABLE 16.1
The molar conductivity of some electrolytes in aqueous solutions at 25°C (S m² mol⁻¹)

Concentration (equivalents/L)	0.0000	0.0005	0.0010	0.0050	0.0100	0.0200	0.0500	0.1000
HCl_4	26.2	422.7	421.4	415.8	412.0	407.2	399.1	391.3
LiCl	115.0	113.2	112.4	109.4	107.3	104.7	100.1	95.9
NaCl	126.5	124.5	123.7	120.7	118.5	115.8	111.1	106.8
KCl	149.9	147.8	147.0	143.6	141.3	138.3	133.4	129.0
NH_4Cl	149.7				141.3	138.3	133.3	128.8
NaI	126.9	125.4	124.3	121.3	119.2	116.7	112.8	108.8
KI	150.4			144.4	142.2	139.5	135.0	131.1
NaOH	247.8	245.6	244.7	240.8	238.0			
$AgNO_3$	133.4	131.4	130.5	127.2	124.8	121.4	115.2	109.1
$0.5CuSO_4$	133.6	121.6	115.3	94.1	83.1	72.2	59.1	50.6
$NaClO_4$	117.5	115.6	114.9	111.8	109.6	107.0	102.4	98.4
$KClO_4$	140.0	138.8	137.9	134.2	131.5	127.9	121.6	115.2

Source: Conway, B. E. 1952. *Electrochemical Data.* New York: Elsevier Publishing Co.

TABLE 16.2
The molar conductivity of some ionic salts near their melting points (S cm² mol⁻¹), arranged in the form of the periodic table

	LiCL		$BeCl_2$			BCl_3		CCl_4		
Λ	164		0.06			0		0		
mp, °C	610		405			−107		−23		
	NaCl		$MgCl_2$			$AlCl_3$		$SiCl_4$		
Λ	135		29			10^{-5}		0		
mp, °C	808		714			192		−70		
	KCl	CuCl	$CaCl_2$	$ZnCl_2$	$ScCl_3$	$GaCl_3$	$TiCl_4$	$GeCl_4$		
Λ	115	86	56	1	15	10^{-7}	0	0		
mp, °C	772	430	772	283	960	78	−25	−50		
	RbCl	AgCl	$SrCl_2$	$CdCl_2$	YCl_3	$InCl_3$	$ZrCl_4$	$SnCl_4$	$NbCl_5$	$MoCl_6$
Λ	83	114	58	51	9.5	14.7		0	10^{-6}	
mp, °C	717	455	875	568	700	586	437	−33	210	
	CsCl	AuCl	$BaCl_2$	$HgCl_2$	$LaCl_3$	$TlCl_3$	$HfCl_4$	$PbCl_2$	$TaCl_5$	WCl_6
Λ	68		65	10^{-3}	39	10^{-3}		2×10^{-5}	10^{-5}	10^{-5}
mp, °C	645		875	277	870	25	432		210	275

Source: Richardson, F. D. 1974. *Physical Chemistry of Melts in Metallurgy*, vol. 2, p.51. London: Academic Press.

An electrochemical cell must have at least one electrical conductor that is not an electron conductor. If electrons were free to pass from one electrode through the system to the other electrode at the same temperature there would be no potential difference between the electrodes, and the system would not be an electrochemical cell.

An electrochemical cell that produces electricity as a result of a spontaneous reaction is called a *galvanic cell*. Batteries and fuel cells are galvanic cells. A cell in which a non-spontaneous reaction is driven by an external source of current is called an *electrolytic cell*. These cells are used for electrowinning of metals from solutions, electroplating of metals, electrorefining of metals and in other applications.

16.5.1 CONTACT POTENTIAL

When two dissimilar electrically conductive substances (metals, semi-conductors or solutions) are brought into physical contact a potential difference is created between them, called the *contact potential*. The microscopic explanation for this is that when they are first contacted, mobile charge carriers (electrons, vacancies or ions) diffuse from one substance to the other to attempt to establish thermodynamic equilibrium, and if there is a net flow of electrons from one substance to the other, one will become negatively charged and the other positively charged, creating a potential difference. Electric potential differences in a cell without current, therefore, exist only at phase boundaries.

There are three main types of contact potential.

- When two dissimilar metals are contacted the local densities of free (or mobile) electrons change so as to form an electrical double layer with an excess of positive charge on one side of the interface and an excess negative charge of the same magnitude on the other side. The electrical double layer creates the contact potential.
- When a metal is immersed in an aqueous environment both oxidation and reduction reactions occur until equilibrium is reached. These reactions create an electrical double layer at the surface which results in an electrical potential difference.
- Ion species with different chemical potentials in two solutions of differing composition, and separated by a porous membrane which permits movement of ions but prevents bulk mixing of the solutions, will diffuse spontaneously across the membrane in the direction of the lower chemical potential. Different ions will diffuse at different rates, resulting in a net charge transfer across the membrane and a potential difference. This called a *liquid junction potential*.

In each of the above cases, the process is self-limiting because the potential difference between the two substances due to the charge separation will increase to a value sufficient to stop further motion of the charge carriers.

An electrochemical cell requires at least two contact potentials. The equilibrium cell potential is the cumulative result of these potential differences.

16.5.2 HALF-CELL AND CELL REACTIONS

When platinum electrodes are placed in acidified water and the leads connected to a power supply, above a certain applied potential difference (1.229 V) hydrogen is liberated at the cathode and oxygen at the anode (Figure 16.4). The reactions at the electrodes can be expressed as follows:

$$\text{Cathode (reduction) reaction:} \quad 2\,H^+(aq) + 2\,e^- = H_2(g)$$

$$\text{Anode (oxidation) reaction:} \quad H_2O(l) = 2\,H^+ + 0.5\,O_2(g) + 2\,e^-$$

These reactions are called the *half-cell reactions* and their sum, namely,

$$H_2O(l) = H_2(g) + 0.5\,O_2(g)$$

is called the *cell reaction*. The cathode reaction consumes electrons and is a reduction reaction while the anode reaction produces electrons and is an oxidation reaction. The net result is equivalent

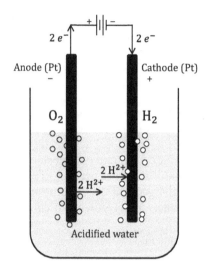

FIGURE 16.4 Illustration of a simple cell for the electrolysis of water.

to a flow of electrons from the cathode to the anode through the water, but in reality the electrons are carried by chemical species.

Since Gibbs energy is a state function, a cell reaction has the same Gibbs energy change as if the reaction had been carried out chemically, for example, in this case, by burning the hydrogen in air followed by reducing the temperature back to that of the cell.

By convention, the above cell is represented schematically as follows:

$$Pt(s) \mid H_2(g) \mid H_2SO_4(aq) \mid O_2(g) \mid Pt(s)$$

The vertical lines indicate boundaries between phases, where there is a contact potential.

16.5.3 GIBBS ENERGY OF CELL REACTIONS

The combined statement of the first and second laws of thermodynamics for a reversible process involving work in addition to that against the external pressure of the system is (Equation 8.10):

$$dG = Vdp - SdT - \delta w'$$

where $\delta w'$ is the additional work. For a process occurring at constant temperature and pressure, therefore

$$dG = -\delta w' \tag{16.4}$$

In an electrochemical process $\delta w'$ is the electrical energy generated by the cell. For a galvanic cell $\delta w'$ is positive, but for an electrolytic cell it is negative as work (in the form of electrical energy) is done on the cell. The work done when a charge of Q Coulombs (C) passes through a potential difference of E Volts is given by:

$$\delta w' = QE \tag{16.5}$$

For a reaction involving 1 mole of one of the reaction species, $Q = zF$ where F is the *Faraday constant* and is the charge carried by Avogadro's number of electrons, and z is the number of electrons involved in the half-cell reactions. F is equal to 96 485 C mol^{-1}. The Gibbs energy change taking place in a reversible cell reaction, therefore, is

$$\Delta G = -z F E \qquad (16.6)$$

If all the components of the cell reaction are in standard state,

$$\Delta G^0 = -z F E^0 \qquad (16.7)$$

It follows that

- *If ΔG for a cell reaction is negative the reaction occurs spontaneously from left to right and work is done by the cell. E in this case is positive as is the case for galvanic cells.*
- *If ΔG for a cell reaction is positive work must be done on the cell for the reaction to occur. E in this case is negative as is the case for electrolytic cells.*

As an example, consider the decomposition of water at 25°C:

$$H_2O(l) = H_2(g) + 1/2\ O_2(g); \Delta G^0 = 237.140\ kJ \qquad (16.8)$$

Since ΔG^0 is positive, the reaction is not spontaneous and must be driven by the application of electrical energy. From Equation 16.7, the necessary applied potential is:

$$E^0 = \frac{237140}{2 \times 96485} = -1.229\ V$$

For the general cell reaction

$$k\,A + l\,B = m\,C + n\,D$$

$$\Delta G = \Delta G^0 + RT \ln \frac{a_C^m \times a_D^n}{a_A^k \times a_B^l}$$

Dividing through by zF, it follows from Equations 16.6 and 16.7 that

$$E = E^0 - \frac{RT}{zF} \ln \frac{a_C^m \times a_D^n}{a_A^k \times a_B^l} \qquad (16.9)$$

This is called the *Nernst equation*. Equation 16.9 enables calculation of the cell potential when components of the cell are not in their standard state. $E = E^0$ when the activities of all reactants and products are unity (in standard state).

Potentials calculated using Equations 16.7 and 16.9 are for reversible reactions, that is, for zero current flow. For the reactions to occur at a finite rate, potentials greater than the reversible potential are necessary for electrolytic cells. Similarly, the potential generated by a galvanic cell drawing current will be less than the calculated reversible potential.

16.5.4 ELECTRODE POTENTIALS

The voltage of an electrochemical cell is the potential difference between the electrodes of each of the half-cells. The potential of a half-cell cannot be measured absolutely, but it can be measured relative to a reference half-cell. For aqueous electrolyte solutions the reference half-cell is the *standard hydrogen electrode* (SHE),

$$Pt(s)|H_2(g, p = 1\ bar)|H^+(aq, a = 1)$$

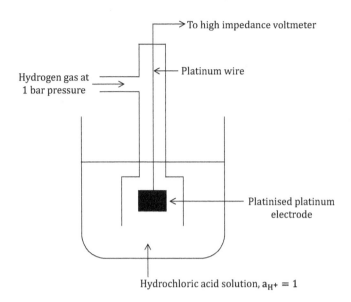

To high impedance voltmeter

Platinum wire

Hydrogen gas at
1 bar pressure

Platinised platinum
electrode

Hydrochloric acid solution, $a_{H^+} = 1$

FIGURE 16.5 Schematic of a hydrogen electrode. Hydrogen is bubbled over the catalytic platinum surface, and the half-cell reaction is between $H_2(g)$ and H^+ ions.

A common form of the hydrogen electrode is shown in Figure 16.5. By convention for aqueous solutions $\Delta_f G^0(H^+, aq, a = 1) = 0$ at all temperatures (Section 12.3), and since $\Delta G^0 = -zFE^0$ the potential of the hydrogen electrode at 1 bar and $a_{H^+} = 1$ is 0 Volts at all temperatures. By combining a standard hydrogen electrode with another electrode to form a cell, the potential of the other electrode will be the voltage of the cell measured at zero current. This voltage is the *standard electrode potential* for the other half-cell reaction.

Electrical potential differences can be measured only by using two pieces of material of the same composition. In practice, these are almost always two pieces of copper attached to the electrodes of the cell. There is a contact potential between the copper and the electrode material, and by convention the magnitude of this is taken to be included in the measured electrode potential value. Standard electrode potentials of half-cell reactions in aqueous solutions have been measured experimentally relative to the hydrogen half-cell, and values are given in Table 16.3 for some common metals.

The cell potential due to a pair of half-cell reactions is obtained by writing the half-cell reactions in such a way that on summing them the cell reaction is obtained. For example, the standard electrode potentials from Table 16.3 of the half-cell reactions for the decomposition of water are:

$$2\,H^+ + 2\,e^- = H_2(g); \quad E^0 = 0\text{ V}$$

and

$$H_2O(l) = 2\,H^+ + 1/2\,O_2(g) + 2\,e^-; \quad E^0 = -1.229\text{ V}$$

The cell potential is the sum of the two half-cell potentials, as written;[*] namely, $0 - 1.229 = -1.229$ V. This is consistent with the value calculated from the Gibbs energy of the decomposition of water using Gibbs energy data for Reaction 16.8.

---

[*] In writing a cell reaction, one electrode reaction is written as a reduction reaction and the other as an oxidation reaction. Their sum gives the cell reaction. If both are written as reduction reactions (as in Table 16.3) then the cell potential is their difference.

TABLE 16.3

Standard electrode potentials at 25°C of some important half-cell reactions in aqueous solutions measured relative to the hydrogen half-cell

Electrode reaction	E^0, Volts	Electrode reaction	E^0, Volts
$Li^+ + e^- = Li$	−3.045	$Sn^{2+} + 2\,e^- = Sn$	−0.136
$K^+ + e^- = K$	−2.925	$Pb^{2+} + 2\,e^- = Pb$	−0.126
$Na^+ + e^- = Na$	−2.714	$Cu(NH_3)_4^{2+} + 2\,e^- = Cu + 4\,NH_3$	−0.12
$Mg^{2+} + 2\,e^- = Mg$	−3.37	$Fe^{3+} + 3\,e^- = Fe$	−0.036
$H_2 + 2\,e^- = 2\,H^-$	−2.25	$2\,H^+ + 2\,e^- = H_2$	0
$Al^{3+} + 3\,e^- = Al$	−1.66	$AgBr + e^- = Ag + Br^-$	0.095
$Zn(CN)_4^{2-} + 2\,e^- = Zn + 4\,CN^-$	−1.26	$HgO + H_2O + 2\,e^- = Hg + 2\,OH^-$	0.098
$ZnO_2^{2-} + 2\,H_2O + 2e^- = Zn + 4\,OH^-$	−1.216	$Sn^{4+} + 2\,e^- = Sn^{2+}$	0.15
$Zn(NH_3)_4^{2+} + 2\,e^- = Zn + 4\,NH_3$	−1.03	$AgCl + e^- = Ag + Cl^-$	0.222
$Sn(OH)_6^{2-} + 2\,e^- = HSnO_2^- + H_2O + 3\,OH^-$	−0.90	$Hg_2Cl_2 + 2\,e^- = 2\,Hg + 2\,Cl^-$	0.2676
$Fe(OH)_2 + 2\,e^- = Fe + 2\,OH^-$	−0.877	$Cu^{2+} + 2\,e^- = Cu$	0.337
$2\,H_2O + 2\,e^- = H_2 + 2\,OH^-$	−0.828	$Ag(NH_3)_2^+ + e^- = Ag + 2\,NH_3$	0.373
$Fe(OH)_3 + 3\,e^- = Fe + 3\,OH^-$	−0.77	$Hg_2SO_4 + 2\,e^- = 2\,Hg + SO_4^{2-}$	0.6151
$Zn^{2+} + 2\,e^- = Zn$	−0.763	$Fe^{3+} + e^- = Fe^{2+}$	0.771
$Ag_2S + 2\,e^- = 2\,Ag + S_2^-$	−0.69	$Ag^+ + e^- = Ag$	0.7991
$Fe^{2+} + 2\,e^- = Fe$	−0.44	$O_2 + 4\,H^+ + 4\,e^- = 2\,H_2O$	1.229
$Bi_2O_3 + 3\,H_2O + 6\,e^- = 2\,Bi + 6\,OH^-$	−0.44	$Cl_2 + 2\,e^- = 2\,Cl^-$	1.36
$PbSO_4 + 2\,e^- = Pb + SO_4^{2-}$	−0.356	$Au^+ + e^- = Au$	1.68
$Ag(CN)_2^- + e^- = Ag + 2\,CN^-$	−0.31	$PbO_2 + SO_4^{2-} + 4\,H^+ + 2\,e^- = PbSO_4 + 2\,H_2O$	1.685
$Ni^{2+} + 2\,e^- = Ni$	−0.25	$O_3 + 2\,H^+ + 2\,e^- = O_2 + H_2O$	2.07
$AgI + e^- = Ag + I^-$	−0.151		

Source: Latimer, W. M. 1952. *The Oxidation States of the Elements and Their Potentials in Aqueous Solution.* New York: Prentice–Hall, Inc.

In practice, the saturated calomel electrode (SCE), based on the reaction between elemental mercury and mercurous chloride (Hg_2Cl_2),

$$Hg(l) \mid Hg_2Cl_2(s) \mid KCl(aq, \text{ saturated}); \quad E^0 = 0.2444 \text{ V at } 25°C$$

was widely used as a secondary reference electrode because it is more robust and easier to use than the hydrogen electrode. However, it has now been largely replaced by the silver chloride electrode for environmental reasons, mercury compounds being highly toxic:

$$Ag(s) \mid AgCl(s) \mid KCl(aq, \text{ saturated}); \quad E^0 = 0.2224 \text{ V at } 25°C$$

16.5.5 TYPES OF ELECTROCHEMICAL CELLS

As noted previously, galvanic cells produce electricity as a result of a spontaneous reaction, and electrolytic cells are those in which a non-spontaneous reaction is driven by an external source of current. Figure 16.6 summarises the main types of electrolytic and galvanic cells.

Electrolytic cells

The cell in Figure 16.4 is an example of a *reduction cell*; ions are reduced from the electrolyte solution. The electrolyte is consumed (reduced) in a reduction cell, but the anode and cathode remain unaltered (unless they are made from reactive materials). For example, under standard state conditions, a copper sulfate solution can be reduced electrolytically (see Figure 16.7) to produce metallic copper (assuming inert electrodes) if the applied potential difference exceeds 0.892 Volts:

$$
\begin{aligned}
\text{Cathode reaction:} \quad & Cu^{2+}(aq) + 2\,e^- = Cu(s) & E^0 = 0.337\text{ V} \\
\text{Anode reaction:} \quad & H_2O(l) = 2\,H^+(aq) + 0.5\,O_2(g) + 2\,e^- & E^0 = -1.229\text{ V} \\
\text{Cell reaction:} \quad & Cu^{2+}(aq) + H_2O(l) = Cu(s) + 0.5\,O_2(g) + 2\,H^+ & E^0 = -0.892
\end{aligned}
$$

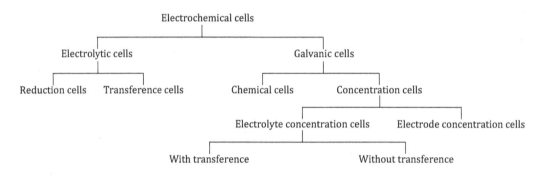

FIGURE 16.6 Simple classification of types of electrochemical cells.

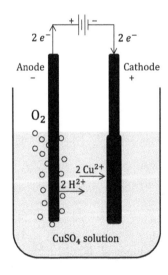

FIGURE 16.7 Example of a reduction cell. When a potential of > 0.892 V is applied, copper is reduced from the copper sulfate solution and deposited on the cathode, oxygen is released at the anode, and the pH of the solution increases due to the release of H⁺ ions.

Copper will be deposited on the cathode, and oxygen will be released at the anode. The copper concentration in the electrolyte solution will progressively decrease while the hydrogen ion concentration will progressively increase. Assuming $p_{O_2} = 1$ bar, the potential required for reduction of the copper at any instant is:

$$E = -0.892 - \frac{RT}{zF} \ln \frac{a_{H^+}^2}{a_{Cu^{2+}}}$$

Thus in a batch process the required potential will increase progressively as copper is depleted from the solution. Commercial processes are continuous, and the solution is continuously replenished to maintain a constant composition (and cell potential).

Now, suppose the electrodes are copper rather than an inert material (Figure 16.8). If the leads are again connected to a power supply, at a potential difference below that required to decompose the water, copper will be transferred from one electrode to the other. The half-cell reactions now would be:

$$\text{Cathode (reduction) reaction:} \quad Cu^{2+}(aq) + 2\,e^- = Cu(s)$$

$$\text{Anode (oxidation) reaction:} \quad Cu(s) = Cu^{2+}(aq) + 2\,e^-$$

The cell reaction is

$$Cu(s) = Cu(s)$$

and all that occurs is that copper is dissolved from the anode and redeposited on the cathode. Schematically, the cell is represented as follows:

$$Cu(s) \,|\, H_2SO_4(aq) \,|\, Cu(s)$$

The standard Gibbs energy change of the reaction is:

$$\Delta G^0 = G_{Cu}^0 - G_{Cu}^0 = 0$$

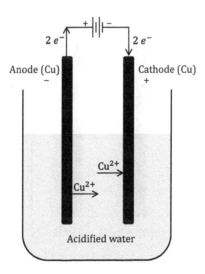

FIGURE 16.8 Illustration of a simple electrolytic transference cell. When a potential is applied, copper is transferred from the anode to cathode via the electrolyte solution.

Hence E^0 is also equal to zero. In practice, an actual potential difference is required to carry out the transfer of the copper at a finite rate.

Cells of the type shown in Figure 16.8 are called *transference cells*. In electrorefining, impure metal is formed to make the anode, and pure metal is produced at the cathode. The electrolyte solution remains unaltered in a transference cell (except for contamination by any soluble impurities released from the anode) while material is transferred from the anode to the cathode.

The above examples are for aqueous solutions, but the electrolyte solution could just as well be a molten salt.

Galvanic cells

It is well known that when a zinc plate is dipped into a copper sulfate solution it becomes coated with a layer of copper due to the redox reaction:

$$Zn(s) + Cu^{2+}(aq) = Cu(s) + Zn^{2+}$$

This reaction can also be carried out as an electrochemical reaction. Consider a cell consisting of copper and zinc electrodes immersed in a solution containing both copper and zinc ions (see Figure 16.9):

$$Cu(s) \,|\, CuSO_4(aq) + ZnSO_4(aq) \,|\, Zn(s)$$

The two half-cell reactions are:

$$\text{Cathode (reduction) reaction:} \quad Cu^{2+}(aq) + 2\,e^- = Cu(s)$$

$$\text{Anode (oxidation) reaction:} \quad Zn(s) = Zn^{2+}(aq) + 2\,e^-$$

and the cell reaction is

$$Zn(s) + Cu^{2+}(aq) = Cu(s) + Zn^{2+}(aq); \quad \Delta G^0 = -212.3 \text{ kJ at } 25°C.$$

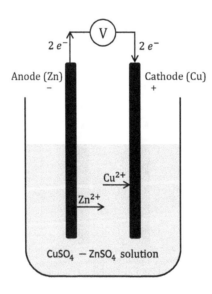

FIGURE 16.9 Illustration of a simple galvanic cell based on the spontaneous reaction: $Zn(s) + CuSO_4(aq) = Cu(s) + ZnSO_4(aq)$.

The Gibbs energy of the reaction is negative so the reaction is spontaneous, and a potential difference will be created, which from Equation 16.7 will be:

$$E^0 = -\frac{\Delta G^0}{zF} = -\frac{-212\,300}{2 \times 96485} = 1.100 \text{ V}$$

This is the reversible potential generated when all the components of the cell reaction are in their standard state. The same result is obtained by summing the standard electrode potentials for the two half-cell reactions:

$$\text{Cathode:} \quad Cu^{2+}(aq) + 2\,e^- = Cu(s); \quad E^0_{Cu^{2+}} = 0.337 \text{ V}$$

$$\text{Anode:} \quad Zn(s) = Zn^{2+}(aq) + 2\,e^-; \quad -E^0_{Zn^{2+}} = 0.763 \text{ V}$$

$$\text{Cell reaction:} \quad Zn(s) + Cu^{2+}(aq) = Cu(s) + Zn^{2+}(aq); \quad \Delta E = 1.100 \text{ V}$$

In this cell, zinc dissolves from the anode and enters the solution and copper will deposit from the solution onto the cathode. Thus, the solution will become enriched in Zn^{2+} ions and depleted in Cu^{2+} ions.

The Gibbs energy (or chemical potential) of the copper and zinc ions are, respectively (Equation 9.48):

$$G_{Cu^{2+}} - G^0_{Cu^{2+}} = RT \ln a_{Cu^{2+}}$$

$$G_{Zn^{2+}} - G^0_{Zn^{2+}} = RT \ln a_{Zn^{2+}}$$

Since, $E = -G/zF$, then dividing the above equations by zF and rearranging yields

$$E_{Cu^{2+}} = E^0_{Cu^{2+}} + \frac{RT}{zF} \ln a_{Cu^{2+}}$$

$$E_{Zn^{2+}} = E^0_{Zn^{2+}} + \frac{RT}{zF} \ln a_{Zn^{2+}}$$

where $E^0_{Cu^{2+}}$ and $E^0_{Zn^{2+}}$ are the standard electrode potentials of copper and zinc (-0.763 V and 0.337 V), respectively. While current is drawn from the cell, the copper ion concentration (and activity) decreases and the zinc ion concentration (and activity) increases. The electrode potential of the zinc increases while that of the copper decreases, as illustrated qualitatively in Figure 16.10. Eventually, the two curves will touch, and the electrode potentials become equal. At this time, the cell potential, given by $E_{Cell} = E_{Cu^{2+}} - E_{Zn^{2+}}$ is equal to zero, and current flow stops (the battery is flat!).

This cell could be driven in reverse by the application of a potential greater than -1.1 V. In this case, zinc would be deposited from the solution on the zinc electrode, and copper would dissolve from the copper electrode into the solution.

When a cell of the above type is not in use, the redox reaction continues due to transport of copper ions to the zinc electrode by normal diffusion rather than aided by an electric field and without the production of an electric current. To prevent this, in practice a porous membrane or diaphragm can be inserted into the cell to form two separate electrode compartments, each containing its own electrolyte (see Figure 16.11). The membrane is made of material which permits ions to pass but prevents bulk mixing of the two solutions and slows the reaction when the cell is not in use. The cell in this case is represented as follows:

$$Cu(s) \,|\, CuSO_4(aq) \,\|\, ZnSO_4(aq) \,|\, Zn(s)$$

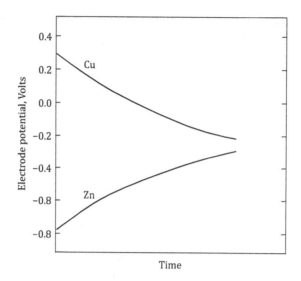

FIGURE 16.10 Typical change in electrode potentials of copper and zinc with time of operation for the galvanic cell in Figure 16.9.

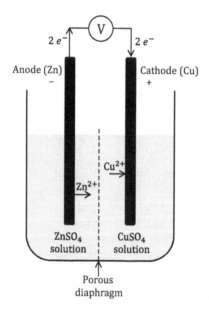

FIGURE 16.11 A galvanic cell with a porous diaphragm to prevent bulk mixing of the two electrolyte solutions but allow the passage of ions.

where the symbol ¦ indicates a junction between two miscible liquids. Cells of this chemistry and design are called Daniell cells.*

In this case, while the battery is in operation, the $ZnSO_4$ solution progressively becomes more concentrated and the $CuSO_4$ solution more dilute, but reaction almost ceases when no

---

* Invented in 1836 by John Frederic Daniell, a British chemist and meteorologist. His cell consisted of a copper pot filled with copper sulfate solution in which was immersed an unglazed earthenware container filled with sulfuric acid and containing a zinc electrode.

current is being drawn. The solutions in the cell can be periodically refreshed to keep the cell functioning.

Because in practice, cells operate in a thermodynamically non-reversible manner, equilibrium within cells is not maintained, and concentration gradients are present. In the above example, concentration gradients will exist across the porous diaphragm when in operation. If the anions and cations diffuse at different rates, one solution near the diaphragm will become negatively charged and the other positively charged resulting in an electrical double layer of positive and negative charges at the junction of the solutions. Thus, a liquid junction potential will be present which will reduce the potential of the cell.

A salt bridge (a liquid form of ionic bridge) is commonly employed to minimise or stabilise the liquid junction potential at the interfaces of two miscible electrolyte solutions. A *salt bridge* is a bridge between the two solutions which consists typically of a saturated solution of potassium chloride or ammonium nitrate (contained in a gel) between the two solutions constituting the junction (see Figure 16.12). The cations and anions in these solutions have very similar diffusion rates, and because they are present in large concentration at the junction they carry almost all the current across the boundary without creating a junction potential. The ions in the salt bridge (such as K^+) diffuse into the solution to replace Cu^{2+} ions and are themselves replaced in the salt bridge by diffusion of Zn^{2+} ions; they play no part in the cathode reaction. The cell in this case is represented as follows:

$$Cu(s) \,|\, CuSO_4(aq) \,\|\, ZnSO_4(aq) \,|\, Zn(s)$$

where the symbol $\|$ represents a liquid junction in which the liquid junction potential is assumed to be eliminated.

The Daniell cell is an example of a *chemical galvanic cell* in which the potential difference generated by the cell is due to a spontaneous chemical reaction. The other type of galvanic cell is the *concentration cell* in which the potential difference generated by the cell is due to a difference in concentration across a boundary. There are two types, electrolyte concentration cells and electrode concentration cells. An example of an *electrolyte concentration* cell is

$$Cu \,|\, CuSO_4 \left(aq, \text{low con}\right) \,|\, CuSO_4 \left(aq, \text{high con}\right) \,|\, Cu$$

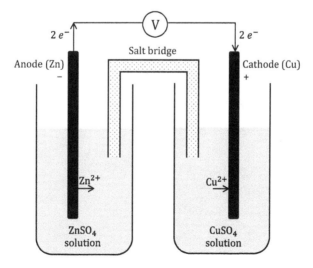

FIGURE 16.12 A galvanic cell with a salt bridge to minimise the junction potential between two electrolyte solutions.

in which copper electrodes are immersed in solutions of copper sulfate of different concentrations separated by a porous diaphragm. The cell reaction is:

$$CuSO_4 \left(aq, high\ con\right) = CuSO_4 \left(aq, low\ con\right)$$

This is a spontaneous reaction, because the concentration gradient across the diaphragm will cause diffusion of ions from the high concentration solution to the low concentration solution, and a potential difference will be created. Since E^0 will be zero, the potential created will be (Equation 16.9):

$$E = -\frac{RT}{2F} \ln \frac{a_{CuSO_4 (aq,\ low\ con)}}{a_{CuSO_4 (aq,\ high\ con)}}$$

Electrolyte concentration cells are commonly the cause of corrosion of metals, when crevices accumulate solutions containing dissolved oxygen or salt at different concentration to the external environment. This is discussed further in Section 17.3.

An example of an *electrode concentration cell* is the widely used oxygen concentration cell, which in practical applications is used to measure the oxygen concentration in gas mixtures. Many different designs are used, and the geometry of the cell depends on the application; one form is illustrated in Figure 16.13. At elevated temperatures, stabilised ZrO_2 partly dissociates to produce oxygen ions which are mobile within the solid lattice. If different oxygen pressures are present on either side of a ZrO_2 element a potential difference is created:

$$Pt(s)\,|\,O_2(g, high\ p_{O_2})\,|\,ZrO_2\,|\,O_2(g, low\ p_{O_2})\,|\,Pt(s)$$

$$Cathode:\quad O_2(g, high\ p_{O_2}) + 4\,e^- = 2\,O^{2-}$$

$$Anode:\quad 2\,O^{2-} = O_2(g, low\ p_{O_2}) + 4\,e^-$$

$$Cell\ reaction:\quad O_2(g, high\ p_{O_2}) = O_2(g, low\ p_{O_2})$$

$$E = -\frac{RT}{4F} \ln \frac{p_{O_2\,(low\ pressure)}}{p_{O_2\,(high\ pressure)}}$$

If the high-pressure side is air,

$$E = -\frac{RT}{4F} \ln \frac{p_{O_2\,(low\ pressure)}}{0.21}$$

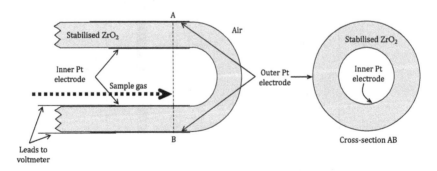

FIGURE 16.13 An oxygen concentration cell utilising stabilised zirconia as the ionic conductor. The platinum leads are to make electrical connection to the zirconia. The potential difference is created by the difference in oxygen pressure inside and outside the zirconia tube.

and the pressure of oxygen in the unknown gas can be calculated from the measured temperature and potential generated by the cell. Stabilised zirconia cells must be operated at sufficiently high temperature that the oxygen ions become mobile (typically 500 to 1000°C).

Because an order of magnitude concentration difference produces less than $60/z$ mV* potential difference at ambient temperature, concentration cells are generally not used for energy production or storage. Their main practical use is in instruments to measure concentrations in solutions or gas mixtures. For solutions, the relationship between the activity of the component of interest and its concentration must be known. If the activity of a component in a solution is known in one half-cell, the other can be calculated from the measured potential difference by means of the Nernst equation.

EXAMPLE 16.1 Calculate cell potential using the Nernst equation

Calculate the potential of the following cell at 25°C:

$$H_2(g), p = 2 \text{ bar} \mid H^+(aq), pH = 1 \parallel Cu^{2+}(aq), a_{Cu^{2+}} = 0.5 \mid Cu(s)$$

SOLUTION

The schematic representation indicates a hydrogen electrode (2 bar pressure) in an acidic solution of $pH = 1$ separated from a copper-bearing solution, with $a_{Cu^{2+}} = 0.5$, by a porous membrane. Since $pH = -\log a_{H^+}$ and $pH = 1$, $a_{H^+} = 0.1$. The half-cell and cell reactions are:

$$\text{Anode:} \quad H_2(g) = 2\,H^+(aq) + 2\,e^-; \quad E^0 = 0 \text{ V}$$

$$\text{Cathode:} \quad Cu^{2+}(aq) + 2\,e^- = Cu(s); \quad E^0 = 0.337 \text{ V}$$

$$\text{Cell reaction:} \quad H_2(g) + Cu^{2+}(aq) = 2\,H^+(aq) + Cu(s); \quad E^0 = 0.337 \text{ V}$$

From the Nernst equation (Equation 16.9),

$$E = E^0 - \frac{RT}{zF} \ln \frac{a_{Cu} \times a_{H^+}^2}{a_{Cu^{2+}} \times p_{H_2}} = 0.337 - \left(\frac{8.314 \times 298.15}{2 \times 96485} \right) \times \ln \left(\frac{1 \times 0.1^2}{0.5 \times 2} \right)$$

$$= 0.337 + 0.059 = 0.396 \text{ Volts}$$

16.5.6 KINETIC EFFECTS

When an electrode is part of a system in which a finite current is being passed or generated, it no longer behaves thermodynamically reversibly, due to the phenomenon of polarisation, and the electrode potential deviates from its reversible value. *Polarisation* is a general term for physical processes by which barriers develop at the interface between an electrode and electrolyte. The difference between the actual potential and the reversible potential is the *overpotential* of the electrode.

There are several causes of polarisation:

Activation polarisation is a result of the activation energy required for an atom of the solid at the anode to transition into an ion in the electrolyte and for the reverse transition at the cathode. The *activation overpotential* is the difference between the reversible electrode potential and the

---

* The reader can verify this using the Nernst equation.

polarised electrode potential. An important example is the discharge of hydrogen gas at a cathode from an aqueous solution by the reaction:

$$2\,H^+ + 2\,e^- = H_2(g); \quad E^0 = 0\ V$$

The *hydrogen overpotential*, as the activation overpotential is called in this case, varies according to the metal on which the hydrogen discharges. For example, at 25°C, the hydrogen overpotential on a platinum electrode is 0 V (by definition), on iron −0.45 V and on zinc −1.13 V in a HCl solution at $a_{H^+} = 1$. The standard electrode potential for the half-cell reaction

$$Zn^{2+} + 2\,e^- = Zn$$

is −0.763 V (Table 16.3). An overpotential of less than −0.763 V on a zinc surface permits the reduction of zinc from aqueous solutions in preference to hydrogen. Since the overpotential for zinc is −1.13 V, zinc can easily be reduced from aqueous solutions.

The activation overpotential of an electrode is not a fixed quantity but a function of the rate of reaction as determined by the current density at the electrode, the relation being given by the *Tafel equation*:

$$\eta_a = a + b \log j$$

where η_a is activation overpotential, j is the current density (Amp m^{-2}) and a and b are constants depending on the temperature and electrode material. At very low current densities η_a will be sufficiently small that hydrogen will be produced in preference to zinc and will continue to form at the same time as zinc at higher current densities.

Concentration polarisation is due to uneven depletion of reagents in the electrolyte when current passes. This causes concentration gradients adjacent to the electrodes and, therefore, a potential drop at each electrode as given by the Nernst equation:

$$E_c = -\frac{RT}{zF} \ln \frac{m_{B^+ \text{(bulk)}}}{m_{B^+ \text{(interface)}}}$$

where $m_{B^+ \text{(bulk)}}$ is the concentration of B^+ in the bulk of the electrolyte solution and $m_{B^+ \text{(interface)}}$ is the concentration of B^+ at the electrode interface. The *concentration overpotential* is the difference between the reversible electrode potential and the polarised electrode potential.

Resistance polarisation is due to the deposition of a reaction layer (often an oxide), or absorption of a gas, on an electrode surface resulting in increased electrical resistance. The *resistance potential* is the difference between the reversible electrode potential and the polarised electrode potential.

The *total overpotential* is the sum of the overpotentials due to the three types of polarisation.

16.5.7 Total cell potential and Ohmic heating

Since $E = IR$ (Equation 16.1), it follows from previous discussion that the total potential of a cell is given by

$$E = E_1 + (R_1 + R_2)I + E_2 + E_3 \text{ Volts} \tag{16.10}$$

where E_1 is the reversible cell potential given by Equation 16.9, R_1 and R_2 are the Ohmic resistances of the electrodes and the electrolyte solution, respectively, I is the current through the cell and E_2

and E_3 are the overpotentials at the cathode and anode, respectively. The electrical energy expended in overcoming the Ohmic resistances is released as heat, and the rate of heat generation is:

$$P = I^2(R_1 + R_2)\,\text{Watts} \tag{16.11}$$

16.5.8 THE LAWS OF ELECTROLYSIS

The quantity of electricity required to deposit or dissolve 1 mole of a substance at an electrode is zF, where, as before, F is the Faraday constant and z is the number of charges involved in the half-cell reactions. The quantity of electricity passing through the cell in time t is It, and, therefore, the mass of substance deposited or dissolved is

$$m = \frac{MIt}{1000\,zF}\,\text{kg} \tag{16.12}$$

where m is the mass (kg) of a substance having an atomic or formula mass M by a current of I (A) passing through the cell for t seconds. The rate of deposition or dissolution (m/t), therefore, depends on the current but not the cell potential. Equation 16.12 is the quantitative statement summarising *Faraday's laws of electrolysis*, discovered empirically by Michael Faraday in 1833. As originally formulated, these laws are:

1. The amount of a substance discharged or dissolved at an electrode is directly proportional to the quantity of electricity passed.
2. If the same quantity of electricity is passed through different electrolytes, the mass of the substances discharged or dissolved is proportional to the chemical equivalent of the substances.

Sometimes, side-reactions may occur; also ohmic heating will occur. As a result, the amount of substance discharged or dissolved at an electrode will be less than predicted by Equation 16.12. This introduces the concept of current efficiency, which is defined as:

$$\text{Current efficiency} = \frac{\text{Actual amount deposited or dissolved} \times 100}{\text{Theoretical amount deposited or dissolved}} \tag{16.13}$$

The energy consumed in an electrical process is given by:

$$w = EIt\,\text{Joules} = \frac{EIt}{3.6 \times 106}\,\text{kW h}$$

Substituting for It in Equation 16.12 and rearranging gives the energy required to deposit or dissolve a given mass of a substance at an electrode:

$$w = \frac{mzEF}{3600 \times M}\,\text{kW h} \tag{16.14}$$

The energy consumption of an electrolytic cell or the energy production of a galvanic cell (per unit mass of substance deposited or dissolved), therefore, depends on the operating potential of the cell and not the current.

16.6   PHASE STABILITY DIAGRAMS

Aqueous *phase stability diagrams*, also known as *Pourbaix diagrams** or $E_H - pH$ diagrams, are maps showing the equilibrium phases in redox systems. These show the redox potential (a measure of the oxidation potential) of the system and the pH (a measure of the activity of hydrogen ions). An example is shown in Figure 16.14 for the $Fe-H_2O$ system.

Pourbaix diagrams are constructed in the same manner as for non-aqueous phase stability diagrams (Section 14.3) except that chemical reactions of the form in Table 16.3 are used. The boundaries between species are represented by lines, and the areas between the lines represent regions of stability. The horizontal axis is pH, and the vertical axis is the potential with respect to the standard hydrogen electrode, E_H (the H stands for hydrogen), calculated using the Nernst equation:

$$E_H = E^0 - \frac{RT}{zF} \ln \frac{a_C^m \times a_D^n}{a_A^k \times a_B^l}$$

To calculate the position of the lines, the activity of the species at equilibrium must be defined. Usually, the activity of a species is approximated as equal to the concentration (for soluble species) or partial pressure (for gases). The same values are used for all species present in the system. For soluble species, the lines are often drawn for molalities of 1 or 10^{-6} mol kg^{-1}. If the diagram involves the equilibrium between a dissolved species and a gas, the pressure is usually set to 1 bar. The lines on the figure are of three types:

- Horizontal lines represent reactions that involve electrons but are independent of pH; neither H^+ nor OH^- ions participate in the reactions, for example,

$$Fe^{2+} + 2\,e^- = Fe$$

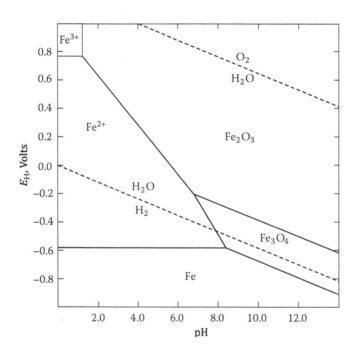

FIGURE 16.14 $E_H - pH$ diagram for the $Fe-H_2O$ system at 25°C; molality of Fe is 10^{-6} mol kg^{-1}. The area between the dotted lines is the region of stability of water.

---

* Named after Marcel Pourbaix (1904–1998), a Russian-born, Belgian chemist.

- Diagonal lines with either a positive or negative slope represent reactions that involve both electrons and H^+ and OH^- ions, for example,

$$2 Fe^{2+} + 2 e^- + OH^- = Fe_2O_3 + H^+$$

- Vertical lines represent reactions that involve either H^+ or OH^- ions but are independent of E_H. In other words, electrons do not participate in the reactions, for example,

$$2 Fe^{3+} + 3 OH^- = Fe_2O_3 + 3 H^+$$

In Figure 16.14 the area between the dashed lines is the region of stability of water. Metallic iron lies outside the area of stability of water; hence iron is unstable in the presence of water. When iron is in contact with water it forms Fe_3O_4 (magnetite) or, in acid solutions, dissolves to form ferrous (Fe^{2+}) ions with evolution of hydrogen. Fe^{2+} ions are oxidised to Fe^{3+} ions in solutions of low pH but only at high values of E_H, that is, under strongly oxidising conditions. Solutions containing Fe^{3+} ions are strongly oxidising and are used in some leaching processes to oxidise other metals present in ores at low valency to make them soluble in acid solutions, for example, oxidising U^{4+} to U^{6+}. In the process of reaction the Fe^{3+} ions are reduced back to Fe^{2+} ions. Fe^{3+} ions can be regenerated by the addition of an oxidising agent such as MnO_2 or $NaClO_3$ and the reaction continued.

There is a tendency for electrons to be removed from species when the potential is high. These conditions may exist near the anode in an electrochemical cell but can also be generated if an oxidising agent (for example, hydrogen peroxide) is present or added to a chemical redox system. In reducing conditions, when the potential is low, the system is able to supply electrons to species by means of a cathode electrode or by the presence or addition of a reducing agent.

$E_H - $ pH diagrams may be used to find conditions suitable for selective leaching of metals from ores or selective precipitation of compounds from solutions. In corrosion science, they may be used to analyse the dissolution and passivation behaviour of different metals in aqueous environments. In geochemistry, they are used to study weathering processes and chemical sedimentation and to investigate the conditions in the past needed to form certain mineral deposits.

16.7 THE USE OF GALVANIC CELLS TO MEASURE THERMODYNAMIC PROPERTIES

Reversible galvanic cells have been widely used to determine thermodynamic properties of reactions and the activities of components of solutions.

1. If a chemical reaction can be performed electrochemically, and a suitable cell can be designed and constructed, the Gibbs energy change of the cell reaction at any temperature can be determined from the measured cell, potential: $\Delta G = -z F E$.
2. The enthalpy change of the cell reaction can be determined from cell potential measurements over a range of temperature. Since $\Delta G = -z F E$,

$$\frac{d\Delta G}{dT} = -z F \left(\frac{dE}{dT} \right)$$

Combining this with Equation 8.13 yields

$$\Delta S = z F \left(\frac{dE}{dT} \right)$$

and, since $\Delta G = \Delta H - T \Delta S$, and $\Delta G = -z F E$,

$$\Delta H = -zFE + zFT\left(\frac{dE}{dT}\right)$$

Thus, if the variation of E with temperature is measured, the enthalpy change can be determined.

3. The Nernst equation can be used to determine the activity (or partial molar Gibbs energy) of a component of a solution. For the general cell reaction $kA + lB = mC + nD$,

$$E = E^0 - \frac{RT}{zF}\ln\frac{a_C^m \times a_D^n}{a_A^k \times a_B^l}$$

If the activity of all the species in the reaction except one is known (measured or controlled), the unknown activity can be calculated from the measured cell potential.

PROBLEMS

Where appropriate use E^0 values from Table 16.3 and $\Delta_r G^0$ values from the NBS tables.

16.1 For the cell

$$Pt(s)\,|\,H_2(g),\ p = 1\ bar\,|\,HCl(aq),\ m = 0.500\ mol\ kg^{-1}\,|\,Cl_2(g),\ p = 1\ bar\,|\,Pt(s)$$

  i.   Write the half-cell and cell reactions.
  ii. Calculate the standard cell potential.
  iii. Calculate the standard Gibbs energy change for the cell reaction at 25°C.
  iv. If the measured cell potential at 25°C is 1.410 V, use the Nernst equation to calculate the mean ionic activity coefficient of HCl in the solution.

16.2 For the cell $Pb(s)\,|\,PbSO_4(s)\,|\,CuSO_4,\ m = 0.01\ mol\ kg^{-1}\,|\,Cu(s)$

  i.   Write the half-cell and cell reactions.
  ii. Calculate the standard cell potential.
  iii. Calculate the standard Gibbs energy change for the cell reaction at 25°C.
  iv. If the measured cell potential at 25°C is 0.5543 V, use the Nernst equation to calculate the mean ionic activity coefficient of $CuSO_4$ in the solution.

16.3 Draw schematic cells for the three types of corrosion concentration cells (Section 17.3), and write the half-cell and cell reactions. Calculate the potential difference created if the system in each case is Fe, water and oxygen. If a salt is present, assume it is NaCl. Assume standard state conditions.

16.4 Calculate the potential of a half-cell consisting of a zinc electrode in equilibrium with zinc ions in solution at 25°C if the activity of the zinc ions is 0.5. Repeat the calculation for a copper electrode in equilibrium with copper ions in solution at 25°C if the activity of copper ions is 0.1. If the two half-cells are connected by means of copper wire, what would be the reversible potential of the cell?

16.5 Calculate the potential of the reaction for the reduction of silver cyanide from solution to form metallic silver, $Ag(CN)_2^- + e^- = Ag + 2\ CN^-$, if the molalities of $Ag(CN)_2^-$ and CN^- are 10^{-5} and 10^{-3} mol kg^{-1}, respectively.

16.6 Magnesium can be produced electrolytically by the reduction of molten magnesium chloride. The overall reaction is $MgCl_2(l) = Mg(l) + Cl_2(g)$ for which $\Delta G^0(1000\ K) = 482.43$ kJ.

  i.   Write the half-cell reactions.
  ii. Calculate the reversible potential required to reduce molten magnesium chloride at 1000 K.

iii. Calculate the minimum energy required to produce 1 kg of molten magnesium, assuming 100% current efficiency.

iv. Are these realistic values of potential and energy for a commercial operation? Explain.

16.7 For the cell $Pt(s) \,|\, H_2(g)$, $p = 1$ bar $|\, HCl(aq) \,|\, AgCl(s) \,|\, Ag(s)$

i. Write the half-cell reactions, and calculate the standard cell potential if the cell reaction is: $H_2(g) + 2\, AgCl(s) = 2\, H^+(aq) + 2\, Cl^-(aq) + 2\, Ag(s)$

ii. If by experimental measurements over a range of temperatures it is found that $dE^0/dT = -6.462 \times 10^{-4}$ V K^{-1}, calculate $\Delta_r G^0$, $\Delta_r S^0$, $\Delta_r H^0$ at 25°C for the cell reaction.

16.8 If the zero-current potential for the cell $Pt(s) \,|\, H_2(g)$, $p = 1$ bar $|\, HCl(aq) \,|\, AgCl(s) \,|\, Ag(s)$ is 0.423 V at 25°C, what is the pH of the electrolyte solution?

16.9 The solid-state electrochemical cell $Pt(s) \,|\, O_2(g) \,|\, ZrO_2 \,|\, Ni(s) \,|\, NiO(s)_2 \,|\, Pt(s)$

is to be used to measure the partial pressure of oxygen in gas mixtures. Derive an equation relating the oxygen partial pressure and temperature of the gas phase to the zero-current potential of the cell. Assume the cell is at the same temperature as the gas. Data:

$$Ni(s) + 0.5\, O_2(g) = NiO(s); \quad \Delta G^0 = -234\ 350 + 85.23\ T \text{ kJ}$$

16.10 In an industrial process, a Ni–Cu alloy containing 0.85 atom fraction of nickel (the anode) is electrorefined in an aqueous solution at 25°C to produce pure nickel (the cathode).

i. Calculate the reversible potential required assuming the alloy is an ideal solid solution.

ii. In practice, both Ni and Cu dissolve from the anode. Solution is withdrawn from the cell, and the copper is removed by cementation using nickel sponge. Calculate the equilibrium ratio of copper to nickel in the purified solution at 25°C. Assume the activities of Cu and Ni ions are equal to their molalities.

iii. The applied cell voltage is 2.0 V and the cathodic current efficiency is 90%. Calculate the energy consumption (in kilowatt-hours per kilogram of refined nickel).

16.11 In the electrowinning of aluminium, the cell reaction is: $Al_2O_3(s) + 1.5\, C(s) = 2\, Al(l) + 1.5\, CO_2(g)$. In a particular operation, the applied voltage is 5.5 V and the current efficiency is 95%.

i. What is the consumption of Al_2O_3 and carbon per kilogram of aluminium produced?

ii. What is the electrical energy required per kilogram of aluminium produced?

16.12 A continuous copper electrowinning process operates at a current density of 300 Amp m^{-2} of cathode and total cell potential cell of 2.2 V.

i. Calculate the current efficiency if the measured energy consumption is 1.89 kWh kg^{-1}.

ii. What mass of copper is deposited on a cathode of surface area 0.95 m^2 after 5 days of operation? Note, the copper is deposited on both sides of the cathode plate.

17 Some applications of electrochemistry

SCOPE

This chapter examines examples of the practical application of the electrochemistry of electrolytic and galvanic cells.

LEARNING OBJECTIVES

1. *Understand the application of thermodynamics to important electrochemical processes.*
2. *Understand how electrolytic cells are used to perform various types of electrolysis: Electrowinning, electrorefining, electroplating, etc.*
3. *Understand the electrochemical nature of corrosion and cementation.*
4. *Understand how galvanic cells in the form of batteries are used to generate and store electrical energy.*
5. *Understand the principles of fuel cells.*

The applications of electrochemisty in everyday life and in industry are numerous and too many to describe here. Rather a selection has been made, and examples of the applications are described as illustrative of the pervasiveness of electrochemical applications.

17.1 ELECTROLYSIS

Electrolysis is the use of a direct current to drive a non-spontaneous redox reaction. Practical applications include the electrolytic production and refining of metals, production of chemicals, electroplating and anodising.

17.1.1 ELECTROWINNING OF METALS

In electrowinning, the metal compound is dissolved in solution or melted (molten salt systems) to form metallic cations, and these are reduced to the atomic state and deposited on the cathode by passage of a current. Metals that lie below hydrogen in Table 16.3 can be reduced from aqueous solutions since water is then stable. However, in practice, many metals above hydrogen can also be reduced due to the hydrogen overpotential effect. Table 17.1 lists common metals which can be produced electrochemically from aqueous solutions. Metals which cannot be produced electrochemically from aqueous solutions can be produced by electrochemical reduction of their molten salts, and a list of the metals in this category is also given in Table 17.1. In most cases the metal chloride has been found to be a suitable salt because chlorides have moderate melting points and high molar conductivities. Other halides can be used, but chlorine is the cheapest of the halogens and, therefore, preferred for the halogenation of metals that occur naturally as oxides using reactions of the type:

$$MO(s) + C(s) + Cl_2(g) = MCl_2(s, l \ org) + CO(g)$$

TABLE 17.1

How the common metals can be produced electrochemically

Metals electropositive with respect to hydrogen (can be reduced from aqueous solutions)		Metals electronegative with respect to hydrogen but which can be reduced from aqueous solutions due to the hydrogen overpotential		Reduction from molten salts	
Au	Cu	Pb	Fe	Nb	Ti
Pt	Bi	Sn	Cr	U	Mg
Ag	Sb	Mo	Zn	V	Na
Hg	W	Ni	Mn	Al	Ca
		Co	Na	Be	
		Cd			

An exception to the use of chlorides is aluminium which is produced from alumina (Al_2O_3) dissolved in cryolite (Na_3AlF_6) to lower its melting point (2345 K). The cryolite is effectively inert and acts as a solvent for the alumina.

EXAMPLE 17.1   Electrowinning of aluminium

Calculate the potential required to reduce aluminium oxide at 1300 K.

SOLUTION

The cell reaction for the reduction of alumina is

$$Al_2O_3(s) = 2\ Al(l) + 1.5\ O_2(g)$$

for which $\Delta G^0 (1300\,\text{K}) = 1262.28\ \text{kJ}$. Therefore,

$$E^0 = -\frac{\Delta G^0}{zF} = -\frac{1\ 262\ 280}{6 \times 96\ 485} = -2.18\ \text{Volts}$$

In practice, graphite electrodes are used, and the cell reaction is actually*

$$Al_2O_3(s) + 1.5\ C(s) = 2\ Al(l) + 1.5\ CO_2(g)$$

for which $\Delta G^0 (1300) = 668.071\ \text{kJ}$. Therefore,

$$E^0 = -\frac{\Delta G^0}{zF} = -\frac{668\ 071}{6 \times 96\ 485} = -1.54\ \text{Volts}$$

COMMENT

An energy advantage has been gained in lowering the cell potential (Equation 16.14), but this is at the expense of consumption of the carbon anode. Since the anode is consumed, any impurities in the carbon are released into the molten cryolite, and, because aluminium is a reactive metal,

---

* At the temperatures of aluminium smelting CO is the stable form of carbon in the presence of graphite. However, in practice CO_2 is formed due to an overpotential effect.

most impurity compounds are reduced and dissolve in the aluminium. This puts a requirement of high purity on the carbon in addition to the electrical and mechanical requirements and makes carbon the most expensive item in the production of aluminium after the cost of the electricity.

The energy required to smelt aluminium is very large due to the stability of Al_2O_3. The difference in enthalpy between two states of a system is the energy required to transform the system from one state to the other at constant temperature and pressure. At 1300 K, $\Delta_r H^0$ for the reaction

$$Al_2O_3(s) + 1.5\, C(s) = 2\, Al(l) + 1.5\, CO_2(g)$$

is 548.75 kJ mol^{-1} of Al. This is positive, meaning energy is required for the process to occur. Since $\Delta_r H^0 = \Delta_r G^0 + T\Delta_r S^0$, the energy has two components. The $\Delta_r G^0$ component is the energy required for the electrochemical reaction (334.03 kJ mol^{-1}), and the $T\Delta_r S^0$ component is the thermal energy required to maintain the temperature at a constant value (214.71 kJ mol^{-1}). In practice, this thermal energy is also provided electrically, through ohmic heating of the molten salt in the reduction cell. Converting the enthalpy value to kilowatt hours gives 5.56 kW h kg^{-1} of Al. This is the best that can be achieved; in practice typical consumption is around 12 kW h kg^{-1} of Al due to inefficiencies in the process.

17.1.2 MANUFACTURE OF CHLORINE

Chlorine, a major industrial chemical, is manufactured by electrolysis of sodium chloride solution. The other main products are sodium hydroxide and hydrogen. The main reactions are:

$$2\, H_2O(l) = 2\, H^+(aq) + 2\, OH^-(aq)$$

$$\text{Cathode:}\quad 2\, H^+(aq) + 2\, e^- = H_2(g)$$

$$\text{Anode:}\quad 2\, Cl^-(aq) = Cl_2(g) + 2\, e^-$$

$$\text{Cell reaction:}\quad 2\, H_2O(l) + 2\, Cl^-(aq) = Cl_2(g) + H_2(g) + 2\, OH^-(aq)$$

As H$^+$ ions are converted to H$_2$, water dissociates forming more H$^+$ and OH$^-$ ions resulting in a gradual build-up of OH$^-$ ions around the cathode and forming a solution of sodium hydroxide. The essential challenge is to have an effective means of separating the anode and cathode reactions so that the Cl$_2$ and NaOH produced cannot react to form sodium hypochlorite (NaClO). This is achieved in modern plants by inserting a cation exchange polymer membrane between the anode and cathode which prevents the passage of gas, anions and water molecules but allows the passage of Na$^+$ ions. The anodes are made of titanium coated with ruthenium dioxide, and the cathodes are nickel.

17.1.3 ELECTROREFINING

In electrorefining, the anode is made of the impure metal, and the solution acts only to carry the metal cations from the anode to the cathode where they are redeposited in the form of pure metal. The impurities in the anode material either dissolve in the electrolyte solution or form precipitates. The thermodynamic state of the metal is nearly identical in both half-cells, and the Gibbs energy of the reaction, therefore, is very small. For example, if the activity of the metal in the impure state is 0.95 (and one in the pure state) then for the refining cell reaction

$$M(s, a = 0.95) = M(s, a = 1)$$

where M is a divalent metal, the cell potential given by the Nernst equation is

$$E = E^0 - \frac{RT}{zF} \ln \frac{1}{0.95}$$

Since $E^0 = 0$ for the reaction $M(s) = M(s)$ E is 0.0007 V at 298 K and 0.002 V at 1000 K. In practice a higher potential is used to achieve a satisfactory rate. Electrochemical refining can be performed in both aqueous and molten salt solutions.

17.1.4 ELECTROPLATING

Electroplating is a process used to reduce dissolved metal cations to form a thin coherent metal coating onto an electrode (electrodeposition). It is used to change the surface properties of an object such as abrasion and wear resistance, corrosion protection or aesthetic qualities (for example, nickel or chromium plating on steel or gold or silver plating on brass jewellery). The part to be plated is the cathode, and the anode is made of the metal to be plated onto the part. Both components are immersed in an electrolyte solution containing one or more dissolved metal salts. The dissolved metal ions are reduced at the cathode and 'plate' onto it, and metal is dissolved from the anode.

The aim is usually to produce a smooth deposit, and generally low ionic strengths are best for this. Smoothing additives are also used which tend to block reactions at high points on the surface where the electric field strength is greatest, and optimum current densities are also necessary. Complexing agents are used to lower ionic strength where the metal to be plated could displace the coating metal from solution, such as copper plating of steel. In that case cyanide complexes are used. Copper is usually applied to steel before plating with nickel and chromium. By using complexes it is also possible to control the ionic strength to allow the plating of two metals at the same time in defined ratios – such as plating with copper and zinc to form a brass coating. In that case the complexing of copper reduces the free ion concentration to low levels compared with zinc.

Electroplating may also be used to build up thickness on undersized parts or to form objects (*electroforming*) by depositing metal onto a model of the part made from a conducting material.

17.1.5 ANODISING

Anodising involves electrochemical oxidation on a solid substrate to form a protective oxide coating. In the anodising of aluminium, the aluminium object forms the anode, and another piece of aluminium forms the cathode. Hydrogen is released at the cathode and oxygen on the surface of the aluminium object. The oxygen reacts with the aluminium to create a layer of aluminium oxide on the surface:

$$\text{Cathode:} \quad 2\,Al(s) + 6\,OH^-(aq) = Al_2O_3(s) + 3\,H_2O + 6\,e^-$$

$$\text{Anode:} \quad 6\,H^+(aq) + 6\,e^- = 3\,H_2(g)$$

$$\text{Cell reaction:} \quad 2\,Al(s) + 6\,OH^-(aq) + 6\,H^+(aq) = Al_2O_3(s) + 3\,H_2O + 3\,H_2(g)$$

Anodising changes the texture of the surface and the crystal structure of the metal near the surface. Anodised aluminium surfaces, for example, are harder than aluminium but have low to moderate wear resistance that can be improved with increasing thickness or by applying suitable sealing substances. Anodic films are generally stronger and more adherent than paint and metal plating but are more brittle.

17.1.6 ENERGY REQUIRED FOR ELECTROLYTIC PROCESSES

The theoretical energy required to carry out an electrolytic process is given by the enthalpy change of the cell reaction. When the cell reaction involves the formation of a gas as it often does for electrowinning (for example, aluminium and magnesium production) the entropy change of the cell reaction is greater than zero, and, since $\Delta G = \Delta H - T\Delta S$, ΔH will be more positive than ΔG. Only energy sufficient to meet the requirements of ΔG has to be supplied electrically, but the balance of energy, $T\Delta S$, has to be supplied or the cell will lose temperature. This can be supplied by burning a fuel or by supplying additional electrical energy for ohmic heating. If the cell is operated adiabatically at constant temperature and all energy is supplied electrically the total energy required will be equal to ΔH.

For a cell reaction for which the entropy change is negative, ΔH is more negative than ΔG, and an amount of heat equal to $T\Delta S$ must be removed to prevent the cell temperature rising (if the system is adiabatic) when electrical energy just sufficient to meet the needs of ΔG is supplied. In electrorefining ΔS will be a very small negative value.

In a real cell there are two additional heat loads which must be taken into account: Thermal energy to raise the temperature of the raw materials to the cell temperature (if it operates above ambient temperature) and to compensate for thermal losses from the cell. In practice, cells are usually run at a higher voltage than required to meet the electrochemical and heat requirements in order to obtain high production rates, and cooling of cells may then be needed to prevent overheating.

17.2 CEMENTATION

Cementation is the reduction of an ion in solution using a metal that has a more negative electrode potential. As is well-known, when a piece of iron is dipped into a copper sulfate solution the iron becomes plated with a deposit of copper metal. The process is electrochemical in nature (see Figure 17.1):

$$\text{Cathode (reduction) reaction:} \quad Cu^{2+}(aq) + 2\,e^- = Cu(s); \quad E^0 = 0.337 \text{ V}$$

$$\text{Anode (oxidation) reaction:} \quad Fe(s) = Fe^{2+}(aq) + 2\,e^-; \quad E^0 = 0.44 \text{ V}$$

$$\text{Cell reaction:} \quad Fe(s) + Cu^{2+}(aq) = Cu(s) + Fe^{2+}(aq); \quad E^0 = 0.777 \text{ V}$$

This reaction has been used commercially to recover copper from copper-bearing solutions using scrap iron as the reductant. Another example is the use of zinc powder to precipitate gold from cyanide solutions in the extraction of gold from ores.

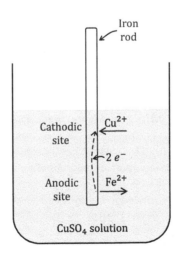

FIGURE 17.1 The mechanism of cementation of copper from solution using iron as the reductant.

17.3   CORROSION

Corrosion is an electrochemical process and can occur in a number of ways. *Galvanic corrosion* can occur when two different metals in contact are immersed in a common electrolyte. The more active metal (the anode) corrodes at a faster rate and the more noble metal (the cathode) corrodes at a slower rate compared with their rates of corrosion if separated. For this reason, zinc is often used as a sacrificial anode for steel structures (for example, galvanised steel). *Concentration cell* corrosion can occur when two or more areas of the same metal object are in contact with different concentrations of the same solution. There are three types:

- Crevice corrosion. In the presence of water, there can be a high concentration of metal ions in crevices under surfaces that are in contact and a low concentration of metal ions in water adjacent to the crevice; there will be a potential difference between the two points. The area of the metal in contact with the high concentration of metal ions (cathode) will be protected, and the area of metal in contact with the low metal ion concentration (anode) will be corroded.
- Differential aeration. Water in contact with a metal surface normally contains oxygen dissolved from the air. A concentration cell can develop at any point where oxygen in the air cannot diffuse freely into the solution thereby creating a difference in oxygen concentration between two points. Corrosion will occur in areas of low oxygen concentration (anode).
- Pitting corrosion. If a metal is protected against corrosion by a tightly adhering passive film and is in contact with salt water, the metal beneath the film will be exposed to corrosive attack in any areas where the film is broken. A potential difference will develop between the large area of the passive film (cathode) and the small area of exposed metal (anode). Rapid pitting of the metal will result.

17.4   BATTERIES

Batteries are galvanic cells used to produce or store electrical energy. There are two types, *primary batteries*, in which the electrode reactions cannot be chemically reversed and which cannot be recharged, and *secondary (or rechargeable) batteries* in which the electrode reactions can be reversed and can be recharged. During recharging, a battery operates as an electrolytic cell.

The best known primary battery is the traditional dry cell battery (~1.55 V), invented by Georges Leclanché in 1866, with a zinc anode casing, a centrally located graphite rod as the cathode and an acidic, water-based electrolyte paste consisting of MnO_2, NH_4Cl, $ZnCl_2$, graphite and starch:

$$\text{Cathode:}\quad 2\,MnO_2(s) + 2\,NH_4^+(aq) + 2\,e^- = Mn_2O_3(s) + 2\,NH_3(aq) + H_2O(l)$$

$$\text{Anode:}\quad Zn(s) = Zn^{2+}(aq) + 2\,e^-$$

$$\text{Cell reaction:}\quad 2\,MnO_2(s) + 2\,NH_4Cl(aq) + Zn(s) = Mn_2O_3(s) + Zn(NH_3)_2Cl_2(s) + H_2O(l)$$

The more efficient alkaline battery (~1.43 V), which has a longer shelf life and more constant output voltage, is a Leclanché cell modified to operate under alkaline conditions by the addition of KOH to the electrolyte:

$$\text{Cathode:}\quad 2\,MnO_2(s) + H_2O(l) + 2\,e^- = Mn_2O_3(s) + 2\,OH^-(aq)$$

$$\text{Anode:}\quad Zn(s) + 2\,OH^-(aq) = ZnO(s) + H_2O(l) + 2\,e^-$$

$$\text{Cell reaction:}\quad Zn(s) + 2\,MnO_2(s) = ZnO(s) + Mn_2O_3(s)$$

The nickel–cadmium battery (~1.4 V) is a water-based cell with a cadmium anode and an oxidised nickel cathode in the form NiO(OH). The battery is designed to maximise the surface area of the electrodes and minimise the distance between them to decrease the internal resistance.

$$\text{Cathode:} \quad 2\,NiO(OH)(s) + 2\,H_2O(l) + 2\,e^- = 2\,Ni(OH)_2(s) + 2\,OH^-(aq)$$

$$\text{Anode:} \quad Cd(s) + 2\,OH^-(aq) = Cd(OH)_2 + 2\,e^-$$

$$\text{Cell reaction:} \quad Cd(s) + 2\,NiO(OH)(s) + 2\,H_2O(l) = Cd(OH)_2 + 2\,Ni(OH)_2(s)$$

The well-known lead-acid battery is a secondary battery which has been used in automobiles for over a hundred years. It consists of high surface area lead plates partially coated with a porous layer of PbO_2 (anode) and $PbSO_4$ (cathode) immersed in dilute sulfuric acid as the electrolyte. The reactions are:

$$\text{Cathode:} \quad PbO_2(l) + HSO_4^-(aq) + 3\,H^+(aq) + 2\,e^- = PbSO_4(s) + 2\,H_2O(l); \quad E^0 = 1.685 \text{ V}$$

$$\text{Anode:} \quad Pb(s) + HSO_4^-(aq) = PbSO_4(s) + H^+(aq) + 2\,e^-; \quad E^0 = 0.356 \text{ V}$$

$$\text{Cell reaction:} \quad Pb(s) + PbO_2(s) + 2\,HSO_4^-(aq) + 2\,H^+(aq) = 2\,PbSO_4(s) + 2\,H_2O(l); \quad E^0 = 2.041 \text{ V}$$

As the cell is discharged, $PbSO_4$ paste forms on the electrodes, sulfuric acid is consumed and water is produced. When an external voltage greater than 2.041 V is applied, the electrode reactions are reversed, and $PbSO_4$ is converted back to metallic lead and PbO_2. Six cells are connected in series to form a conventional car battery of 12 Volts.

The popular lithium-ion battery is a rechargeable battery in which lithium ions move from the anode to the cathode during discharge and back when charging. The batteries have a lithium compound as the cathode and porous graphite as the anode. Both electrodes allow lithium ions to move in and out of their structures by processes called insertion (intercalation)* and extraction (deintercalation), respectively. The cathode is typically lithium cobalt oxide (lithium cobaltate), lithium manganese oxide (lithium manganate), lithium iron phosphate, lithium nickel manganese cobalt (NMC) or lithium nickel cobalt aluminium oxide (NCA). The electrolyte consists of lithium salts in an organic solvent. The reactions, for a lithium cobaltate cathode are:

$$\text{Cathode:} \quad CoO_2 + Li^+ + e^- = LiCoO_2$$

$$\text{Anode:} \quad LiC_6 = C_6 + Li^+ + e^-$$

$$\text{Cell reaction:} \quad CoO_2 + LiC_6 = C_6 + LiCoO_2$$

17.5 FUEL CELLS

Fuel cells are batteries that are continuously supplied with reactants. The energy available from simple fuels such as hydrogen, methane and methanol is utilised to directly provide electrical energy. The hydrogen and carbon in the fuel combine with oxygen in the air to produce water and carbon dioxide. The catalysed reactions for a hydrogen fuel cell are:

$$\text{Cathode:} \quad O_2(g) + 4\,H^+ + 4\,e^- = 2\,H_2O(g)$$

$$\text{Anode:} \quad 2\,H_2 = 4\,H^+ + 4\,e^-$$

$$\text{Cell reaction:} \quad 2\,H_2(g) + O_2(g) = 2\,H_2O(g)$$

---

* Intercalation is the reversible inclusion or insertion of a molecule or ion into a material with a layered atomic structure.

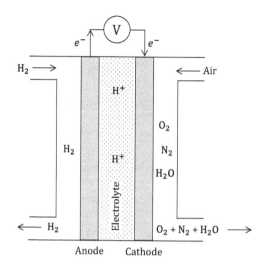

FIGURE 17.2 Schematic of a proton-exchange membrane fuel cell.

There are several types of fuel cells, but all consist of an anode, a cathode and an electrolyte that allows positively charged hydrogen ions (protons) to move between the two half-cells. In one form (see Figure 17.2), a proton-conducting polymer membrane* contains the electrolyte solution that separates the anode and cathode. On the anode side, hydrogen diffuses to the anode catalyst where it dissociates into protons and electrons. The protons pass through the membrane, then diffuse through the electrolyte to the cathode. The electrons travel through an external circuit because the membrane is electrically insulating. Oxygen molecules react on the cathode catalyst with the electrons (from the external circuit) and protons to form water.

Fuel cells are classified by the type of electrolyte used and by the difference in start-up time which can range from a second for proton exchange membrane fuel cells to around 10 minutes for solid oxide fuel cells, which operate at elevated temperatures. Fuel cells convert the energy in fuel to electricity much more efficiently than can be achieved by burning the fuel and raising steam to drive a generator since they are not limited by the efficiency of thermal energy conversion to mechanical energy (see Section 7.2.3 and Example 7.1). The energy efficiency of a fuel cell is typically 50–60% (see Example 8.7).

Fuel cells can be operated in reverse, for example to generate hydrogen and oxygen from steam. However, fuel cells operated backwards generally do not make very efficient systems unless they are purpose-built to do so. The elevated temperature and the heat content of the steam significantly reduce the electrical energy used compared to that required for the electrolysis of liquid water.

---

* Typically Nafion™, a sulfonated tetrafluoroethylene-based fluoropolymer-copolymer, is used.

Answers to problems

2.3	3.125, 2.743, 0.626, 0.401, 0.360, 2.171
2.4	2.30, 1.47, 20.25, 55.49, 1.71
2.5	4.69×10^{24}, 4.57×10^{23}
2.6	(a) i. 25% SiO_2, 75% $Ca(OH)_2$; ii. 19.2% SiO_2, 57.7% $Ca(OH)_2$, 23.1% $MgCO_3$; iii. 15.6% SiO_2, 46.9% $Ca(OH)_2$, 18.8% $MgCO_3$, 18.8% $BaSO_4$
	(b) i. $x_{SiO_2} = 0.291$, $x_{Ca(OH)_2} = 0.709$; ii. $x_{SiO_2} = 0.233$, $x_{Ca(OH)_2} = 0.567$, $x_{MgCO_3} = 0.199$; iii. $x_{SiO_2} = 0.218$, $x_{Ca(OH)_2} = 0.529$, $x_{MgCO_3} = 0.186$, $x_{BaSO_4} = 0.067$
3.1	985 mL
3.2	247.9 kPa
3.3	1.977 moles
3.4	−211.8°C
3.5	30.41

3.6

	vol%	x	p
H_2	15.4	0.154	0.277
O_2	23.1	0.231	0.415
Ar	42.3	0.423	0.762

3.7	2.45 g
3.8	0.5 atm
3.9	$p = 6.5$ bar; $p_{N_2} = 5.0$ bar; $p_{O_2} = 1.5$ bar
3.10	$p = 130.2$ bar; $p_{H_2} = 123.3$ bar; $p_{N_2} = 3.54$ bar; $p_{CO_2} = 3.38$ bar
3.11	(a) 3.44×10^6 Pa; (b) 3.57×10^6 Pa
3.12	(a) 721.7°C; (b) 679.9°C
4.2	1.60 kJ
4.3	50.7 J; 253.3 J
4.4	206.2 kJ
4.5	−287.07 kJ
4.7	605.4 kJ mol^{-1}
4.8	−2043.17 kJ
4.9	$\Delta_r H^0 (298.15) = -890.3$ kJ; $\Delta_r H^0 (1200) = -801.6$ kJ
	$\Delta_r H^0 (298.15) = -91.9$ kJ; $\Delta_r H^0 (1200) = -111.6$ kJ
	$\Delta_r H^0 (298.15) = -215.6$ kJ; $\Delta_r H^0 (1200) = -233.5$ kJ
4.10	$\Delta_{vap} H^0 = 40.65$ kJ mol^{-1}
	$\Delta_{vap} U^0 = 37.55$ kJ mol^{-1}
4.11	31,031 J g^{-1}
4.12	$\Delta_{fus} H^0 = 4803$ J mol^{-1}
5.1	$H^0(CaO, 900) = 605.195$ kJ mol^{-1} (data from Table 5.1) $= 605.193$ kJ mol^{-1} (data from Table 5.3)
	$T_{fus} = 3200$ K; $\Delta_{fus} H_0 (CaO) = 97.50$ kJ mol^{-1} (data from Table 5.1)

5.4

T (K):	1500 K	1600 K	1700 K
$H(T) - [H(298)]$ (kJ mol^{-1}):	30.65	33.52	86.60

6.1	6.90×10^5 kJ; 305.5 kg water
6.2	753.6 kJ
6.3	~3200 K
6.4	$\Delta_r H^0 (298.15) = -2219$ kJ; i. 2098 K; ii. 2204 K
6.5	$\Delta_r H^0 (298.15) = -442.1$ kJ; excess heat: 187.0 kJ mol^{-1}

7.1   9.38 J mol^{-1} K^{-1}; 23.50 J mol^{-1} K^{-1}; 87.40 J mol^{-1} K^{-1}

7.2 -57.43 J

7.3   11.53 J; 18.70 J; 0.0 J

7.5   10.7%; 44.9%

7.6 -13.81 J mol^{-1} K^{-1}

7.7   39.19 J

7.8   i. 90.11 J; ii. -203.0 J; iii. 22.8 J

7.9 $\Delta_r S^0(298.15) = -242.8$ J mol^{-1} K^{-1}; $\Delta_r S^0(1200) = -0.7$ J mol^{-1} K^{-1}

   $\Delta_r S^0(298.15) = -198.1$ J mol^{-1} K^{-1}; $\Delta_r S^0(1200) = -235.2$ J mol^{-1} K^{-1}

   $\Delta_r S^0(298.15) = -6.8$ J mol^{-1} K^{-1}; $\Delta_r S^0(1200) = -43.0$ J mol^{-1} K^{-1}

8.2 $\Delta_r G^0(298.15) = -817.9$ J mol^{-1}; $\Delta_r G^0(1200) = -800.8$ J mol^{-1}

   $\Delta_r G^0(298.15) = -32.8$ J mol^{-1}; $\Delta_r G^0(1200) = 170.6$ J mol^{-1}

   $\Delta_r G^0(298.15) = -213.6$ J mol^{-1}; $\Delta_r G^0(1200) = -181.9$ J mol^{-1}

8.4   0.227 kW h; efficiency $= 92\%$

9.1   23.76 Pa

9.2   145.6 g

9.3 2.42×10^{-6} atm; 2225 K

9.4   0.04 mL mol^{-1}

9.6   iv. -21.0 kJ mol^{-1}

9.7 $\gamma_{Al} = 0.74$; $a_{Al} = 0.45$

9.11   Yes

9.12   33.016 kJ

9.13 $a_{Si}^H = 0.55$; $\gamma_{Si}^H = 1.09$

10.1 6.37×10^{-9} J; 1.26×10^{28}; 5.28

10.3   0.0061 g kg^{-1} O_2; 0.055 g kg^{-1} N_2

10.4   i. Oxygen dissolves in atomic form; ii. 0.129 mass%

10.5   ~725°C

10.6   863 K

10.7   i. $\Delta H^0 \approx 110.5$ kJ; $\Delta S^0 \approx 240$ kJ

   ii. 183°C

   iii. -48°C

10.8 $K = 13.2$; $\Delta_r G_0 = -15.50$ kJ; $\Delta_r H^0 = 155.3$ kJ; $\Delta_r S^0 = 244.6$ J

10.9   ii. 1180 K; iii. $\Delta_{vap} H^0 = 111$ kJ

10.10   i. 2420 K; ii. 30.5 Pa; iii. 8800 Pa; iv. 8350 Pa

10.11   0.0048 mass%

10.12   No solid solution: Reaction proceeds until all Zn and FeS is consumed.

   Solid solution: $x_{FeS} = 0.083$; $x_{ZnS} = 0.917$; $n_{Fe} = 0.917$; $n_{Zn} = 0.0083$

10.13   8.71 g Si recovered; 31% efficiency of recovery; equilibrium gas composition (by volume): 47.3% HCl, 23.7% $SiCl_4$, 29.0% HCl_3Si

11.1   Decomposition commences at 147°C and becomes rapid at 188°C.

11.2 4.87×10^{-5}; 4.54×10^{-6}

11.3   33.2 vol%

11.4 $p_{CO} < 0.34$ bar

11.6 $a_{Al} = 0.16$; $x_{Al} = 0.13$

12.1

$\Delta_r H^0$ kJ mol^{-1}	$\Delta_r S^0$ J mol^{-1}	$\Delta_r G^0$ kJ mol^{-1}	K
-153.8	-32.23	-144.39	1.98×10^{25}
-223.18	-207.14	-161.324	1.838×10^{28}
-44.4	-161.43	3.628	0.2314

12.5

	$a_{\pm}$	a_e
NaCl	0.146	0.0213
Na_2SO_4	0.114	0.00149

$CaCl_2$	0.152	0.00354
$MgSO_4$	0.026	0.000676

12.6

	$a_\pm$	a_e
NaCl	0.253	0.0639
Na_2SO_4	0.150	0.00336

12.7 3.32×10^{-4} mol kg^{-1}

12.8

	m_{Ag^+}, mol kg^{-1}	$m_{S^{2-}}$, mol kg^{-1}	$m_{Ag(CN)_2^-}$, mol kg^{-1}
water	6.0×10^{-17}	3.0×10^{-17}	–
0.15 mol kg^{-1} NaCN	3.38×10^{-23}	9.46×10^{-5}	1.89×10^{-4}

12.9 5.50×10^{-9} mol kg^{-1}

12.10 0.034 mol kg^{-1}

12.11 1.8 mol kg^{-1}

12.12

	$\gamma_{SO_4^{2-}}$	$a_{SO_4^{2-}}$
D–H limiting law	0.317	0.00634
Extended D–H	0.419	0.00839

12.13 $I_m = 0.06$ mol kg^{-1}; $\gamma_\pm = 0.662$; $m_\pm = 0.0317$ mol kg^{-1}; $a_\pm = 0.0210$

12.14 0.222

12.15 3.19×10^{-4} mol kg^{-1}

12.16 $\alpha = 0.3494$; $m_{H^+} = 0.01398$ mol kg^{-1}; $m_{H_2PO_4^-} = 0.01398$ mol kg^{-1}; $m_{H_3PO_4} = 0.02602$ mol kg^{-1}; pH = 1.885

12.17 $\alpha = 0.3913$; $m_{H^+} = 0.01565$ mol kg^{-1}; $m_{H_2PO_4^-} = 0.01565$ mol kg^{-1}; $m_{H_3PO_4} = 0.02435$ mol kg^{-1}, pH = 1.869

12.18

	Unit activity coefficients	D–H limiting law
α	0.1179	0.1275
m_{H^+} mol kg^{-1}	0.004715	0.00510
m_{F^-} mol kg^{-1}	0.004715	0.00510
m_{HF} mol kg^{-1}	0.03528	0.03490
pH	2.326	2.329

12.19 $a_{Cu^{2+}} = 0.04682$; $a_{Cl^-} = 0.2351$; $a_{CuCl_2} = 0.003766$

13.1 337 K; 156 000 Pa

13.2 3.17×10^4 bar; 5.02×10^4 bar

13.3 $\Delta_{sub}H = 31.2$ kJ mol^{-1}; $\Delta_{vap}H = 22.1$ kJ mol^{-1}; $\Delta_{fus}H = 9.1$ kJ mol^{-1}

13.5 28.4 kJ mol^{-1}

13.6 0.011 K

13.7 1.85

13.8 42.8 kJ mol^{-1}

13.12 Mg, Zn and Pb are amenable.

14.1 First reaction: $P = 2$; $C = 1$; $F = 1$
Second reaction: $P = 3$; $C = 2$; $F = 1$

14.4 $r = 4$; $s = 7$; $P = 4$; $F = 1$

14.5 $p_{FeCl_2} = 4.42 \times 10^{-5}$ bar; $p_{FeCl_3} = 0.16$ bar

14.7 $x_{Mn} = 0.0011$; $p_{O_2} = 5.12 \times 10^{-9}$ bar; $x_{Mn} = 0.024$

14.8 Ca and Mg can be removed in the slag; Zn and Cd can be removed as gases.

15.1 At 1 bar: vol% $N_2O_4 = 5.4$; vol% $NO_2 = 94.6$
At 10 bar: vol% $N_2O_4 = 30.0$; vol% $NO_2 = 70.0$

15.3 iv. vol% $H_2 = 39$; vol% $N_2 = 50$; vol% $NH_3 = 11$
v. vol% $H_2 = 44$; vol% $N_2 = 50$; vol% $NH_3 = 6$
vi. vol% $H_2 = 60$; vol% $N_2 = 25$; vol% $NH_3 = 15$

15.4 0.7 vol% SO_2, 9.7 vol% O_2, 82.0 vol% N_2, 7.6 vol% SO_3

15.5   i. 0.0 g Zn; 8.54 g Fe; 14.90 g ZnS; 1.55 g FeS

ii. 8.56 g alloy (1.22 wt% Zn, 98.78 wt% Fe); 14.75 g ZnS; 1.69 g FeS

iii. 8.63 g alloy (6.44 wt% Zn, 93.56 wt% Fe); 16.37 g NiS–FeS solid solution (85.99 wt% ZnS, 14.01 wt% FeS)

15.6   i. 21.02 g Ni; 0.0 g Fe; 73.25 g NiO; 25.73 g FeO

ii. 21.02 g alloy (99.87 wt% Ni, 0.13 wt% Fe); 73.29 g NiO; 25.69 g FeO

iii. 21.02 g alloy (99.95 wt% Ni, 0.05 wt% Fe); 98.99 g NiO–FeO solid solution (73.31 wt% NiO, 25.98 wt% FeO)

iv. 21.01 g alloy (99.02 wt% Ni, 0.98 wt% Fe); 99.00 g NiO–FeO solid olution (74.26 wt% NiO, 25.72 wt%)

16.1   ii. 1.36 V; iii. −262.4 kJ; iv. 0.756

16.2   ii. 0.463 V; iii −89.345 kJ iv. 0.47

16.4   −0.772 V; 0.307 V; 1.079 V

16.5   −0.251 V

16.6   ii. −2.500 V; iii 19.84 MJ kg^{-1} or 5.512 kW h kg^{-1}

16.7   ii. $\Delta_r G^0 = 42.84$ kJ ; $\Delta_r S^0 = -124.7$ J K^{-1}; $\Delta_r H^0 = -80.02$ kJ

16.8   1.76

16.10 i. −0.0021 V; ii. 5.1×10^{-20}; iii. 2.03 kW h kg^{-1}

16.11 i. 1.89 kg Al$_2$O$_3$ and 0.333 kg C per kg Al; ii. 17.24 kW h kg^{-1}

16.12 i. 98.2%; ii. 79.6 kg

Index